Advanced Level Physical Chemistry

A. HOLDERNESS

A new edition of the *Physical* section of *Inorganic and Physical Chemistry* revised by

J. LAZONBY, PH.D.
Lecturer in Chemistry, University of York

HEINEMANN EDUCATIONAL BOOKS
LONDON

Heinemann Educational Books

LONDON EDINBURGH MELBOURNE AUCKLAND TORONTO
HONG KONG SINGAPORE KUALA LUMPUR
IBADAN NAIROBI JOHANNESBURG
LUSAKA NEW DELHI

ISBN 0 435 65434 9

First published as part of *Inorganic and Physical Chemistry*, 1961
Reprinted 1962
Second Edition 1963
Reprinted 1964 (with corrections)
Reprinted 1966 (with corrections)
Reprinted 1967 (with corrections)
Third Edition (*Physical Chemistry* section only) 1976

Published by Heinemann Educational Books Ltd
48 Charles Street, London W1X 8AH

Text set in 10/11 pt. Monotype Times New Roman, printed by letterpress, and bound in Great Britain at The Pitman Press, Bath

Preface

Advanced Level Physical Chemistry is a new edition of the Physical section of *Inorganic and Physical Chemistry.* At the time of going to press the Inorganic section is in preparation and will be published at a later date under the title *Advanced Level Inorganic Chemistry.*

In addition to attention to units and nomenclature, the revision of the Physical section has required a regrouping, retitling, and rewriting of some of the original chapters in order to produce a more coherent treatment which is consistent with modern Advanced Level Syllabuses. New material has been introduced, particularly in the parts of the book which deal with electron arrangements, crystal structures, and energy changes.

The policy on units has been to adopt SI units entirely, except in such cases as the use of the convenient mmHg alongside $N\ m^{-2}$ for pressure. In making decisions about nomenclature, the recommendations of the Association for Science Education and the Examining Boards have been taken into account.

Thanks are due to Mr Martyn Berry, Mr J. P. Chippendale, Dr John Dyson, and Mrs Freda Stevens for their helpful advice on the content and editorial work on the manuscript and proofs.

J. L.

1976

Contents

1. Classical Atomic and Molecular Theories

Elements, Compounds, and Mixtures

Late in the eighteenth century, sufficient chemical knowledge had accumulated to allow scientists to recognize elements and compounds as different species of substances. At this period an element was defined as follows:

An **element** is a substance which cannot be decomposed into simpler substances by any known chemical process.

At the same period, the following definition was given for a chemical compound:

A **compound** is a material of uniform composition throughout its bulk and containing two or more elements in a state of chemical combination.

It was clearly recognized that, during chemical combination, the elements concerned lost their own characteristic chemical properties and that the compound possessed a new set of chemical properties of its own. For example, sodium is a very reactive metallic element which attacks water violently at ordinary temperature. Chlorine is a poisonous, greenish-yellow, gaseous element. The 'chemical combination' of these two produces sodium chloride (common salt), which is a white solid, non-poisonous, and soluble in water without reaction with it. Scientists of the early nineteenth century had no idea of the real nature of 'chemical combination', and, in fact, its interpretation in essentially electrical terms has been possible only in the twentieth century. This interpretation is developed to some degree in Chapter 7.

The two most important characteristics of compounds are:

1. All pure samples of a given compound are identical in composition by mass. (This statement disregards the occurrence of isotopy. For a discussion of this phenomenon, see page 29.)
2. Chemical action is usually needed to separate the constituent elements from a compound. For example, to obtain hydrogen (element) from its

compound water, the chemical action of sodium on water or of heated magnesium or iron on steam is required.

$$2Na(s) + 2H_2O(l) \rightarrow 2NaOH(aq) + H_2(g)$$
$$Mg(s) + H_2O(g) \rightarrow MgO(s) + H_2(g)$$
$$3Fe(s) + 4H_2O(g) \rightleftharpoons Fe_3O_4(s) + 4H_2(g)$$

Mixtures can be made from elements or compounds or both. In mixtures, the constituents merely lie mingled together. Mixtures of given constituents can usually be made widely variable in composition and the various constituents can always be recovered by *physical* action alone. For example, iron filings and sulphur can be mixed in any desired proportion and the sulphur can be recovered by solution in carbon disulphide, filtration, and evaporation of the solvent. This is a succession of purely physical processes. In addition, mixing is usually a quiet process; by contrast, chemical combination is often accompanied by large energy changes which exhibit themselves as the evolution of heat, explosion, or flame.

Laws of Chemical Combination by Mass

Around the beginning of the nineteenth century, two laws, which were based entirely on the results of experiments, were known to govern chemical combination.

1. *The Law of Conservation of Mass*

This states (in its chemical aspect) that:

> The total products of a given chemical reaction have the same mass as the total reactants involved in the reaction.

The most convincing experimental evidence for this law was supplied by Landolt in 1906.

Solutions of pairs of chemical reagents were introduced separately into the two arms of a Landolt tube (Figure 1.1). The tube was sealed off by

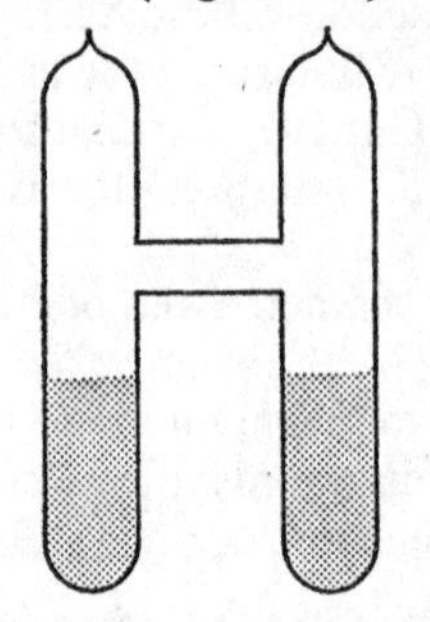

Figure 1.1. Landolt tube

heating and its mass was found at room temperature. The reaction was induced by inverting the tube and allowing the solutions to mix. The mass of the tube containing the products was then determined. Allowance had to be made for two main possibilities of error which arose from the slight evolution of heat during the reaction. This heat caused loss of moisture from the surface of the tube and expansion of the tube, so that it displaced a greater volume of air. Both these factors tended to decrease the mass of the tube. The errors were eliminated if sufficient time was allowed for the tube to regain its original state. Landolt investigated fifteen chemical reactions, of which two typical cases were:

$$Ag_2SO_4(aq) + 2FeSO_4(aq) \rightarrow 2Ag(s) + Fe_2(SO_4)_3(aq)$$
$$5HI(aq) + HIO_3(aq) \rightarrow 3H_2O(l) + 3I_2(aq)$$

He found that the change of mass during the reactions did not exceed one part in ten million, which was well within the limit of experimental error.

It should be noted that, according to modern ideas, emission of energy involves a loss of mass according to the famous Einstein equation, $E = mc^2$, where E is the energy, m the mass, and c the velocity of light (in appropriate units). Consequently, an emission of energy during chemical action can cause loss of mass and so interfere with the above law. In practice, however, the loss of mass is so very small as to be neglible. Even so the true conservation is one of mass and energy together.

2. *The Law of Constant Composition (or Definite Proportions)*

All pure samples of a given chemical compound contain the same elements in the same proportions by mass.

This experimental law was put forward by Proust in 1799. It was strongly disputed by Berthollet, but convincing experimental evidence such as that obtained by Stas finally led to its acceptance. In one particular case, Stas prepared silver chloride by heating silver in chlorine, and by dissolving silver in nitric acid and precipitating the chloride by ammonium chloride and by hydrochloric acid. In the three products, 100 parts by mass of silver combined with 32.8425, 32.842 and 32.8475 parts by mass of chlorine.

In more recent years, the phenomenon of *isotopy* has been recognized and is relevant to this law. Isotopy is the existence of atoms of the same chemical properties but different relative atomic mass. For example, lead associated with uranium minerals has a relative atomic mass of 206, but lead associated with thorium minerals has a relative atomic mass of 208. It is obvious that compounds of these isotopes with another element cannot have a constant composition. This is an exceptional case. Usually, elements encountered in chemical practice contain isotopes in *fixed proportions* and so act as if their relative atomic masses are completely constant.

Berthollide compounds

Deviations from the Law of Constant Composition corresponding to Berthollet's ideas have been studied in solids since about 1930. In a solid, AB_n, the ideal situation is to have every particle (ion or atom) of A placed in its appropriate point in the solid lattice (page 86) and every such point occupied; and similarly for B. Such a perfect arrangement can be expected only at 0 K(kelvin), where thermal vibration of particles is absent. At other temperatures lattice defects may occur, arising from such vibrations. These defects may be produced in three ways: (a) by interchange of particles of A and B, (b) by migration of particles of A or B away from their correct locations in the lattice to *interstitial* (see page 90) positions (these are called *Frenkel defects*), and (c) when locations are left vacant by migration of particles of A or B to the surface of the material (called *Schottky defects*).

Type (a) is very unlikely to occur in *ionic* compounds, e.g. KCl, because of energy factors involved, but is found in metallic alloys. At ordinary temperatures, ionic compounds, e.g. KCl, CaO, show no observable lattice defects and conform perfectly to the Law of Constant Composition. As the melting point is approached, however, defects may become more appreciable. For example, potassium chloride at 1000 K has about 0.004 per cent of lattice vacancies.

Iron(II) sulphide, ideally FeS, has been shown to vary between $FeS_{1.00}$ and $FeS_{1.14}$, or in reverse, $Fe_{1.00}S$ and $Fe_{0.88}S$. This variation is caused by lattice deficiency in iron, that is, $Fe_{1.00}S$ to $Fe_{0.88}S$ is the more correct representation. This is known to be so because the material *decreases* in density (by loss of Fe) as the proportion of sulphur increases. If sulphur were added interstitially to true FeS, an increase in density would be expected. The relative loss of iron is compensated electronically by the conversion of Fe^{2+} cations to Fe^{3+}, so that, in a sense, the product has the composition of a continuously variable mixture of FeS and Fe_2S_3. Similarly, iron(II) oxide has been shown to vary between $FeO_{1.06}$ and $FeO_{1.19}$ at 1700 K. As in the case of iron(II) sulphide, this arises from the omission of some iron from the lattice and the conversion of some Fe^{2+} to Fe^{3+} to compensate.

The Atomic Theory of Dalton

Against this chemical background of the early nineteenth century, John Dalton in 1808 was able to produce the set of postulates known collectively as the Atomic Theory. Atomic ideas have, however, a very long history. Dalton's fame rests on the fact that he converted atomic ideas from 'ineffectual thoughts casually entertained' (Whitehead) into a set of definite ideas which could be (a) used to explain known experimental observations and (b) used to make predictions which could be subjected to experimental test. These ideas, which constitute the classical **Atomic Theory**, are:

1. All elements are made up of small particles called atoms.
   An **atom** is defined as the smallest particle of an element which can take part in a chemical reaction.
2. Atoms cannot be created or destroyed.
3. Atoms are indivisible.
4. Atoms of the same element are exactly alike in every way, notably in mass.
5. Atoms combine together in small whole numbers.

Some of these ideas have had to be restated in the light of more recent discoveries.

The Atomic Theory clearly explains the two experimental Laws of Chemical Combination, which were known at that time:

1. **Law of Conservation of Mass** (page 2). The Atomic Theory stated that atoms cannot be created or destroyed. From this postulate it follows that all the atoms present at the beginning of a chemical reaction should be present at the end of it, with no gain or loss of material. That is, the total mass of the products of a chemical reaction should be equal to the total mass of the reagents used.
2. **Law of Constant Composition** (page 3). The Atomic Theory postulated that all the atoms of the same element are exactly alike. From this the conclusion can be drawn that pure samples of a given chemical compound, containing atoms of the same elements in the same proportions, must be identical in composition by mass.

Two more Laws of Chemical Combination can be deduced from the Atomic Theory.

3. The **Atomic Theory** postulated that all atoms of the same element are exactly alike and that the atoms combine together in small whole numbers. If the second postulate is true, compounds between the elements A and B must be of the type AB, A_2B, AB_2, A_2B_3, etc. If all the atoms of A are exactly alike, the amount 'A' (one atom of A) must be a fixed amount. The amounts of B which combine with this fixed amount A in these respective compounds are B, B/2, 2B, 3B/2, etc. These are in the proportions 2B:B:4B:3B. If all the atoms of B are alike, these must be simple, whole-number proportions. This conclusion can be tested by analysing sets of compounds such as copper(I) and copper(II) oxides, mercury(I) and mercury(II) chlorides, and the three oxides of lead, and is found to be true. It is expressed in the form known as the **Law of Multiple Proportions:**

   If two elements, A and B, combine to form more than one compound, the various masses of B, which combine with a fixed mass of A, are in a simple whole-number ratio.

Dalton himself was the first person to verify this law experimentally in about 1804.

4. The second conclusion from the above two postulates describes the more complicated relations of a number of different elements, C, D, E, F, etc., with an element, A. If atoms combine in small whole numbers, the compounds formed by these elements with A and with each other must be of the simple types AC, A_2C, AC_2, AD, AD_2, CD, CD_3, EF_3, etc. From this it follows that, if all atoms of the respective elements are exactly alike, the proportions in which C, D, E, F, etc. combine with a fixed mass of A are also the proportions in which C, D, E, F, etc. combine with each other, or are simple multiples of those proportions. This is known as the **Law of Reciprocal Proportions**:

   The masses of elements C, D, E, F, etc. which combine with (or displace) a fixed mass of element A, are the masses of C, D, E, F, etc. which combine with (or displace) each other, or are simple multiples or submultiples of these masses.

   The following figures illustrate the law. Let the fixed mass of oxygen be 8 g. The masses of other elements which combine with the 8 g of oxygen are zinc, 32.5 g; sulphur, 8 g (in SO_2); hydrogen, 1 g. The law says that, if zinc, sulphur or hydrogen combine with (or displace) each other, they must do so in the proportions of 32.5:8:1 or some simple multiples of these. Actually, 32.5 g of zinc combine with 16 g of sulphur, 1 g of hydrogen combines with 16 g of sulphur, and 1 g of hydrogen is displaced by 32.5 g of zinc, as the law requires.

   The most convincing experimental verification of this law was published by Berzelius in 1812.

Dalton's Atomic Theory could not, of course, be directly verified because the atoms were quite inaccessible to the measuring instruments of the day. However, the theory provided a satisfactory explanation of the known experimental laws and led to the prediction of others, which were successfully tested by experiment.

Combining masses

When carrying out an investigation into how two elements are involved in a particular reaction, quite often the easiest experimental data which can be obtained is the mass of one element which combines with a certain mass of the other element. These masses are called **combining masses** or, when referred directly or indirectly to the common standard of one part by mass of hydrogen, they are called **equivalent masses**. (It is sometimes more convenient to use either oxygen or chlorine as the standard when determining the equivalent mass of an element, in which case 8 parts by mass of oxygen or 35.5 parts by mass of chlorine must be used as these are the masses which combine with 1 part by mass of hydrogen.)

Relative atomic masses

Dalton used the combining masses of elements to determine values for their relative atomic masses. He tried to compare the mass of one atom of each element with the mass of one atom of hydrogen. (Hydrogen was chosen as the reference element as its atom has a smaller mass than atoms of other elements.) In order to make this comparison, it was necessary for Dalton to assume that atoms of different elements combine in the simplest possible ratios.

For example, when it was found that 8 g of oxygen combined with 1 g of hydrogen, it was assumed that 1 atom of oxygen combined with 1 atom of hydrogen and that these combining masses represented equal numbers of atoms. Therefore, the atomic mass of oxygen relative to hydrogen was found to be 8. Obviously, to obtain the correct value, it was necessary to know that the formula of water is H_2O and that 1 g of hydrogen represents twice as many atoms as 8 g of oxygen, from which it can be deduced that the atomic mass of oxygen relative to hydrogen is 16.

The lack of knowledge of the ratios in which atoms combine led to the determination of conflicting results for the relative atomic masses of elements.

It was not until Cannizzaro (1858) realized the full significance of the contributions of Gay-Lussac and Avogadro to the molecular theory of gases that this problem, for elements such as carbon, nitrogen, and oxygen, was overcome.

Molecular Theory of Gases

The development of the molecular theory began with the following experimental laws.

Boyle's Law (1662)

> The volume of a fixed mass of gas is inversely proportional to its pressure, temperature remaining constant.

Charles' Law (1787)

> The volume of a given mass of gas is directly proportional to its absolute temperature, pressure remaining constant.

Gay-Lussac's Law of Gaseous Volumes (1808)

> When gases react, they do so in volumes which bear a simple ratio to one another and to the volume of the product if it is a gas, temperature and pressure remaining constant.

This law is illustrated by experimental observations, such as that at constant temperature and pressure:

2 volumes of hydrogen combine with 1 volume of oxygen to give 2 volumes of steam;

2 volumes of ammonia are formed from 3 volumes of hydrogen and 1 volume of nitrogen.

A notable feature of these laws is that they apply (with very minor variations) to all gases, irrespective of whether the gases are elements or compounds and, also, of their chemical nature, which can be very variable. There must be some explanation of this universal similarity of gaseous behaviour. Under the influence of the atomic ideas which pervaded chemistry in the early nineteenth century, Berzelius (and others) put forward the hypothesis that the factor common to all gases is the occurrence of equal numbers of atoms in equal volumes of all gases at constant temperature and pressure. This idea proved unacceptable for reasons of which the following is typical.

By experiment, 1 volume of hydrogen combines with 1 volume of chlorine to give 2 volumes of hydrogen chloride at constant temperature and pressure. If the idea of Berzelius is accepted, this statement can be rendered in the form:

n atoms of hydrogen combine with n atoms of chlorine to give $2n$ 'compound atoms' of hydrogen chloride.

This simplifies to:

1 atom of hydrogen combines with 1 atom of chlorine to give 2 'compound atoms' of hydrogen chloride.

It is obvious that one atom of hydrogen cannot contribute to two 'compound atoms' of hydrogen chloride unless the hydrogen atom can be split. But Dalton's Atomic Theory required the atom to be indivisible so the Berzelius idea was abandoned.

It was replaced in 1811 by the very famous and important idea usually known as **Avogadro's Hypothesis**. This states that:

Equal volumes of all gases at the same temperature and pressure contain the same number of molecules.

It is important here to be clear about the nature of the atom and the molecule. They are defined in the following way:

An **atom** of an element is the smallest particle of the element which can take part in a chemical reaction.

A **molecule** of an element or compound is the smallest particle of it which can exist separately.

The two are not necessarily the same particle in the case of any given gas and, in particular, the molecule may be polyatomic and so capable of splitting into individual atoms as required. This recognition of the occurrence of the two different particles, the atom and the molecule, removed the difficulty which destroyed the atomic idea of Berzelius discussed above.

The following is an example of the way in which Avogadro's Hypothesis provides a satisfactory explanation of Gay-Lussac's Law.

By experiment:
1 volume of hydrogen combines with 1 volume of chlorine to give 2 volumes of hydrogen chloride.

Applying Avogadro's Hypothesis we can say:
n molecules of hydrogen combine with n molecules of chlorine to give $2n$ molecules of hydrogen chloride.

Simplifying:
1 molecule of hydrogen combines with 1 molecule of chlorine to give 2 molecules of hydrogen chloride.

1 molecule of hydrogen chloride must contain $\frac{1}{2}$ molecule of hydrogen; therefore 1 molecule of hydrogen must contain at least 2 atoms of hydrogen.

Chemical evidence such as the fact that hydrochloric acid only forms one series of salts (unlike sulphuric acid which forms two), indicates that hydrogen chloride contains 1 atom of hydrogen per molecule and, therefore, 1 molecule of hydrogen must contain 2 atoms.

By a somewhat similar argument it can be shown that the molecule of chlorine contains 2 atoms, and the reaction between hydrogen and chlorine may be diagrammatically represented by Figure 1.2 at constant temperature and pressure.

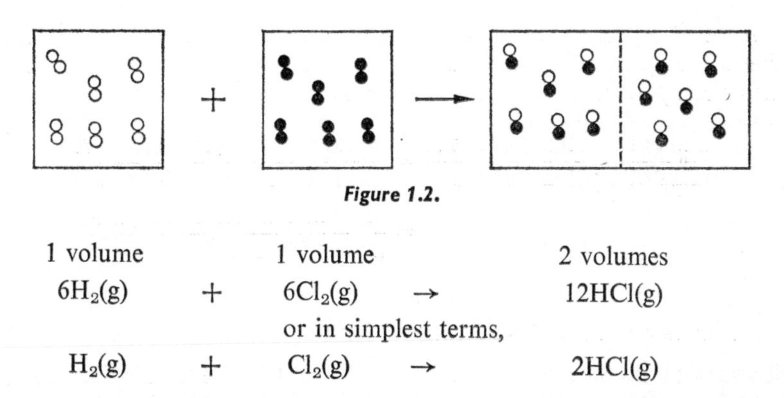

Figure 1.2.

1 volume		1 volume		2 volumes
$6H_2(g)$	+	$6Cl_2(g)$	→	$12HCl(g)$
		or in simplest terms,		
$H_2(g)$	+	$Cl_2(g)$	→	$2HCl(g)$

Thus it can be seen that the application of Avogadro's Hypothesis has produced a theory for this reaction which is consistent with the experimental evidence.

Since their molecules contain 2 atoms each, hydrogen and chlorine are said to be *diatomic*, the term atomicity being defined as follows:

The **atomicity** of an element is the number of atoms contained in 1 molecule of the element.

The importance of Avogadro's Hypothesis

The importance of the hypothesis lies in the fact that, since it asserts that equal volumes of gases contain equal numbers of molecules, it enables us to change over directly from a statement about volumes of gases to the same statement about molecules of gases. Every time we make a statement about one volume of a gas, we are also making a statement about a certain number of molecules of it, and that number, by Avogadro's Hypothesis, is always the same, no matter what the gas may be. Consequently we can change over at will, in any statement about gases, from volumes to molecules and *vice versa*, if the temperature and pressure are constant.

This means that by applying the hypothesis to volume measurements of gases, we can probe right to the heart of a chemical reaction, to the actual molecules themselves. It is an enormous step to change directly from an experimental statement like:

2 volumes of hydrogen combine with 1 volume of oxygen giving 2 volumes of steam (temperature and pressure constant)

to:

2 molecules of hydrogen combine with 1 molecule of oxygen giving 2 molecules of steam.

Another aspect of the link between volumes and molecules of gases is that if the masses of equal volumes of two gases are compared, then it is also a comparison of the masses of one molecule of each of the gases, i.e.

$$\frac{\text{mass of 1 volume of gas A}}{\text{mass of 1 volume of gas B}} = \frac{\text{mass of } n \text{ molecules of gas A}}{\text{mass of } n \text{ molecules of gas B}}$$

$$= \frac{\text{mass of 1 molecule of gas A}}{\text{mass of 1 molecule of gas B}}$$

Cannizzaro and Relative Atomic Masses

It was not until the work of Cannizzaro (1858) that the full significance of the above statements was realized and that Avogadro's Hypothesis was generally accepted. Cannizzaro recognized the possibility of comparing the molecular masses of gases by comparing their densities. Using hydrogen as the reference element, he deduced a relationship between the density of a gas, relative to hydrogen (relative vapour density), and its molecular mass relative to the mass of one atom of hydrogen (relative molecular mass).

$$\text{Relative vapour density} = \frac{\text{mass of a volume of gas}}{\text{mass of equal volume of hydrogen}}$$

at constant temperature and pressure.

Applying Avogadro's Hypothesis, we can say directly:

$$\text{relative vapour density} = \frac{\text{mass of 1 molecule of gas}}{\text{mass of 1 molecule of hydrogen}}$$

But hydrogen is diatomic, and so

$$\text{relative vapour density} = \frac{\text{mass of 1 molecule of gas}}{\text{mass of 2 atoms of hydrogen}}$$

$$2 \times \text{relative vapour density} = \frac{\text{mass of 1 molecule of gas}}{\text{mass of 1 atom of hydrogen}}$$

$$= \text{relative molecular mass of the gas}$$

i.e. **the relative molecular mass of a gas is twice its relative vapour density.** Therefore, whenever it is possible to determine experimentally the relative vapour density of a gas (see page 125 for experimental methods), it is possible to determine its relative molecular mass. Cannizzaro used the relative molecular masses of gases, determined in this manner, in his very important method for the determination of relative atomic masses. He realized that if he selected a sufficiently large number of compounds of a particular element, then it was likely that at least one of those compounds would contain only one atom of the element per molecule. He then employed the following steps in order to determine the relative atomic mass of the element.

1. He determined the relative vapour density of each compound containing the element.
2. He doubled the relative vapour density and thus obtained the relative molecular mass of each compound.
3. He analysed each compound for the percentage of the element by mass.
4. Using the results from 2 and 3 he calculated the mass of the element in one relative molecular mass of each compound.

The table on page 12 illustrates application of the method to carbon. The figures in the last column correspond to the presence of one, two, three or more carbon atoms per molecule. The lowest mass is 12 and the others are multiples of 12. Now it is obvious that if the list contains any compound containing only one carbon atom per molecule, that compound will be the first, methane, because in this the mass of carbon is the least. If, therefore, the molecule of methane does actually contain only one carbon atom, the relative atomic mass of carbon is 12. This process has been applied to a very large number of carbon compounds, and the mass of carbon in the relative molecular mass has always been found to be 12, or a multiple of 12, but never less. From this it is concluded that the least mass of carbon there can ever be in the relative molecular mass of one of its compounds is 12, that this mass corresponds to the presence of one carbon atom, and that the relative atomic mass of carbon is 12.

The following table illustrates the application of the method to carbon.

Compound	*Relative vapour density by experiment)*	*Relative molecular mass (=2 × vapour density)*	*Percentage of carbon by mass (by experiment)*	*Mass of carbon in the molecular mass*
Methane	8	16	75.0	$75.0 \times \frac{16}{100} = 12$
Ethane	15	30	80.0	$80.0 \times \frac{30}{100} = 24$
Propane	22	44	81.8	$81.8 \times \frac{44}{100} = 36$
Ethene (ethylene)	14	28	85.7	$85.7 \times \frac{28}{100} = 24$
Ethyne (acetylene)	13	26	92.3	$92.3 \times \frac{26}{100} = 24$

The method can be applied to determine the relative atomic masses of any element forming a large number of gaseous or easily vaporized compounds. It thus became possible, for the first time, to determine reasonably accurate values for the relative atomic masses of such important elements as oxygen, carbon, nitrogen, and sulphur. In effect, the work of Gay-Lussac, Avogadro, and Cannizzaro had overcome the problem of not knowing the ratios in which atoms of elements such as these combine. It was the lack of knowledge of these ratios which had caused so much trouble to Dalton and others in the early part of the nineteenth century (see page 7).

When comparing the combining ratios of different elements it was found to be useful to employ hydrogen as the reference element. Thus if an element of symbol M combined with hydrogen to give a compound of formula MH_n, then n was known as the combining power or valency of the element. If the relative atomic mass of M is A, it follows that in this compound A g of M must combine with n g of hydrogen (since the relative atomic mass of hydrogen is, by definition, 1). Therefore 1 g of hydrogen must combine with A/n g of M and this value is, by definition, the equivalent mass of M. This leads to the relationship

$$\text{relative atomic mass} = \text{equivalent mass} \times \text{valency}$$

or

$$A = A/n \times n$$

Thus if the valency of an element can be determined, the above relationship can be used to convert the equivalent mass into the relative atomic mass. However the relationship must be applied with caution, as certain elements can exhibit more than one combining power. Therefore it is essential that the combining power used is the same as that exhibited by the element in the compound from which its combining mass is determined.

Other Methods for the Determination of Relative Atomic Masses

1. *Using Dulong and Petit's Law*

The above relationship was used, along with the experimental law of Dulong and Petit (1820), to determine the relative atomic masses of certain metallic elements. Dulong and Petit observed that a relationship existed between the relative atomic masses and specific heat capacities of most solid metallic elements. The **Dulong and Petit's Law**, stated in a modern form, is:

> For solid elements near room temperature, relative atomic mass × specific heat capacity is approximately constant at 27 J K^{-1}.

This law has been used to determine relative atomic masses of metals in the following way.

Consider indium as a newly discovered metal. Its equivalent mass was found to be 37.8; its specific heat capacity (by experiment) was 0.24 J g^{-1} K^{-1}. By Dulong and Petit's Law,

$$\text{relative atomic mass} \times 0.24 = 27 \text{ (approx.)}$$

i.e.

$$\text{relative atomic mass} = \frac{27}{0.24} = 110 \text{ (approx.)}$$

But

$$\text{relative atomic mass} = \text{equivalent mass} \times \text{valency}$$

$$\therefore \text{valency} = \frac{110}{37.8} = 3 \text{ (must be an integer)}$$

$$\begin{aligned}\text{accurate relative atomic mass} &= \text{equivalent mass} \times \text{valency}\\ &= 37.8 \times 3\\ &= 113.4\end{aligned}$$

Any divergence of the valency from an integer arises from the fact that the law of Dulong and Petit is only approximate and can only give an approximate value for a relative atomic mass.

2. *Using the Law of Isomorphism*

The phenomenon of isomorphism has proved useful for determining relative atomic masses and, more particularly, for correcting them. As first put forward in 1819 by Mitscherlich, the **Law of Isomorphism** stated:

> Substances of similar chemical character, having equal numbers of atoms combined in a similar way, show identity of crystalline form.

It is now known that this statement is too precise. No two substances have exactly the same crystalline form. In cases of general similarity of crystalline form, there are always minute angular differences. The law is, however, quite useful if it is taken in the looser form of an assertion such as:

> Substances of corresponding chemical composition often show similarity in crystalline form.

That is, such substances are often isomorphous, showing at least two of the following characteristics: a general similarity of crystalline shape, formation of mixed crystals, formation of overgrowths on each others' crystals.

Isomorphism has proved useful in cases such as the following. In the early nineteenth century silver was taken to be divalent, its sulphide was taken to be AgS, and its relative atomic mass to be 216 (H = 1). Dumas showed copper(I) sulphide, Cu_2S, and silver sulphide to be isomorphous. The formula of silver sulphide was then corrected to Ag_2S and the relative atomic mass of silver to 108.

The atomic mass of selenium was determined from the fact that potassium sulphate and potassium selenate are isomorphous. By analogy with the sulphate, K_2SO_4, the selenate must be represented as K_2SeO_4. The percentage of selenium in the selenate was found to be 35.75 per cent. From this,

$$\frac{\mathrm{Se}}{\mathrm{K_2SeO_4}} \times 100 = 35.75$$

Inserting the relative atomic masses K = 39 and O = 16, we have

$$\frac{\mathrm{Se}}{\mathrm{Se} + 142} \times 100 = 35.75$$

Solution of this equation yields the result Se = 79.

3. *Using a mass spectrometer*

A mass spectrometer provides the most accurate method for determining relative atomic masses and its use has now superseded all other methods.

After some initial experiments by Thompson, the first mass spectrometer was constructed by Aston in 1919. Aston's mass spectrograph and more modern versions of the instrument are discussed in some detail on page 30.

A summary of the Development of the Methods for the Determination of Relative Atomic Masses

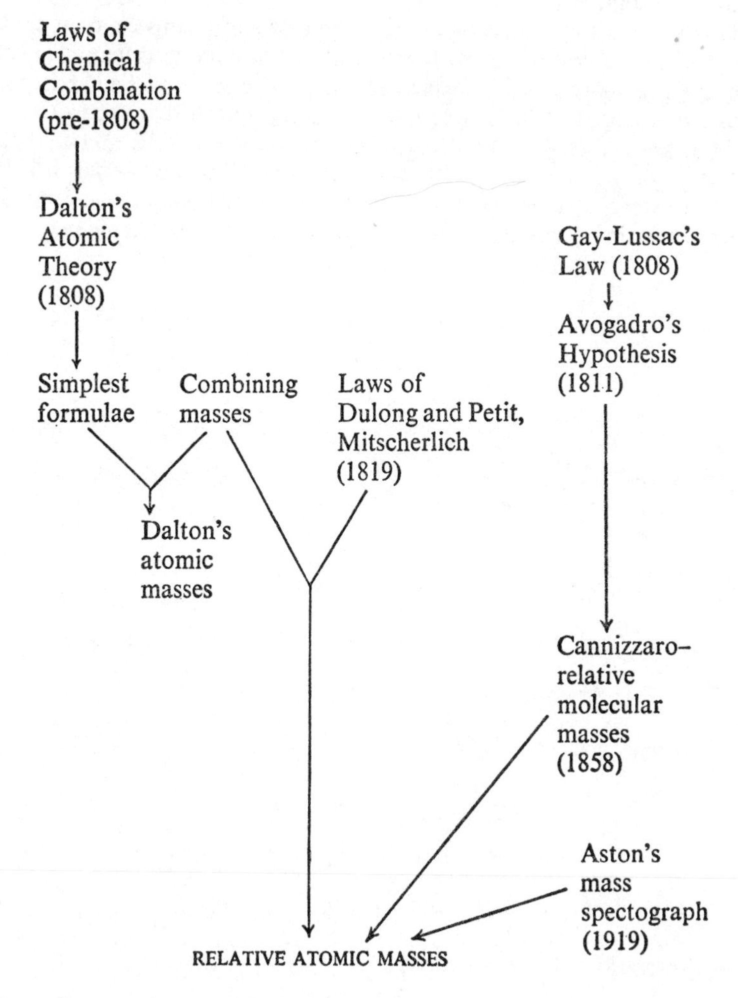

Lists of approximate and accurate values of relative atomic masses are given on page 308.

Standards for Relative Atomic Masses

The accepted reference element for the determination of relative atomic masses has varied. In the early days, hydrogen, being the element with atoms of least mass, seemed to be the obvious choice. Towards the end of the nineteenth century, the methods used for determining relative atomic masses largely depended on finding the combining masses of elements. Therefore, as more elements combine directly with oxygen than with hydrogen, it was decided to fix the atomic mass of oxygen at 16 and to use this as the standard.

Finally, with the realization, during the first half of this century, that a lot of elements consist of mixtures of atoms of slightly different masses (isotopes, see page 29), it became necessary to select a particular isotope as the standard. In 1961 it was internationally agreed that the isotope of carbon, known as ^{12}C, should have an atomic mass of 12 and that this isotope should always be used as the standard. ^{12}C was chosen as it is the most convenient standard, from an experimental point of view, to use when determining relative atomic masses with a mass spectrometer.

2. Moles and Molarity

The availability of a reliable set of relative atomic masses has enabled chemists to use reacting masses to establish formulae for compounds and equations for reactions. Also, once an equation is known, it is possible to predict reacting masses and the masses of products. In their modern form, these calculations involve the unit amount of substance known as a **mole.**

The approximate relative atomic mass of oxygen is 16 on the H = 1 scale, which means that each atom of oxygen is 16 as heavy as atoms of hydrogen (ignoring, for the moment, the existence of isotopes). It follows from this that any masses of these elements, which are in the ratio of 16:1, must contain equal numbers of atoms. For example, 16 g of oxygen must contain the same number of atoms as 1 g of hydrogen. By similar arguments, it is possible to see that the relative atomic mass of any element, expressed in grams (a g-atom), will contain the same number of atoms as 1 g of hydrogen.

This constant number of atoms is called the **Avogadro Constant** and is given the symbol L. The amount of an element containing the Avogadro Constant number of atoms is called 1 mole of atoms of the element.

In addition to atoms, chemical reactions can involve molecules of elements, molecules of compounds, ions, or electrons. It is therefore useful to extend the definition of a mole to cover these other species. For example, 1 mole of oxygen molecules (O_2) contains 2 moles of oxygen atoms and has a mass of 32 g.

The modern **definition of a mole** on the $^{12}C = 12$ scale is:

1 mole is the amount of a substance containing the same number of particles (atoms, molecules, ions etc.) as there are atoms present in 12 g of the isotope of carbon, ^{12}C.

This number, which is the accepted value for the Avogadro Constant, is 6.02×10^{23}, e.g. 6.02×10^{23} molecules of carbon dioxide (CO_2) contain 6.02×10^{23} atoms of carbon (1 mole–mass 12 g) and $2 \times 6.02 \times 10^{23}$ atoms of oxygen (2 moles–mass 32 g). Therefore 1 mole of carbon dioxide has a mass of $12 + 32 = 44$ g.

The mass of 1 mole of a substance is thus the formula mass of the substance.

The following table gives a few simple examples of the relationships between *masses* and *numbers of moles.*

Formula	*Mass of 1 mole*	*Mass of substance involved*	*Number of moles involved*
Na (atoms)	23 g	57.5 g	2.5
Cl_2 (molecules)	73 g	7.3 g	0.1
H_2O (molecules)	18 g	0.9 g	0.05
NaCl (formula units)	58.5 g	11.7 g	0.2
CO_3^{2-} (ions)	60 g	1.8 g	0.03

When referring to 1 mole of a compound which exists in the form of a giant lattice (see page 86), it is misleading to refer to 1 mole of molecules, as individual molecules (represented, for example, by the formula NaCl) do not exist. A more satisfactory term is 1 mole of 'formula units'.

A large proportion of chemical reactions involve aqueous solutions of reagents and it is useful to express the concentration of these solutions in terms of the number of moles of solute dissolved in 1 dm^3 of solution. This is known as the **molarity** of the solution.

The simple examples in the following table will serve to illustrate the relationships between mass dissolved, volume of solution, and molarity of solution.

Formula	*Mass of 1 mole*	*Mass of solute dissolved*	*Volume of solution*	*Molarity of solution*
NaOH	40 g	4.0 g	1000 cm^3	$\frac{4.0}{40} = 0.1$ M
H_2SO_4	98 g	4.9 g	250 cm^3	$\frac{4.9 \times 4}{98} = 0.2$ M
Na_2CO_3	106 g	15.9 g	2000 cm^3	$\frac{15.9}{2} \times \frac{1}{106} = 0.075$ M

Due to the direct relationship between moles and numbers of particles (atoms, molecules, or ions), the mole concept is of fundamental importance. It is obviously of more interest to know the numbers of particles involved in a reaction, rather than just the reacting masses. The following worked examples illustrate some of the more common uses of the mole in quantitative chemistry.

To Calculate the Formula of a Compound from its Composition by Mass

Hydrated magnesium sulphate contains 9.8 per cent magnesium, 13.0 per cent sulphur, 26.0 per cent oxygen, and 51.2 per cent water of crystallization. (Relative atomic masses, Mg = 24, S = 32, O = 16, H = 1.)

	Mg	S	O	H_2O
Percentage by mass	9.8	13.0	26.0	51.2
Number of moles	$\frac{9.8}{24}$	$\frac{13.0}{32}$	$\frac{26.0}{16}$	$\frac{51.2}{18}$
	= 0.408	= 0.406	= 1.63	= 2.84
Divide each by the smallest	$\frac{0.408}{0.406}$	$\frac{0.406}{0.406}$	$\frac{1.63}{0.406}$	$\frac{2.84}{0.406}$
	= 1	= 1	= 4	= 7

The formula of hydrated magnesium sulphate is $MgSO_4,7H_2O$. This calculation gives the *empirical formula* of the compound, i.e. the simplest formula which expresses its composition by mass. In some cases the molecular formula of a compound is not the same as its empirical formula. This problem mainly arises with organic compounds. To decide this point it is necessary to find the relative molecular mass of the compound. For example, the empirical formula of ethane can be shown to be CH_3 and, as its relative molecular mass is found to be 30, its molecular formula is C_2H_6.

To Determine the Equation for a Reaction

By titration, 25.0 cm^3 of a 0.1 M solution of sodium hydroxide was found to react completely with 28.2 cm^3 of a solution of sulphamic acid (NH_3SO_3) containing 8.54 g of the acid per dm^3 of solution. (Relative atomic masses, N = 14, H = 1, S = 32, O = 16.)

To find the number of moles of NaOH in 25.0 cm^3 of solution:

1000 cm^3 of solution contains 0.1 moles

$$25.0 \text{ cm}^3 \text{ of solution contains } 0.1 \times \frac{25}{1000} = 0.0025 \text{ moles}$$

To find the number of moles of sulphamic acid in 28.2 cm^3 of the solution:

The mass of 1 mole of the acid is $14 + 3 + 32 + 48 = 97$ g

$$1000 \text{ cm}^3 \text{ of the acid contains } \frac{8.54}{97} \text{ moles}$$

$$28.2\ cm^3\ \text{contains}\quad \frac{8.54}{97} \times \frac{28.2}{1000}$$

$$= 0.00248\ \text{moles}$$

From this it can be seen that 1 mole of sulphamic acid will react completely with 1 mole of sodium hydroxide and that the left-hand side of the equation is

$$NH_3SO_3(aq) + NaOH(aq) \rightarrow$$

The products are likely to be a salt plus water and the complete equation for the reaction is probably

$$NH_3SO_3(aq) + NaOH(aq) \rightarrow NaNH_2SO_3(aq) + H_2O(l)$$

Gravimetric Calculation From an Equation

Calculate the mass of nitric acid necessary to react with 10 g of lead(II) oxide and the mass of lead(II) nitrate formed in the process. (Relative atomic masses, Pb = 207, N = 14, O = 16, H = 1.)

The equation for the reaction is

$$PbO(s) + 2HNO_3(aq) \rightarrow Pb(NO_3)_2(aq) + H_2O(l)$$

From the equation,

1. 1 mole of lead(II) oxide reacts with 2 moles of nitric acid.
   The mass of 1 mole of lead(II) oxide = 207 + 16 = 223 g
   The mass of 1 mole of nitric acid = 1 + 14 + 48 = 63 g
   Therefore 223 g of lead(II) oxide reacts with 2 × 63 g of nitric acid
   and 10 g of lead(II) oxide reacts with $2 \times 63 \times \frac{10}{223}\ g = 5.65\ g$
2. 1 mole of lead(II) oxide produces 1 mole of lead(II) nitrate.
   The mass of 1 mole of lead(II) nitrate = 207 + 28 + 96 = 331 g
   Therefore 223 g of lead(II) oxide produces 331 g of lead(II) nitrate
   and 10 g of lead(II) oxide produces $331 \times \frac{10}{223}\ g = 14.8\ g$

The examples in this chapter illustrate the fundamental importance of the mole as a unit amount of substance. As previously stated, chemists are interested in the numbers of particles involved in reactions, and their way of counting particles is to convert the masses of solids, or the volumes of solutions of known concentration, or the volumes of gases at known temperatures and pressures, into numbers of moles.

3. Fundamental Particles and Their Arrangement in Atoms

Dalton regarded atoms as ultimate particles of matter, discrete and indivisible, but by the end of the nineteenth century this idea had been abandoned. It is now believed that atoms are made up of certain fundamental particles. In order to account for the properties of matter which are of interest to a chemist, it is only necessary to consider three fundamental particles – the **electron**, the **proton**, and the **neutron**.

The Electron

In the late nineteenth century a great deal of work was carried out on the effects of electrical discharges through gases at very low pressures of the order of 1 mmHg (133 N m^{-2}) or less (see Figure 3.1). One of the results of this work was the discovery of 'cathode rays'. These rays were observed as a faintly luminous beam which caused a fluorescence when it hit the glass wall of the discharge tube.

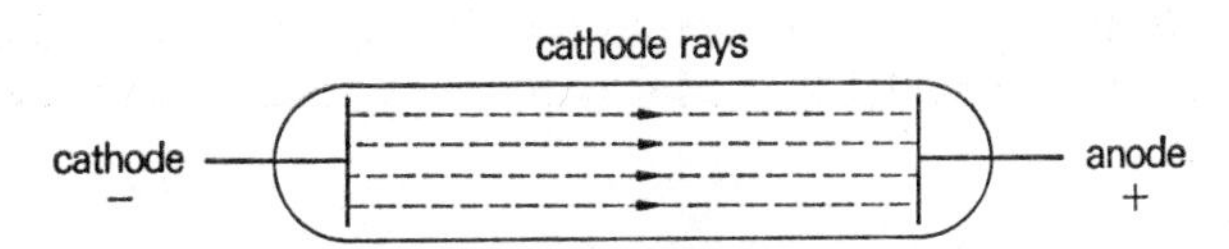

Figure 3.1. Production of cathode rays in a discharge tube

Cathode rays were investigated and found to have the following properties:

1. They emerge from the cathode at right angles, travel in straight lines, and cast a shadow of an object placed in their path.
2. They are deflected by an electrostatic field away from the negative plate, proving that the rays carry a negative charge.
3. They are deflected by magnetic fields.
4. They can cause a small paddle wheel to rotate by exerting a mechanical pressure.

From a consideration of these facts, it was concluded that cathode rays consist of a stream of negatively charged particles, to which the name **electrons** was given. In 1897, J. J. Thomson, by measuring the angles through which the rays were deflected by known magnetic and electric fields, succeeded in determining a value for the ratio of charge to mass (e/m) for the electron. It was found that the ratio e/m was constant irrespective of the nature of the cathode, the nature of the residual gas, or the voltages employed. Thus it was considered that the electron is a fundamental particle common to all atoms.

About 1910, Millikan determined the charge on the electron by observing oil drops which were charged by the capture of gaseous positive ions. These ions were produced by a beam of X-rays (see page 26) which removed electrons from air molecules. By recording the movement of the charged oil drops in a known electric field, he was able to establish that the smallest charge on an oil drop was 1.6×10^{-19} coulomb and that all other charges were simple multiples of this value. This charge, 1.6×10^{-19} coulomb, was taken to be the charge on the electron, and together with the e/m value it showed that the mass of an electron is about 1/1840 of the mass of a hydrogen atom.

The electron, however, cannot be regarded as a particle without qualification, as it can exhibit properties of a wave-like character; for example, a stream of electrons can be diffracted by crystals, which act as diffraction gratings (see page 64).

The Proton

As early as 1886 it was realized that as negative particles were observed moving away from the cathode in a discharge tube, it was likely that positive particles were moving towards the cathode. These were observed and investigated by allowing them to pass through a perforated cathode. The arrangement required to produce a beam of positive 'rays' is shown in Figure 3.2.

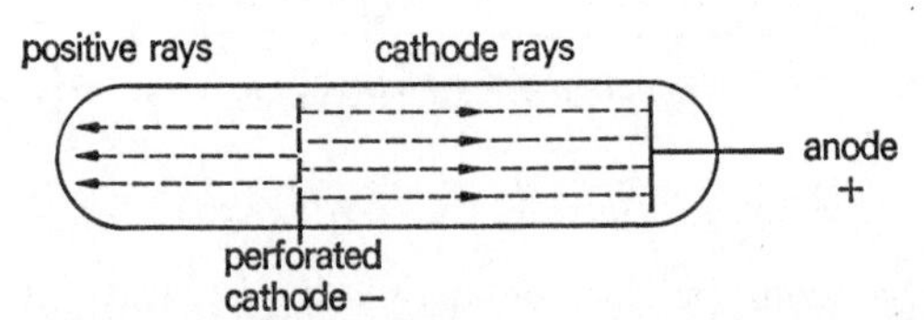

Figure 3.2. Production of positive rays in a discharge tube

The e/m ratio for positive rays was determined by Thomson in 1912 and it was found that the values obtained did depend on the residual gas. It was concluded from this that the rays were positive ions formed when the highly energetic electrons travelling from the cathode collided with the molecules of gas left in the discharge tube. These collisions resulted in the breakdown of some molecules into atoms and the removal of one or more electrons to form positive ions.

The largest *e*/*m* value was obtained with hydrogen as the residual gas, which indicated that the hydrogen ion had the smallest mass. The masses of ions from other gases were found to be approximately whole number multiples of the mass of the hydrogen ion. Thus the hydrogen ion, formed by the loss of one electron from a hydrogen atom, $H - e^- \rightarrow H^+$, was thought to be a fundamental particle present in all atoms and it was called **a proton.**

The Neutron

The existence of a neutral fundamental particle was predicted by Rutherford in 1920, but it was not until 1932 that the neutron was discovered. The experimental work which led to the discovery involved the use of radioactive materials (see page 37). A radioactive substance contains atoms which are unstable and emit radiation spontaneously. One type of radiation consists of alpha-particles. Rutherford showed that alpha-particles consist of helium ions, He^{2+}, of relative mass 4.

In 1932 it was discovered that several of the lighter atoms, and in particular the atoms of beryllium and boron, give out a very penetrating radiation if they are subjected to the action of alpha-particles emitted from polonium. Chadwick put forward the view that the radiation consisted of fast-moving particles comparable in mass with hydrogen atoms and possessing no electrical charge. This last factor accounts for their penetrating power, as they are little affected by either positively or negatively charged particles in the atoms which they encounter. The name **neutron** was given to this type of particle.

The origin of the neutron in the above process appears to be that the alpha-particles of relative mass 4 are captured by boron atoms of relative mass 11. These are converted to nitrogen of relative mass 14 with emission of a neutron of relative mass 1. Using modern nomenclature (see page 30) this nuclear reaction may be represented by the equation

$$^{4}_{2}He + ^{11}_{5}B \rightarrow ^{14}_{7}N + ^{1}_{0}n$$

The characteristics of the three fundamental particles are summarized in the following tables:

	Approx. relative mass	*Relative charge*
Electron	$\frac{1}{1840}$	−1
Proton	1	+1
Neutron	1	no charge

The Distribution of Particles in the Atom

As early as 1906, Rutherford noticed that if alpha-particles pass first through a very thin sheet of metal (about one-millionth of a centimetre thick) and then on to a photographic plate, the effect produced on the plate is not sharply defined but is diffuse at the edges. This was attributed to 'scattering' of the alpha-particles by some action of the metallic atoms in the sheet through which they had passed. Three years later Geiger and Marsden, under the guidance of Rutherford, studied this scattering more carefully and showed that while most of the alpha-particles were only slightly affected by a sheet of gold of thickness 6×10^{-5} cm, about one particle in eight thousand was scattered through an angle of 90° or more from its original direction. Rutherford later commented that, considering the high velocity and mass of the alpha-particle and the thickness of the metal used, this result was 'about as credible as if you fired a fifteen-inch shell at a piece of tissue paper and it came back and hit you'.

In 1911, on the basis of Geiger and Marsden's work, Rutherford put forward his nuclear theory of the atom. He showed that it was extremely improbable that the large deflections of alpha-particles could arise from a succession of small deflections in the same direction. These large deflections require the assumption of a region in the atom which fulfils the following conditions:

1. It must contain a positive charge because the positive alpha-particles are repelled by it.
2. It must be relatively massive and highly charged because the deflection of the alpha-particles is so great.
3. It must occupy a very small space because most alpha-particles are little affected and very few are repelled to any marked extent.

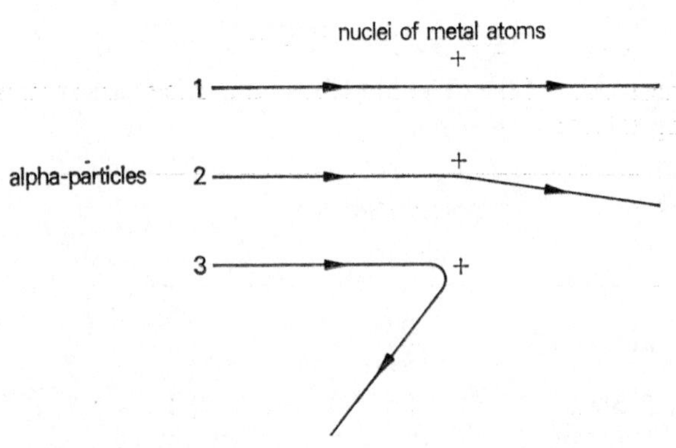

Figure 3.3 Scattering of alpha-particles by atomic nuclei

This is the postulate of the atomic *nucleus*, very small relative to the whole bulk of the atom, massive and positively charged. The general idea is illustrated in Figure 3.3. In the figure, alpha-particle 2 suffers slight scattering by distant effects of the nuclei. Particle 3 is one of the small minority which suffer a large deflection by close encounter with a nucleus.

Rutherford was able to derive an equation relating the scattering to the thickness of the metal, the charge on the nuclei of the metallic atoms, and the velocity of the alpha-particles, but the war of 1914 intervened and it was not until its close that the equation could be accurately tested.

The Wilson Cloud-chamber

In 1912, Wilson, using his cloud-chamber, obtained evidence which supported Rutherford's nuclear theory. The cloud-chamber consisted of a cylindrical glass container in which was air saturated with water vapour (or, better, with water vapour and alcohol vapour). The base of this cylinder (in later models) consisted of a thin rubber diaphragm fixed at its edges and kept taut by air pressure on the side away from the cylinder. By automatic devices it was arranged that a valve could operate to reduce the air pressure supporting the diaphragm so that rapid expansion occurred in the cloud chamber while, immediately afterwards, alpha-particles were allowed to enter the chamber. They ionized the gases through which they passed and the ions acted as centres of condensation for drops of liquid. By strong lateral illumination, two cameras (at right-angles to each other) then photographed these drops, which appeared as a continuous trail on the photographic plate. In this way, the tracks of alpha-particles could be accurately catalogued and studied in three dimensions.

Wilson found that the great majority of such tracks were straight lines, but a few showed forking into a longer and a shorter arm (see Figure 3.4).

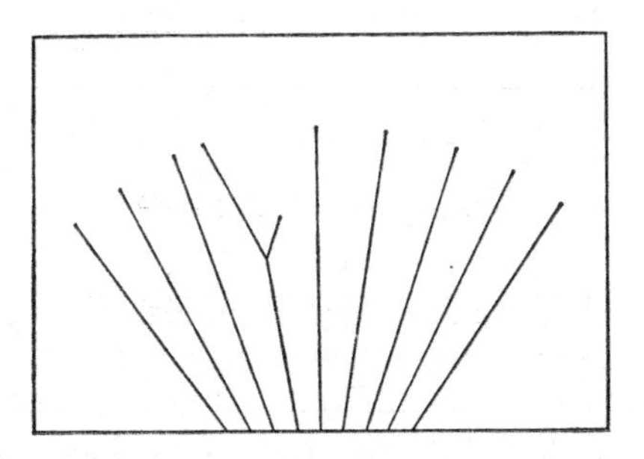

Figure 3.4. ***Tracks of alpha-particles***

This supported Rutherford's nuclear theory. For the straight tracks were those of the great majority of alpha-particles which passed through the gases, ionizing atoms by striking out electrons, but encountering no nuclei because of the minute size of the nuclei. The small minority of alpha-particles showing forked tracks underwent close encounter with a nucleus and were deflected into a changed track (long arm) while the ionized atom (or molecule) of the gas also showed an ionizing track (short arm).

Moseley's Experiments and Atomic Number

The work of von Laue at Zurich and the Braggs at Leeds had recently shown that crystals could act as diffraction gratings for X-rays. Moseley, working in Rutherford's laboratory at Manchester University in 1913, devised an apparatus in which cathode rays (i.e. electrons in very rapid motion) were allowed to fall on an anti-cathode containing the element under investigation. The element then produced its characteristic X-ray spectrum, containing (generally) two strong lines. Moseley's first set of results was obtained for the elements calcium to copper in the Periodic Table of Elements. He was able to show that the square roots of the characteristic X-ray frequencies of these elements were proportional to the atomic numbers of the elements, i.e. their ordinal number in the Periodic Table. The effect is called the 'Moseley stair-case' (see Figure 3.5).

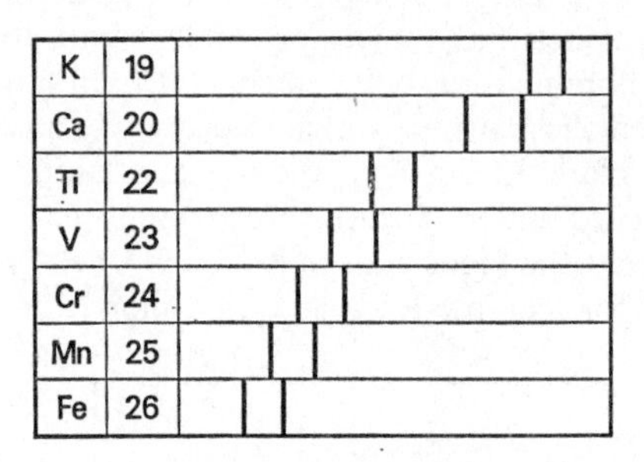

Figure 3.5. X-ray spectra

In Moseley's words, 'we have here a proof that there is in the atom a fundamental quantity that increases by regular steps as we pass from one element to the next. This quantity can only be the charge on the atomic nucleus.' He went on to argue that, on average, atomic masses increase by about two units from element to element in the Periodic Table. So, the number of charges on the nucleus being about half the relative atomic mass, it is very probable that in passing from one element to the next, one proton is added to the nucleus. Moseley was able to assert on these grounds that all possible elements up to aluminium were known and that, between aluminium and gold, only four elements remained to be discovered.

After the war Chadwick (1920) returned to the alpha-particle scattering experiments and was able to show, with a probable error of between 1 and 2 per cent, that the nuclei of copper, silver, and platinum carried positive charges of 29.3, 46.3, and 77.4 units respectively. The atomic numbers of these elements in the Periodic Table are 29, 47, and 78, leaving little doubt that the number of protons in the nucleus of these atoms is the same as the ordinal number of the element in the Periodic Table. Similar results were later obtained by various workers for aluminium, magnesium, and gold.

This number – the ordinal number of the element in the Periodic Table, the number of protons in the atomic nucleus and, since the atom is electrically neutral, the number of electrons outside the nucleus – is called the *atomic number* of the element and is usually denoted by the symbol Z. For lighter elements, the atomic number of an element is about half its relative atomic mass, as shown by the following table.

	Relative atomic mass	*Atomic number*
Oxygen	16	8
Sodium	23	11
Aluminium	27	13

It was to account for this difference that Rutherford in 1920 predicted the existence of a neutral fundamental particle. As previously described (page 23) this particle, the neutron, was discovered in 1932 and was found to have a relative mass almost equal to that of a proton. Thus, the nucleus of an atom of relative atomic mass A and atomic number Z contains Z protons and $(A - Z)$ neutrons.

The general result so far may be summarized in the following way. All atoms contain a nucleus which is very small, both absolutely and relative to the size of the complete atom. The nucleus is made up of protons equal in number to the ordinal number of the element in the Periodic Table, and of neutrons (except the hydrogen nucleus which is a single proton). The nucleus is positively charged by the protons present. For the lighter atoms the numbers of protons and neutrons are about equal, but in the heavier atoms the number of neutrons increases more rapidly than the number of protons. For example, if the relative atomic mass of uranium is taken as 238, its atoms must contain 146 neutrons and 92 protons. A few typical nuclei are given in the table on page 28.

	Atomic number	Nucleus		Relative atomic mass
		Protons	Neutrons	
Fluorine	9	9	10	19
Sodium	11	11	12	23
Aluminium	13	13	14	27

Atoms are neutral and so the number of electrons around the nucleus must be equal to the number of protons in the nucleus. The more detailed arrangement of the electrons around the nucleus will be discussed in Chapter 5.

Since the proton and the neutron both have a relative mass very close to unity, it would be expected that the relative atomic masses of all elements should be integral (or very nearly so). This is in fact the case for a large number of elements. The exceptions, like chlorine (relative atomic mass 35.5), arise from the fact that the naturally occurring element consists of a mixture of atoms of different masses. Some chlorine atoms have a relative atomic mass of 35 and some, 37. This phenomenon, called *isotopy*, which arises from differing numbers of neutrons in the two nuclei, will be discussed in the next chapter.

4. The Nucleus

Isotopy

Atomic nuclei are made up of protons and neutrons. The protons must be accompanied by an equal number of extra-nuclear electrons to produce an electrically neutral atom, but a neutron contributes no charge and requires no such balancing by an electron. It is, therefore, theoretically possible for two or more atoms to exist possessing the same number of protons on the nucleus, the same number and arrangement of electrons outside the nucleus, and, therefore, the same chemical properties (see Chapter 5), but having different numbers of neutrons in the nucleus and therefore different relative atomic mass.* Such cases are very common and the phenomenon is known as **isotopy**. It may be formally defined as follows:

> **Isotopy** is the occurrence of atoms with different relative atomic mass but the same atomic number, and chemical properties which are identical (or nearly so).

The idea of isotopy may be illustrated by the element chlorine, which has two principal isotopes, ^{35}Cl and ^{37}Cl.

^{35}Cl			^{37}Cl		
Nucleus: Protons	17	35	Protons	17	37
Neutrons	18		Neutrons	20	
Atomic number	17		Atomic number	17	
Electrons	17		Electrons	17	

The identical atomic numbers and electron arrangements produce the same chemical properties in the two isotopes, but the difference of two neutrons in the nucleus produces a difference of two units of relative atomic mass. The ordinary element, chlorine, has a relative atomic mass of 35.45, showing that it contains about 75 per cent of $^{35}_{17}Cl$ and 25 per cent of $^{37}_{17}Cl$ by mass.

* Or mass numbers (see below).

The numerical convention used in connection with isotopes is the following. Consider the expression $^{a}_{b}X$. X is the symbol representing one atom of the isotopic element; *a* and *b* are figures written in front of the symbol. The upper figure, *a*, is called the *mass number* of the isotope and is the sum of the number of protons and the number of neutrons in the nucleus of the atom. The lower figure, *b*, is called the **atomic number** of the element, X, and is the number of protons in the nucleus.

The relative atomic mass of the naturally occurring element is the weighted mean of the mass numbers of the isotopes present, i.e. a mean value which takes into account the relative abundance of each isotope in the naturally occurring element.

Most elements exhibit isotopy. A few do not, having only a single type of atom, notable examples being fluorine, sodium and aluminium. The record number appears in tin, which has ten isotopes of mass numbers from 112 to 124 (atomic number 50). Many elements have more than six isotopic forms. It is a notable fact that elements of odd atomic number never show more than two isotopes, while elements of even atomic number show much more extensive isotopy, e.g.

Element	*Atomic number*	*Isotopes*
Ag	47	Two (107, 109)
Cd	48	Eight (106–116)
In	49	Two (113, 115)
Sn	50	Ten (112–124)
Sb	51	Two (121, 123)
Te	52	Eight (120–130)

Evidence for the existence of isotopes

1. *Evidence from the mass spectrometer*

In 1919 Aston developed his mass spectrograph with which he analysed the positive rays from neon and found that there were two ions present with relative masses of 20 and 22 respectively. The modern mass spectrometer has been developed from Aston's mass spectrograph and its essential components are indicated in Figure 4.1.

The apparatus must be evacuated in order to avoid interference by the air molecules. A vaporized sample of the element is introduced at A and is subjected to a beam of electrons from the hot filament B. The electron beam removes one or more electrons from the atoms, thus producing positive ions. The positive ions are accelerated to the same velocity through a slit by an electric field, C, and then deflected by a magnetic

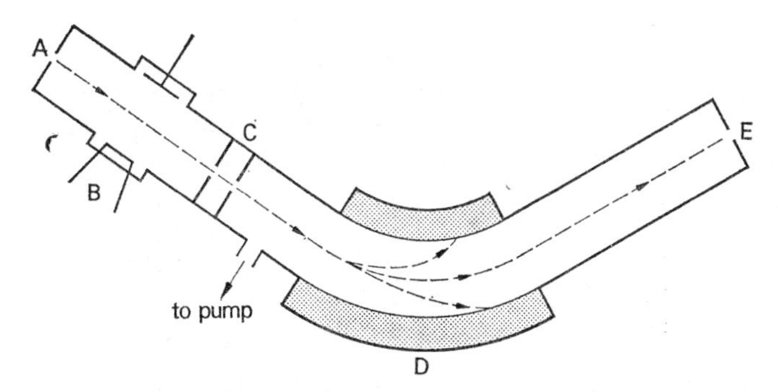

Figure 4.1. *The essential components of a modern mass spectrometer*

field, D. The angle of deflection of each ion depends on its ratio of charge to mass. If all the ions have single positive charges, the angles of deflection will be directly related to their mass numbers.

The strength of the magnetic field is varied so that each type of ion, in turn, is detected by a collector, E. When the ions strike the collector they produce an electric current, the magnitude of which is proportional to the abundance of the isotope which gives rise to the ions. The detector currents are recorded as a series of lines or peaks which is called a mass spectrum. Each line or peak represents one isotope and the height of the line or peak is proportional to the relative abundance of the isotope.

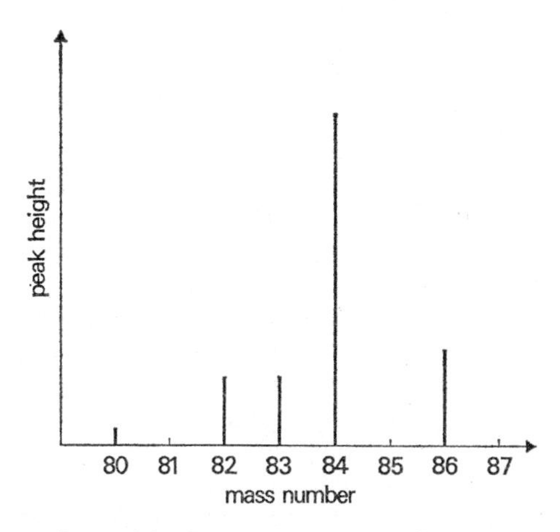

Figure 4.2. *The mass spectrum of krypton*

The mass spectrum of naturally occurring krypton, Figure 4.2, provides the information shown in the table on page 32.

Mass numbers of isotopes detected	Percentage abundance, (taking total line length as 100 per cent)	Relative mass of isotopes in 100 atoms of mixture
80	3.0	240
82	11.5	943
83	11.5	945.5
84	57.0	4788
86	17.0	1462
		Total 8387.5

The mean value per atom is therefore $\frac{8387.5}{100} = 83.87$

This weighted mean is the relative atomic mass of the naturally occurring element which, by this method, can be determined to a very high degree of accuracy.

2. *Evidence of isotopy from diffusion*

When experiments by Thomson, which were forerunners of the mass spectrograph, showed the probability of isotopy in neon ($^{20}_{10}Ne$ and $^{22}_{10}Ne$) Aston attempted their separation by diffusion experiments, based on **Graham's Law.** This states that:

> The rates of diffusion of gases are inversely proportional to the square roots of their densities (under comparable conditions).

The lighter neon isotope should, therefore, have a more rapid diffusion rate than the heavier, in the proportion of $\sqrt{22}:\sqrt{20}$. Aston was able to separate neon (obtained from air) into two fractions differing in mass number by more than 0.13 unit, which is greater than any possible experimental error. Similar diffusion experiments on hydrogen chloride (chief constituents $H^{35}_{17}Cl$ and $H^{37}_{17}Cl$) have produced similar results.

A similar diffusion technique was used on an enormous scale in 1944–5 to separate the fissionable isotope of uranium, $^{235}_{92}U$, from the isotope of mass number 238. The gaseous compound UF_6 was employed in the diffusions, and after separation was heated to generate the pure 235-isotope. This was then applied in the atomic bomb.

3. *Evidence from lead associated with radioactivity*

Radioactivity is a phenomenon resulting from spontaneous changes in the nuclei of the heaviest atoms (page 37). There are two main radioactive series, starting from uranium and thorium, respectively. Lead is the end-product of each and is generated by changes briefly summarized as:

$$\underset{(238)}{\text{Uranium}} \xrightarrow[\text{particles gives}]{\text{by loss of 3 alpha-}} \underset{(226)}{\text{Radium}} \xrightarrow[\text{alpha-particles gives}]{\text{which by loss of 5}} \underset{(206)}{\text{Lead}}$$

$$\underset{(232)}{\text{Thorium}} \xrightarrow{\text{by loss of 6 alpha-particles gives}} \underset{(208)}{\text{Lead}}$$

Each alpha-particle (He^{2+}) emitted represents a loss of four units of atomic mass so that, as the above figures indicate, lead associated in nature with uranium minerals would be expected to have a mass number of 206, while lead from thorium minerals would be expected to have a mass number of 208. Actual results obtained experimentally are:

Uranium minerals	*Mass number of lead*
Cleveite (Norway)	206.08
Pitchblende (W. Africa)	206.05
Thorium minerals	
Thorite (Ceylon)	207.8
Thorite (Norway)	207.9

These samples of lead differ in mass number by almost the two required units. They are indistinguishable chemically, having the same atomic number, 82, but differ in that the heavier isotope has two extra neutrons in its nucleus.

Such cases in nature are very rare. In a general way the relative atomic masses of elements are constant, which shows that the isotopic mixtures are constant in composition. Some differences, usually very slight, have been reported, however. For example, atmospheric oxygen has been shown to contain 3 per cent more of the isotope $^{18}_{8}O$ than oxygen from the waters of Lake Michigan.

Limitations of Isotopy

It will perhaps have been noted that the variation of mass number in isotopes of a given element is not very great, being rarely more than eight units. The reason for this is that the addition of neutrons to a given nucleus soon produces instability and this sets a limit to isotopy. Consequently, the order of relative atomic masses for the elements is the same as the order of atomic numbers in most cases. The important exceptions are Ar—K, Co—Ni, Te—I, in which the order of relative atomic mass reverses the order of atomic number. This similarity of order for most elements has been important in the development of chemical classification because it enabled Mendeléeff to devise the Periodic Table on the basis of relative atomic mass though its true basis is atomic number.

Isotopy of hydrogen

Hydrogen is known to exist in three isotopic forms, though only one of them occurs in considerable quantity in ordinary hydrogen gas. The main constituent of hydrogen is made up of atoms containing one proton as nucleus and one electron. This isotope is denoted by the symbol ^1_1H, and is called protium.

Ordinary hydrogen also contains a small proportion (about 1 in 4500) of an isotope having, as nucleus, one proton and one neutron, with one electron outside the nucleus. This isotope's atom has about twice the mass of the lighter atom. The chemical properties of the heavier isotope resemble those of ^1_1H (though its reaction rates are different), because of the similar electron structure. The effect of the added neutron is, however, relatively great in such simple atoms and the compounds of the heavy isotope, notably the oxide, differ appreciably from the compounds of protium. For this reason, it has been considered desirable to give the heavy isotope an individual name, *deuterium*, and a separate symbol, D.

The third isotope, *tritium*, symbol T, possesses a nucleus consisting of one proton and two neutrons and has one electron outside this nucleus. It can be produced by interaction between neutrons of low energy and the isotope of lithium, ^6_3Li. The products are tritium and helium.

These three isotopes of hydrogen are represented diagrammatically as follows:

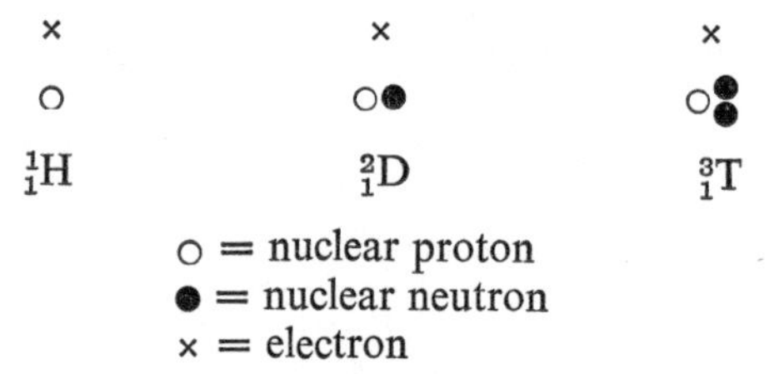

Production of deuterium oxide, 'heavy water', D_2O, and deuterium, D_2

During electrolysis of ordinary water, which contains about 0.02 per cent of deuterium oxide, hydrogen and deuterium may both discharge at the cathode.

$$2H^+(aq) + 2e^- \rightarrow H_2(g) \qquad 2D^+(aq) + 2e^- \rightarrow D_2(g)$$

There is, however, a preferential discharge of the light isotope, with the result that deuterium oxide concentrates in the residue. This circumstance is utilized in the following way. A 0.5 M solution of potassium hydroxide in water is electrolysed using nickel electrodes till the concentration of the alkali reaches about 5 M. The potassium hydroxide is then neutralized by passage of carbon dioxide and the water is distilled off. The processes are repeated on the distillate and, after seven such stages, reasonably pure deuterium oxide is obtained. The gas evolved in the later stages contains increasing proportions of deuterium and is burnt with oxygen, the product being returned to the earlier cells. Deuterium itself can be obtained at the cathode by electrolysis of the deuterium oxide with potassium hydroxide as electrolyte and nickel electrodes.

This electrolytic process for producing deuterium oxide has one great disadvantage—the enormous consumption of electric power, which is in the region of 40 000 ampere-hours per gramme of deuterium oxide.

Properties of deuterium, D_2

In a general way, the chemical properties of deuterium resemble those of hydrogen (^1_1H) but deuterium usually reacts at a different rate. Its explosion with oxygen is less violent. Quite a number of deutero-analogues of ordinary hydrogen compounds have been prepared. By passing benzene vapour and deuterium over nickel at 400 K the hydrogen atoms of benzene are replaced by deuterium atoms.

$$C_6H_6(g) + 3D_2(g) \rightleftharpoons C_6D_6(g) + 3H_2(g)$$

Similar replacement in carboxylic acids has also been observed.

Deuterium, hydrogen and water also engage in the reversible reaction:

$$H_2(g) + D_2O(g) \rightleftharpoons H_2O(g) + D_2(g)$$

Under similar conditions the reverse reaction is about three times as rapid as the forward reaction, so that the deuterium concentration in the water tends to rise.

By the reaction between deuterium oxide and calcium carbide, dideutero-ethyne has been prepared and used in further syntheses.

$$CaC_2(s) + 2D_2O(l) \rightarrow C_2D_2(g) + Ca(OD)_2(s)$$

Properties of deuterium oxide, 'heavy water', D_2O

This liquid is colourless like ordinary water, but differs appreciably from it as the following table shows.

	Water	*Deuterium oxide*
Boiling point	373.0 K	374.4 K
Freezing point	273.0 K	276.8 K
Maximum density	1.0 g cm^{-3}	1.108 g cm^{-3}

Deuterium oxide exhibits the same behaviour as water in showing a maximum density at a certain temperature and lower densities both above and below this temperature. For ordinary water the temperature of maximum density is 277 K and for deuterium oxide 284 K.

Separation of isotopes

Separation of isotopes is usually difficult because of the almost complete identity of their chemical behaviour. The most successful separation methods are physical.

1. *Separation by positive-ray methods*

This has already been discussed in principle on page 30. The mass spectrograph separates isotopes and causes them to register the separate lines of the mass spectrum. By actual collection of the particles, the isotopes may be obtained separated. This was done on a minute scale for the isotopes $^{6}_{3}Li$ and $^{7}_{3}Li$ in 1934 and was later applied to the two uranium isotopes, $^{235}_{92}U$ and $^{238}_{92}U$.

2. *Separation by diffusion*

This was discussed earlier (page 32) in connection with the isotopes of neon, of chlorine (combined in hydrogen chloride), and of uranium (using UF_6).

3. *Evaporation methods*

Lighter atoms tend to evaporate more readily than heavier atoms from a mixture of the two. In similar conditions, the rate of evaporation is inversely proportional to the square-root of the mass of the atom. Using this principle, Brönsted and Hevesy allowed mercury to evaporate at very low pressure and at 320 K. The vapour was condensed on a glass surface cooled by liquid air so that the mercury solidified. After a time the solid distillate was removed, melted, and refractionated in a similar way. Repetition of the process produced mercury differing by nearly 0.2 unit in atomic mass from ordinary mercury. That is, a partial separation of isotopes was accomplished.

4. *Separation by thermal diffusion*

Enskog (1911) showed that, if a mixture of gaseous molecules is enclosed in a tube with its ends at different temperatures, heavier molecules tend to concentrate at the cooler end. Clusius and Dickel applied this observation to isotopes. They employed a vertical tube over 30 m long, with its sides cooled by water and an electrically heated wire running down the middle. In this way, convection currents were made to assist the thermal separation and the isotopes $^{35}_{17}Cl$ and $^{37}_{17}Cl$ were almost completely separated.

Mass defect, binding energy, and nuclear stability

The mass spectrometer can be used to determine the relative masses of nuclei to such a high degree of accuracy that it has been possible to detect that the relative mass of a nucleus is always slightly less than the sum of the relative masses of its constituents. This loss in mass represents the binding energy of the nucleus (energy and mass being related by the equation, $E = mc^2$, page 3).

The nature of the forces holding the nucleus together is not fully understood, but an interesting observation is that only neutron–proton ratios which lie within fairly narrow limits give rise to stable nuclei (see

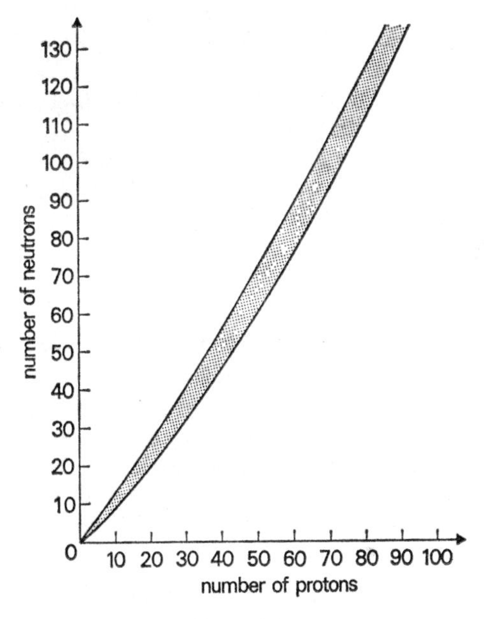

Figure 4.3.

Figure 4.3). Just outside these limits, nuclei may exist, but they spontaneously disintegrate until stable nuclei are formed. This spontaneous disintegration of unstable nuclei gives rise to radioactivity.

Radioactivity

Radioactivity was first noticed by Becquerel (1896) in uranium salts. Rays emitted by these salts were found to affect a photographic plate in the same way as light, to ionize gases, causing the discharge of a gold-leaf electroscope, and to produce phosphorescence in certain materials, e.g. zinc sulphide. Marie and Pierre Curie, working in Paris about 1900, detected more intense radioactivity in a new element (named by them *polonium*), and later they isolated another very powerfully radioactive element, which they named *radium*.

Kinds of radiation

The radiation emitted by uranium was found to be composite and of three types, originally named *alpha-*, *beta-* and *gamma-rays* before the true nature of each was recognised.

The **alpha-rays** were shown to undergo deflection in an electrostatic field away from the positively charged plate and so must carry positive charge themselves. They are now known to be helium ions, He^{2+}, of relative mass 4.

The **beta-rays** underwent a deflection opposite to that of alpha-particles in the electric field and so must carry negative charge. By determination of the ratio charge/mass, the particles were shown to be electrons. Beta-radiation is simply an electron stream.

The **gamma-radiation** was unaffected by electrical fields and was found to consist of electromagnetic waves of very high frequency, i.e. very short wavelength (about 5×10^{-11} cm).

Results of radioactive changes

The general principles of radioactive change are the following:

Alpha-particle emission. When a nucleus emits an alpha-particle, it loses two protons and two neutrons, i.e. four units of mass. The lost neutrons do not affect the chemical nature of the element. The two lost protons reduce the atomic number of the element by 2 and cause the product to take up a position two groups *lower* in the Periodic Table. Thus, $^{226}_{88}Ra$ in Group II loses an alpha-particle and produces $^{222}_{86}Rn$ in Group 0.

$$^{226}_{88}Ra \rightarrow {}^{222}_{86}Rn + {}^{4}_{2}He$$

Beta-particle emission. Beta-disintegration can be considered as equivalent to the splitting of a neutron in the nucleus into an electron, which is then emitted as a beta-particle, and a proton, which the nucleus retains. The electron is so light that loss of it leaves the relative atomic mass of the residue virtually unchanged. The extra proton increases the atomic number of the element by one unit and it takes up a position one group *higher* in the Periodic Table. For example, the radioactive isotope of lead, $^{210}_{82}Pb$, in Group IV, loses a beta-particle (electron) to give a radioactive isotope of bismuth, $^{210}_{83}Bi$, in Group V.

$$^{210}_{82}Pb \rightarrow {}^{210}_{83}Bi + {}^{0}_{-1}e$$

Radioactive changes involve the nucleus alone and are hence called **nuclear reactions.** As can be seen above, nuclear reactions can be represented by equations. These equations, unlike equations for chemical reactions, involve the formation of isotopes of different elements. The equations are balanced by ensuring that the total relative masses (upper figures) on each side of the equation are equal and that the total relative charges (lower figures) on each side of the equation are equal.

Decay series

When the product of a radioactive disintegration is itself unstable, then a further radioactive disintegration will occur. Thus a whole series of spontaneous radioactive disintegrations can occur (see the thorium decay series below), each step involving the loss of either an alpha- or a beta-particle or very occasionally both, until eventually a stable isotope is formed.

Thorium decay series

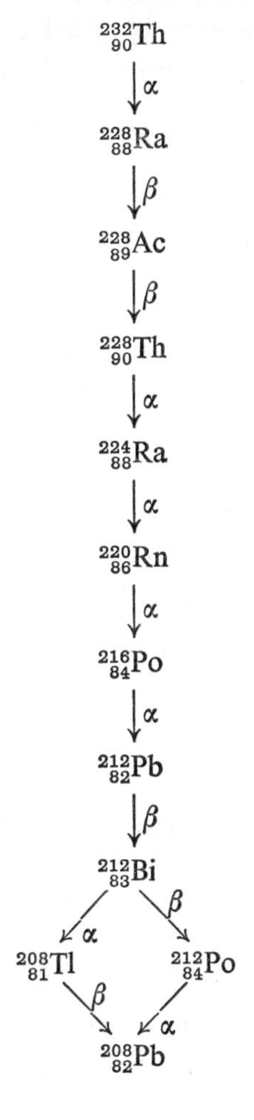

It is now recognized that there are several series of radioactive changes, named from the elements in which they now originate in nature, e.g. the *actinium* series, the *uranium* series (which contains radium), and the *thorium* series. All these series show many and complicated changes, ending in some isotope of lead (atomic number 82), e.g. the thorium series produces $^{208}_{82}Pb$ and the uranium series, $^{206}_{82}Pb$. Another series, based upon *neptunium*, has bismuth as its end-product.

Half-life

Radioactive change follows an exponential law, that is, the rate of decay of a given material is directly proportional, at any instant, to the amount of material present. This means that, kinetically, it is a first-order reaction and it can be expressed by the formula (see page 258)

$$kt = \ln \frac{N_0}{N}$$

where N_0 is the original number of atoms of the isotope,
N is the number of atoms of the isotope left after time t,
k is the radioactive decay constant for the isotope.

The time for half the nuclei to disintegrate (i.e. when $N = \frac{1}{2}N_0$) is represented by $t_{\frac{1}{2}}$ and is known as the half-life of the isotope. Substituting in the above equation,

$$t_{\frac{1}{2}} = \frac{\ln 2}{k}$$

Hence the half-life of a particular isotope is constant and is independent of the amount of starting material. It is therefore customary to characterize radioactive isotopes by quoting their half-lives. For example, the half-life of radium is 1620 years, i.e. 1 g of radium held in 1972 will have decreased in amount to 0.5 g by about the year 3592. Half-lives range from fractions of a second to millions of years.

Nuclear Energy

The energy changes associated with nuclear reactions are very much larger than those associated with chemical reactions. For example, when radium loses an alpha-particle to form radon, 4.2×10^{11} joules are evolved per mole of radium.

$$^{226}_{88}Ra \rightarrow {}^{222}_{86}Rn + {}^{4}_{2}He$$

The possibility of utilizing nuclear energy changes in a humanly controlled way has existed since Rutherford brought about the first artificial transmutation of one element into another. This was accomplished (1919) by the action of swift alpha-particles on nitrogen gas. In this process the alpha-particle was captured by the nucleus of $^{14}_{7}N$ (a gain of 4 units of mass) with subsequent emission of a proton (loss of one unit of mass) and the production of the isotope of oxygen, $^{17}_{8}O$.

$$^{4}_{2}He + ^{14}_{7}N \rightarrow ^{17}_{8}O + ^{1}_{1}H$$

The chief process which has so far yielded energy from nuclear sources is the process of **nuclear fission**. Its discovery in 1939 is usually associated with the German nuclear scientist, Hahn. He found that the nucleus of the uranium isotope, $^{235}_{92}U$, could absorb a neutron and then break up into two roughly equal parts with mass numbers lying between 72 and 162. Neutrons are also emitted. During this fission there is a loss in mass of about 0.2 units per mole of $^{235}_{92}U$. This mass is converted into energy according to the equation $E = mc^2$, where E is the energy in joules, m is the mass in kg, and c is the velocity of light in m s^{-1}.

$$\begin{aligned} E &= 0.2 \times 10^{-3}\,(3 \times 10^{8})^2 \\ &= 1.8 \times 10^{13}\ \text{J mol}^{-1} \end{aligned}$$

So the heat change associated with the nuclear fission of $^{235}_{92}U$ is of the order of 2×10^{10} kJ mol^{-1}. Relative to energy changes during ordinary chemical reactions, such as the burning of carbon in which about 4×10^{2} kJ mol^{-1} are evolved, this output of energy is stupendous.

Atomic bomb

This process of nuclear fission of $^{235}_{92}U$ can be made into a 'chain-reaction' in suitable conditions. This is because, while fission occurs as a *result* of

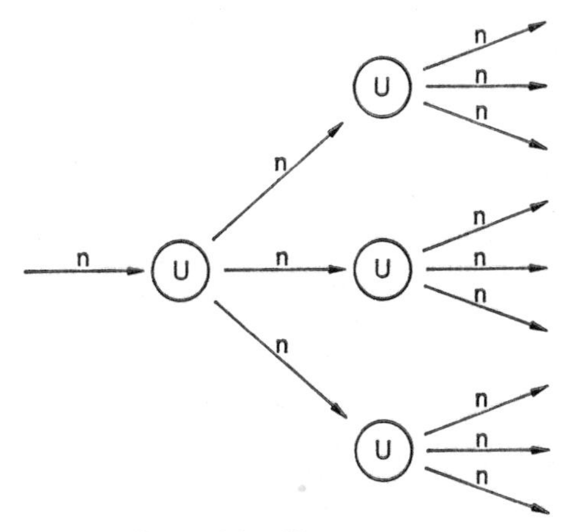

Figure 4.4. Chain-reaction

neutron absorption by the uranium nucleus, the process also *emits* neutrons. These are available for absorption for new fissions, and so on. If the mass of uranium is small, the neutrons may escape from it at such a rate that the chain-reaction cannot be maintained but, above a certain critical mass, the uranium absorbs the neutrons so as to produce very rapid nuclear fission of at least a very considerable proportion of it. The energy liberated then produces a terrific explosion, with a temperature of several million degrees Kelvin for a short period after its occurrence.

The essential ideas relating to the production of the atomic bomb were (a) the isolation of a sufficient amount of the isotope $^{235}_{92}U$ from natural uranium, (b) the invention of a device which would keep the uranium apart in quantities below the critical mass until the decisive moment and then shoot them rapidly together, and (c) the provision of a 'tamping' device to reflect escaping neutrons back into the fissionable material. A source of neutrons (such as beryllium activated by alpha-particles) is also required to initiate the fission. The isotope of plutonium, $^{239}_{94}Pu$, will produce a similar chain-reaction by fission and was used in the second atomic bomb of 1945.

Hydrogen bomb

The nuclear reactions occurring in this bomb involve the fusion of hydrogen isotopes into helium. These reactions are activated by the heat from a smaller uranium or plutonium fission bomb. It is possible to construct more powerful fusion bombs as the restriction of critical mass, which is important in the case of a fission bomb, does not apply in this case. Energy is released in this case because a helium atom has a mass slightly less than that of the hydrogen isotopes which combine to form it.

Nuclear fusions only occur at very high temperatures and hence they are known as thermonuclear reactions. Stellar energy is due to thermonuclear reactions which produce temperatures of several millions of degrees. The sun is estimated to contain enough hydrogen to continue radiating at its present rate for several thousand million years.

Nuclear reactors

The problem of adapting nuclear energy for useful purposes is that of releasing the same energy as in the atomic bomb, but steadily over a relatively long period and at a temperature convenient for making use of the energy.

$^{235}_{92}U$ forms only 0.7 per cent of natural uranium and is difficult and costly to purify. It is much more economical if natural uranium, containing 99.3 per cent of $^{238}_{92}U$, can be used. This isotope can absorb fast neutrons but not slow neutrons. $^{235}_{92}U$ can absorb either of the two types and undergo fission. Fast neutrons can be slowed down, sufficiently to affect $^{235}_{92}U$ only, by passage through graphite in which they collide with the carbon nuclei and lose energy. They are then rejected by $^{238}_{92}U$ but are available for $^{235}_{92}U$ fission, which will occur rapidly enough to maintain the chain-reaction.

Calder Hall, which came into operation in 1956, was the first nuclear reactor to produce power on an industrial scale. In this type of reactor uranium fuel rods in metal containers are arranged between blocks of pure graphite. The reaction is controlled by lowering boron steel rods into this atomic pile. The control rods absorb neutrons and hence slow down the reaction. For safety, the pile is surrounded by several feet of concrete to absorb stray radiation. A circulation of carbon dioxide at about 7 atmospheres pressure absorbs the liberated heat energy. The gas leaves the reactor at a temperature in the region of 600 K. The heat is conveyed to boilers which generate steam and operate alternators for the production of electrical power.

A more recent development has been the reactor at Dounreay. The fuel used in this case is $^{239}_{94}Pu$ which will undergo fission when bombarded with fast neutrons, hence it is not necessary to include the bulky graphite moderator. Another advantage of the Dounreay reactor is that, by allowing some of the neutrons from the plutonium fission to bombard $^{238}_{92}U$, more plutonium can be formed. Thus, by the series of nuclear reactions given below, this reactor produces its own fuel and is known as a **breeder reactor,**

$$^{238}_{92}U + ^{1}_{0}n \rightarrow ^{239}_{92}U \rightarrow ^{239}_{93}Np + ^{0}_{-1}e$$

$$^{239}_{93}Np \rightarrow ^{239}_{94}Pu + ^{0}_{-1}e$$

Attempts are being made to control nuclear fusion which, if it proves possible, will have the advantage that the fuel, isotopes of hydrogen, is more readily available than the heavy metals needed for the fission reactors.

Artificial Elements

The most complex atomic nucleus known on earth up to about 1940 was that of uranium (atomic number 92). Since that time more complex nuclei have been produced. The first two of these, neptunium and plutonium, as indicated in the equations, have been built up from $^{238}_{92}U$.

By using a variety of projectiles (even as heavy as $^{16}_{8}O$) and targets, this series of elements has been built up as far as kurchatovium which has an atomic number of 104.

'Artificial' production of radioactive isotopes

The development of methods for the transmutation of elements has also proved useful in that it is now possible to prepare radioactive isotopes which do not exist in nature. This means that when selecting a radioactive isotope for a particular purpose in industry, medicine or some type of research, there is now a much larger variety of isotopes to choose from.

For example, if sodium chloride is subjected to the action of an intense concentration of neutrons in a uranium pile for a few days, a certain proportion of the $^{23}_{11}Na$ will absorb one neutron per atom to form a radioactive sodium isotope, $^{24}_{11}Na$. In similar conditions, sulphur, $^{32}_{16}S$, will absorb one neutron per atom, emit a proton, and so finish as a radioactive isotope of phosphorus, $^{32}_{15}P$.

$$^{32}_{16}S + ^{1}_{0}n \rightarrow ^{32}_{15}P + ^{1}_{1}p$$

Uses of radioactive isotopes

The uses of radioactive isotopes may be categorized as follows:

1. those in which the effects of radiations are used,
2. those in which the fact that the radiations are easily detected are put to use.

The first category may be subdivided into:

(a) *Curative uses*

Cancerous growths may sometimes be eradicated by exposure to gamma-rays from radioactive materials. Radium has been used for this purpose but was scarce and expensive. More recently radioactive cobalt has been used and is a much cheaper and more easily available material. It can be made by irradiating ordinary cobalt, containing $^{59}_{27}Co$, with neutrons to produce $^{60}_{27}Co$. This radioactive cobalt is placed near the centre of a lead sphere which absorbs unwanted radiation. The wanted radiation is allowed to escape down a very narrow hole in the lead and directed at the tissue to be treated.

Radioactive phosphorus has been used in the treatment of leukaemia and radioactive iodine in disease of the thyroid gland.

(b) *Sterilization of medical equipment*

Radiation is also used for the sterilization of such items as disposable hypodermic syringes. The syringes are sterilized after they have been packed and they are not unpacked until immediately before they are to be used.

The second category may be subdivided into:

(i) *Tracer techniques*

The radioactive isotope of an element does not differ chemically from the 'ordinary' atoms, therefore the radioisotope may be used as a marker, tracing what occurs in some process under review. A biological example is the tracing of the course of photosynthesis by using carbon dioxide containing the radioactive $^{14}_{6}C$ instead of the ordinary $^{12}_{6}C$.

In organic chemistry the labelling of a particular atom in a compound, by preparing the compound with a radioactive isotope in that position, can be used to trace the reaction path and hence help to elucidate the mechanism of the reaction. An example in inorganic chemistry is the determination of the solubility of a sparingly soluble salt. A saturated solution is prepared using the salt containing a known proportion of radioactive isotope. The concentration of this solution is determined from the intensity of the radiation. The analysis of such a dilute solution would not be possible by more conventional means.

In industry the tracer technique can be used to detect the leaks in water mains; or, when the grade of oil is changed in a pipeline, radioactive material is added to the first of the new flow and is readily detected at the receiving end.

(ii) *Absorption of radiation*

The thickness of a sheet of metal can be continuously checked by measuring the absorption of radiation which is directed through it. A similar technique can be used to detect that containers are full of a particular product.

(iii) *Dating*

Carbon-14 dating depends upon the fact that the percentage of $^{14}_{6}C$ in the atmosphere, and hence in all living things, is almost constant as it is continuously regenerated by the action of cosmic radiation on nitrogen. From the moment death occurs the $^{14}_{6}C$ will not be replenished from the atmosphere and the $^{14}_{6}C$ which is present will decay at a definite rate. The half-life of this isotope is 5600 years, and by measuring the proportion of $^{14}_{6}C$ remaining it is possible to put an approximate date on the time of death.

(iv) *Radiography*

Faults in welded metal plates and pipes can be detected by a radiography examination. The advantage of this method is that the source of radiation can be very small compared to the equipment required to produce the more conventional X-rays.

5. Electrons in the Atom

The atom is an electrically neutral structure. It is, therefore, obvious that it must contain electrons equal in number to the protons in the nucleus. The question is, how are the electrons distributed? An indication of the arrangement of the electrons around the nucleus is provided by atomic emission spectra.

Atomic Emission Spectra

When atoms of an element are heated or subjected to an electrical discharge, it is observed that the element emits electromagnetic radiation. This is caused by the atoms of the element absorbing some energy and then emitting it in the form of electromagnetic radiation. When this radiation is passed through a spectrometer, it is observed that, unlike sunlight, it is not a continuous spectrum. The spectrum obtained consists of a number of definite lines where each line corresponds to a definite wavelength of radiation. In the visible part of the spectrum this is observed as a number of lines, each of different colour.

The fact that it is a line spectrum and not continuous indicates that only certain energy absorptions and emissions are possible. This can only be accounted for by the existence of definite electron energy levels within an atom, each absorption being due to the movement (jumping) of an electron from one energy level to a higher one. When the electron reverts to a lower energy level, energy is emitted in the form of electromagnetic

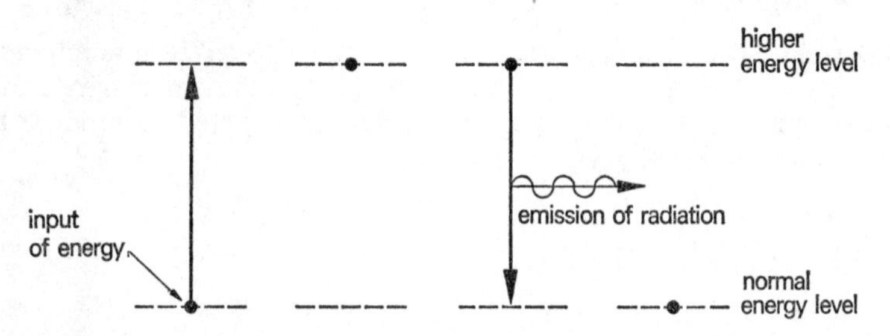

Figure 5.1. Diagrammatic representation of absorption and emission of energy by an electron

radiation and this radiation will have a definite wavelength corresponding to the definite energy drop. The discrete amounts of energy which can be absorbed or emitted by an atom are called **quanta** (see Figure 5.1).

The energy and frequency of the radiation are related by Planck's equation:

$$E = h\nu$$

where E is the energy, ν is the frequency, and h is Planck's constant.

Hydrogen spectrum

Each element has a characteristic pattern of lines in its atomic emission spectrum. Hydrogen, having only one electron per atom, is the simplest example to consider. Its spectrum consists of several series of lines all of which can be accounted for by postulating the existence of energy levels as indicated in Figure 5.2. The possible energy levels are denoted by increasing values of n. The lowest energy level, or **ground state,** is described as the $n = 1$ state.

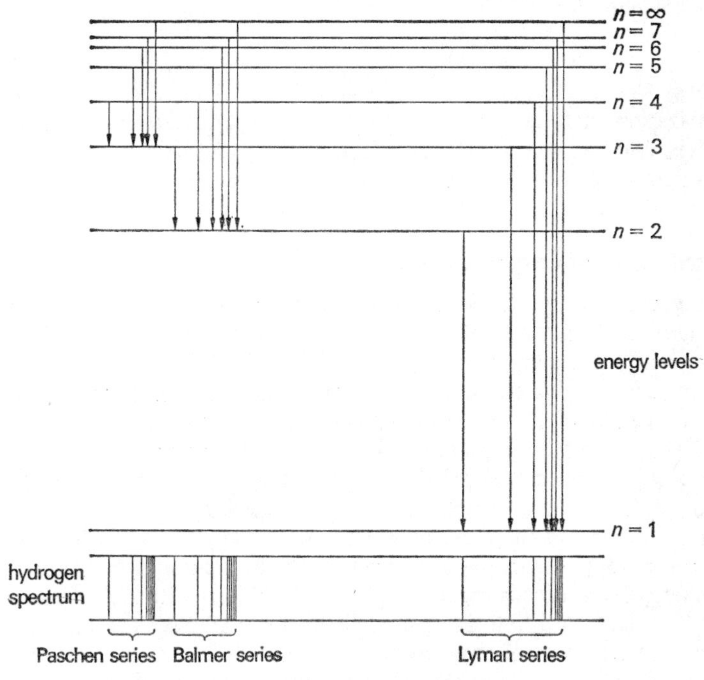

Figure 5.2. The energy 'drops' which produce part of the hydrogen atomic emission spectrum

Each series is produced by the reverting of electrons to a particular energy level from all the energy levels above it. Thus the Lyman series is due to electrons reverting back to the lowest energy level, $n = 1$. The Balmer series, which occurs in the visible part of the spectrum, is due to electrons reverting to the $n = 2$ state; the Paschen series to the $n = 3$ state, and so on.

Convergence frequency and ionization energy

The spaces between the successive lines in each series gradually become smaller as the frequency increases, indicating that the higher energy levels in the atom gradually become closer together until eventually they merge. If an electron can be made to jump beyond this highest energy level, it will not return and the atom will have ionized. The energy required to remove an electron from its ground state ($n = 1$) to beyond the highest energy level is the ionization energy for that electron in that atom. Therefore, in the case of hydrogen, the convergence frequency for the Lyman series must correspond to the ionization energy of hydrogen.

The convergence frequency of the Lyman series is of the order of 3.3×10^{15} Hz. Taking Planck's constant as 4×10^{-13} k Js mol^{-1}, and using the equation $E = h\nu$, the ionization energy is given by

$$E = 4 \times 10^{-13} \times 3.3 \times 10^{15}$$
$$= 1.3 \times 10^{3} \text{ kJ mol}^{-1}$$

Ionization energies, derived from spectroscopic data, are very important as their interpretation in terms of electron energy levels enable chemists to rationalize a lot of the chemical behaviour of elements.

Distribution of Energy Levels

Atomic emission spectra indicate that there are only certain energy levels available in an atom. Ionization energies provide an indication of how these energy levels are distributed.

Consider first successive ionization energies, i.e. the energy required to remove the first electron, then the second electron, then the third electron, and so on, from atoms of a particular element. A graph of the logarithm of successive ionization energies of potassium against the number of electrons removed (Figure 5.3) shows that, as expected, each electron is more difficult to remove than the previous one. However, in addition to this general trend, there are some distinct breaks in the graph, indicating that the energy levels are arranged in groups. These start with 2 electrons, with fairly similar ionization energies, which are both near the nucleus and difficult to remove, followed by 8 electrons with fairly similar ionization energies (but distinctly less then the previous 2), followed by another group of 8 and finally by a single electron which, compared to all the other

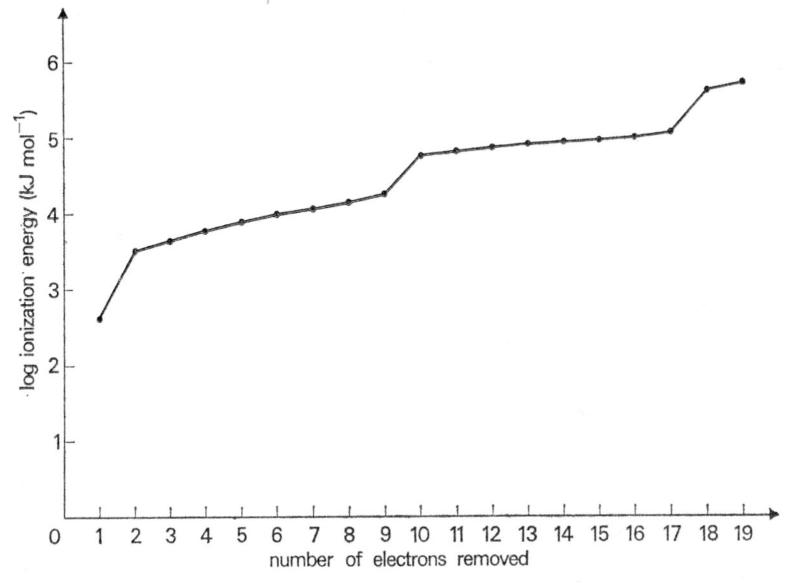

Figure 5.3. Lg of successive ionization energies of potassium

electrons, is very easily removed. These groups of energy levels are referred to as **electron energy shells**. Each shell is denoted by a particular value of n (principal quantum number) or by a letter. For example, the nineteen potassium electrons are arranged as follows:

Shell	K $n = 1$	L $n = 2$	M $n = 3$	N $n = 4$
Number of electrons	2	8	8	1

The electrons of the first twenty elements arranged in energy shells are given on page 50.

A point which will be taken up later (Chapter 6) is that elements which are chemically similar are found to have the same number of electrons in their outer shells.

A more detailed insight into the distribution of energy levels is obtained by plotting a graph of the first ionization energies of the elements against their atomic numbers (see Figure 5.4, on page 51).

For the first twenty elements the general shape of the graph follows the 2, 8, 8, 2 pattern of the principal electron energy shells. In addition the 8 electrons of the $n = 2$ shell and the 8 electrons of the $n = 3$ shell appear to be subdivided into 2, 3, 3 arrangements. A more advanced analysis of spectra shows that there are only two energy levels, one holding 2 electrons

Elements of the first twenty elements arranged in energy shells

	1	2	3	4
1 H	1			
2 He	2			
3 Li	2	1		
4 Be	2	2		
5 B	2	3		
6 C	2	4		
7 N	2	5		
8 O	2	6		
9 F	2	7		
10 Ne	2	8		
11 Na	2	8	1	
12 Mg	2	8	2	
13 Al	2	8	3	
14 Si	2	8	4	
15 P	2	8	5	
16 S	2	8	6	
17 Cl	2	8	7	
18 Ar	2	8	8	
19 K	2	8	8	1
20 Ca	2	8	8	2

and the other 6, and the extra break in the graph occurs at the point where the 6 level is half full, which is a particularly stable arrangement. In order to account for the graph beyond the first twenty elements it is necessary to include subshells, one type holding 10 electrons (e.g. atomic numbers 21–30) and one holding 14 electrons (e.g. atomic numbers 58–71). These subsidiary energy shells are denoted by the azimuthal (or subsidiary) quantum number, l, which can have values of 0, 1, 2 etc. up to a maximum of $n - 1$. They are also known by the letters s, p, d, and f which are the initial letters of the names given to certain spectral lines.

Subshell	$l = 0$ s	$l = 1$ p	$l = 2$ d	$l = 3$ f
Maximum number of electrons	2	6	10	14

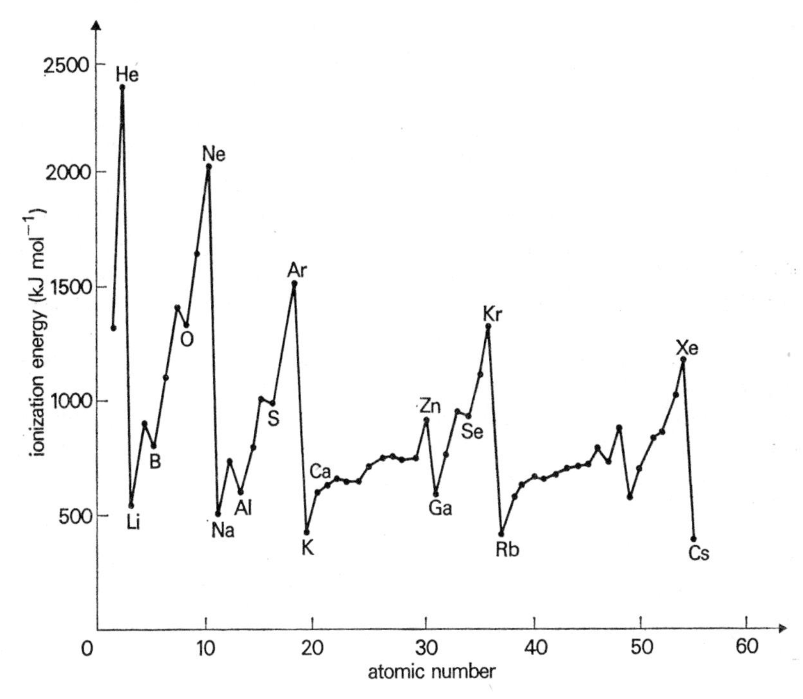

Figure 5.4. *Variation of first ionization energies with atomic number*

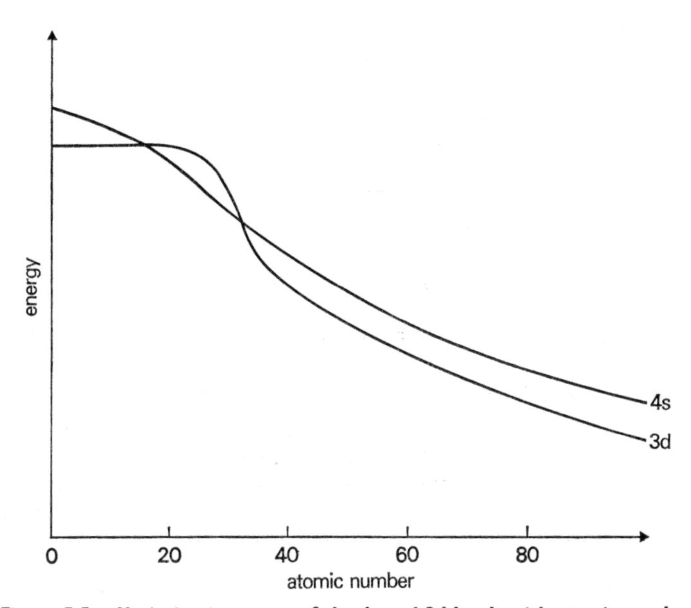

Figure 5.5. *Variation in energy of the 4s and 3d levels with atomic number*

It is therefore possible to characterize an electron by its principal quantum energy shell and its subsidiary quantum energy shell. Thus an electron in the s subshell of the third shell is known as a 3s electron.

The first 2 electrons of all elements are in the 1s energy level, the next 2 in the 2s, the next 6 in the 2p and so on. The 2s electrons of, say, sodium will be in a different environment than the 2s electrons of, say, uranium; it might therefore be expected that the absolute energy levels of these electrons will be different. It is of particular interest to consider the variation in energy of the 4s and 3d electrons with atomic number (Figure 5.5).

It can be seen from Figure 5.5 that the 4s is at a lower energy level than the 3d in the atomic number region 20. The electron shells are filled in order of increasing energy and therefore the two outer electrons of calcium are in the 4s shell and not the 3d shell. The next ten elements after calcium have their extra electrons added to the 3d shell, thus giving rise to the d-block or transition elements.

The outer electron configurations of chromium, $3d^5\ 4s^1$, and copper, $3d^{10}\ 4s^1$, are anomalous and indicate that there is a special stability associated with the half filled and completely filled d subshell.

The relative energies of the subshells for the first thirty-six elements are given in Figure 5.6.

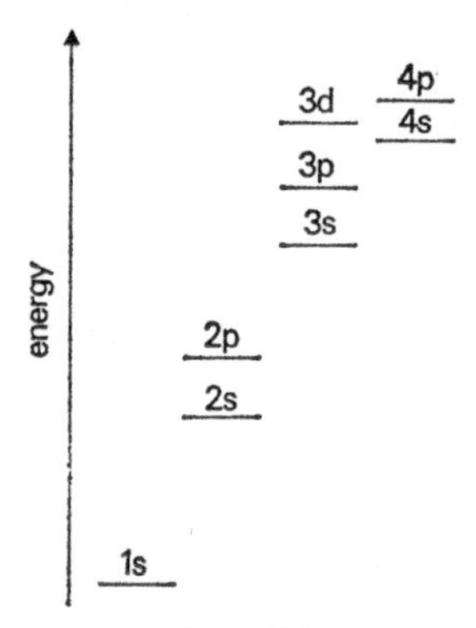

Figure 5.6.

The subshells are filled in order of increasing energy according to the **Aufbau Principle.**

Figure 5.6 can be used to determine the electron arrangement in an atom, a superscript being used to denote the number of electrons in a particular subshell. For example, chlorine has an atomic number of 17 and therefore 17 electrons around its nucleus. These electrons will be arranged as follows:

$$1s^2\ 2s^2\ 2p^6\ 3s^2\ 3p^5$$

Iron has an atomic number of 26 and its electron arrangement is

$$1s^2\ 2s^2\ 2p^6\ 3s^2\ 3p^6\ 3d^6\ 4s^2$$

An alternative way of remembering the order in which the subshells are filled is provided by Figure 5.7.

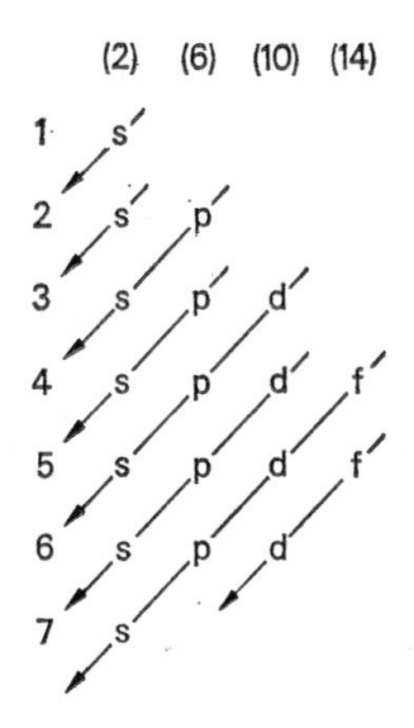

Figure 5.7.

Electron spin

On closer examination, some apparently single spectral lines are found to be double lines. These *doublets* are accounted for by electrons having spin and there being two directions of spin. Two electrons with opposite spin can form a stable pair of electrons. Thus the 2 electrons in a 1s subshell can be represented by [↑↓], the 6 electrons in a 2p subshell will exist as three pairs and can be represented by [↑↓][↑↓][↑↓], the 10 electrons in a 3d subshell will exist as five pairs and can be represented by [↑↓][↑↓][↑↓][↑↓][↑↓] and the 14 electrons in a 4f subshell will exist as seven pairs and can be represented by [↑↓][↑↓][↑↓][↑↓][↑↓][↑↓][↑↓]. The electronic configuration of, for example, fluorine may be written as

1s	2s	2p
[↑↓]	[↑↓]	[↑↓][↑↓][↑]

Within a particular subshell electronic configurations are always written with the maximum number of unpaired electrons (Hund's Rule). For example, oxygen (atomic number 8) is written as

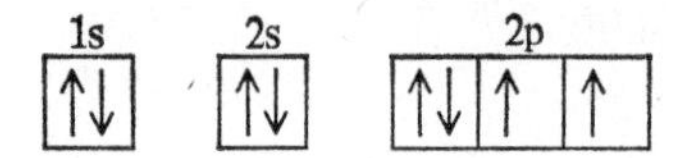

and not as

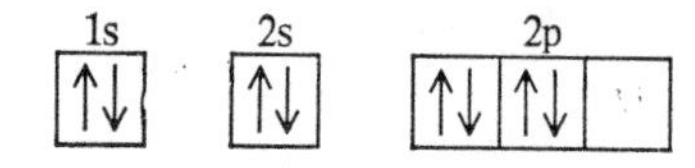

Quantum Numbers

In describing the energy state of a particular electron two quantum numbers have so far been used:

1. The **principal quantum number, *n*,** where $n = 1, 2, 3$ etc. and denotes the principal quantum energy shell in which the electron is found.

2. The **azimuthal** or **subsidiary quantum number, *l*,** which can have values of $l = 0$, $l = 1$, $l = 2$ up to $l = n - 1$. This means that when $n = 3$, l can have values of 0, 1, and 2. That is, there are three types of subshell (s, p, and d) in the third principal shell.

In order to describe completely the energy state of an electron two additional quantum numbers are required.

3. The **magnetic quantum number, *m*,** which can have values from $+l$ to $-l$. Thus, when $l = 2$ (d subshell), m can have five values, $+2$, $+1$, 0, -1, -2. The use of this quantum number stems from work on the effect of magnetic fields on spectral lines which gave an indication of the number of energy levels within a particular subshell. Thus when $l = 2$, m can have five values; therefore there are five energy levels in this subshell.

4. The **spin quantum number, *s*,** which can have values of $+\frac{1}{2}$ or $-\frac{1}{2}$ and describes the fact that each energy level in a subshell can hold two electrons of opposite spin. This means that for the $l = 2$ subshell there are five energy levels (five values of m) each of which can hold 2 electrons giving a total of 10 electrons for a d subshell.

The quantum numbers of the electrons in the first three principal energy shells are as follows.

	n	l	m	s	
First shell	1	0	0	$+\frac{1}{2}$	1s subshell
	1	0	0	$-\frac{1}{2}$	
	2	0	0	$+\frac{1}{2}$	2s subshell
	2	0	0	$-\frac{1}{2}$	
	2	1	+1	$+\frac{1}{2}$	
Second shell	2	1	+1	$-\frac{1}{2}$	
	2	1	0	$+\frac{1}{2}$	
	2	1	0	$-\frac{1}{2}$	2p subshell
	2	1	−1	$+\frac{1}{2}$	
	2	1	−1	$-\frac{1}{2}$	
	3	0	0	$+\frac{1}{2}$	3s subshell
	3	0	0	$-\frac{1}{2}$	
	3	1	+1	$+\frac{1}{2}$	
	3	1	+1	$-\frac{1}{2}$	
	3	1	0	$+\frac{1}{2}$	3p subshell
	3	1	0	$-\frac{1}{2}$	
	3	1	−1	$+\frac{1}{2}$	
	3	1	−1	$-\frac{1}{2}$	
Third shell	3	2	+2	$+\frac{1}{2}$	
	3	2	+2	$-\frac{1}{2}$	
	3	2	+1	$+\frac{1}{2}$	
	3	2	+1	$-\frac{1}{2}$	
	3	2	0	$+\frac{1}{2}$	3d subshell
	3	2	0	$-\frac{1}{2}$	
	3	2	−1	$+\frac{1}{2}$	
	3	2	−1	$-\frac{1}{2}$	
	3	2	−2	$+\frac{1}{2}$	
	3	2	−2	$-\frac{1}{2}$	

It can be seen from this that no two electrons in an atom can have all four quantum numbers with identical values (Pauli Exclusion Principle). For example, the two electrons in the 3s subshell have quantum numbers

$$n = 3,\ l = 0,\ m = 0,\ s = +\tfrac{1}{2}$$
$$n = 3,\ l = 0,\ m = 0,\ s = -\tfrac{1}{2}$$

Atomic Orbitals

In the early days (Bohr 1914), the definite energy levels indicated by atomic spectra were visualized as definite electron orbits. At a later stage it was realized that, due to its extremely small mass, it would not be possible to determine experimentally the exact path and velocity of an electron in an atom (Heisenberg 1927). However, by considering the wave properties of an electron, it has proved possible to calculate the probability of finding an electron in a particular position.

If the probability of finding the 1s electron of hydrogen at a particular distance from the nucleus is plotted as a radial charge density against the

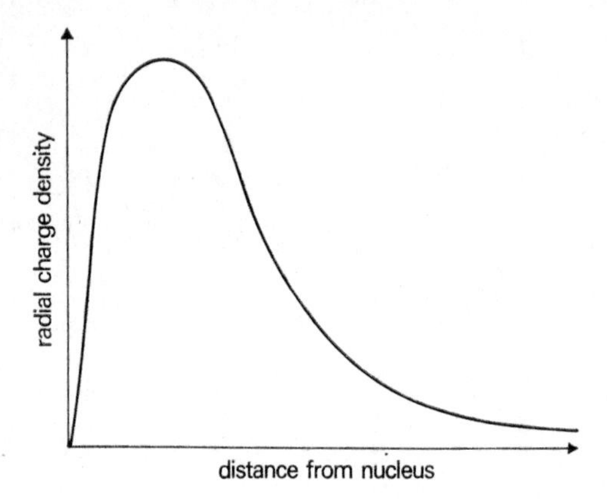

Figure 5.8.

distance from the nucleus (Figure 5.8), it can be seen that the most likely place to find the electron will be within a spherical shape fairly close to the nucleus. This shape can be thought of as the volume within which there is, say, a 99 per cent chance of finding the electron. The volume is called an atomic orbital.

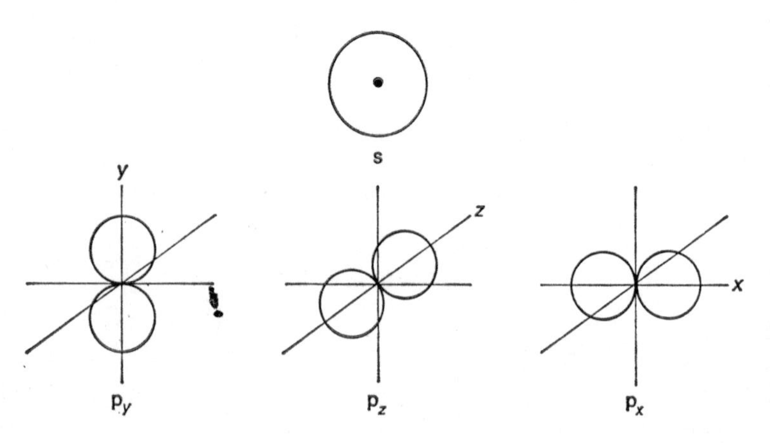

***Figure 5.9.** Two-dimensional representations of the shapes of s, p_z, p_y, and p_z orbitals*

The *s* atomic orbitals are spherical and hold a maximum of 2 electrons. The 6 electrons of a p subshell are found to be in three orbitals, each holding 2 electrons. The p orbitals, which resemble dumb-bells in shape, are on axes at 90° to each other and are referred to as the p_x, p_y, and p_z orbitals (Figure 5.9).

6. The Periodic Classification of the Elements

This very important scheme of classification was first put forward by the Russian chemist, Mendeléeff, in 1869. It had, however, certain partial forerunners in the form of *Döbereiner's Triads* and *Newlands' Law of Octaves.*

Döbereiner's Triads

In 1829, Döbereiner drew attention to the existence of sets of *three* chemical elements (hence *triads*), the members of which resemble each other closely in chemical properties and have one of two relations among their atomic masses – either the atomic masses of all three are similar, or the middle one is close to the arithmetic mean of the other two. Actual cases are:

Cl 35.5	Ca 40.1	Fe 55.85 Atomic masses
Br 80 (A.M. 81.25)	Sr 87.6 (A.M. 88.75)	Co 58.94 similar
I 127	Ba 137.4	Ni 58.69

These triads are actually selections from much broader similarities, but are of historical interest because they represent the earliest known attempt to trace an interdependence between the properties of elements and their atomic masses.

Newlands' Law of Octaves

This idea was put forward in 1867. Newlands developed the idea that, if the elements are written in order of increasing atomic masses (as below), a Law of Octaves can be discerned. By the operation of this law, the first, eighth, and fifteenth elements show chemical similarity, as do the second, ninth, and sixteenth, and the third, tenth, and seventeenth, and so on. Written as in the table below (which includes the known elements of Newlands' time), the law should place chemically similar elements in the same vertical columns. Notice that no noble gases appear; they remained unknown till the 1890s.

H	Li	Be	B	C	N	O
F	Na	Mg	Al	Si	P	S
Cl	K	Ca	Cr	Ti	Mn	Fe

Similarities such as those of F and Cl; Li, Na, and K; C and Si; N and P, are obvious, but the scheme soon broke down, e.g. Fe is not suitable for inclusion with O and S.

Here Newlands had the first glimmerings of the idea that has produced the Periodic Classification, but he did not develop it adequately. In particular, he failed to allow for the imperfect chemical knowledge of the mid-nineteenth century. It is enough to point out that the third horizontal period should have read:

Cl K Ca (Sc) Ti (V) (Cr) Mn Fe

The bracketed elements were unknown or misplaced in the Newlands table.

The Periodic Classification of the Elements (Mendeléeff, 1869)

This classification was based on the Periodic Law which, as given by Mendeléeff, stated that the properties of the elements are periodic functions of their relative atomic masses. That is, if the elements are arranged in order of increasing atomic mass, elements with similar properties occur at regular intervals. Instead of arranging the elements in the form of a continuous list, Mendeléeff placed the elements in horizontal rows (**periods**) so that elements with similar properties appeared in the same vertical column (**group**). The noble gases and a few other elements, such as gallium and germanium, were not discovered until later and were therefore absent from the early Periodic Table.

On the basis of relative atomic mass, two anomalies appeared in the early table. Cobalt (58.94) and nickel (58.69), and tellurium (127.6) and iodine (126.9), had to have their atomic mass order reversed to bring them into correct placing on chemical grounds. The strict order of atomic mass would have separated iodine from the other halogens into Group VI, while tellurium would have been separated from the closely-related elements sulphur and selenium, and placed in Group VII with the halogens. Another anomaly arose in the 1890s when the newly discovered argon (39.9) had to be put *before* potassium (39.1) to keep argon with the other noble gases and potassium with the alkali metals. These cases showed clearly that atomic mass is not really the true basis of the table.

The Periodic Table
(Mendeléeff's arrangement)

	Group								
Period	0	I A B	II A B	III A B	IV A B	V A B	VI A B	VII A B	VIII
1		H 1							
2 (*Short*)	He 2	Li 3	Be 4	B 5	C 6	N 7	O 8	F 9	
3 (*Short*)	Ne 10	Na 11	Mg 12	Al 13	Si 14	P 15	S 16	Cl 17	
4 (*Long*)	Ar 18 39.9	K 19 39.1 Cu 29	Ca 20 Zn 30	Se 21 Ca 31	Ti 22 Ge 32	V 23 As 33	Cr 24 Se 34	Mn 25 Br 35	Fe Co Ni 26 27 28 58.9 58.7
5 (*Long*)	Kr 36	Rb 37 Ag 47	Sr 38 Cd 48	Y 39 In 49	Zr 40 Sn 50	Nb 41 Sb 51	Mo 42 Te 52 127.6	Tc 43 I 53 126.9	Ru Rh Pd 44 45 46
6 (*Long*)	Xe 54	Cs 55 Au 79	Ba 56 Hg 80	La 57 58–71 Rare earths Tl 81	Hf 72 Pb 82	Ta 73 Bi 83	W 74 Po 84	Re 75 At 85	Os Ir Pt 76 77 78
7	Rn 86	Fr 87	Ra 88	Ac 89	Th 90	Pa 91	U 92		

The Rare Earths	Ce 58	Pr 59	Nd 60	Pm 61	Sm 62	Eu 63	Gd 64	Tb 65	Dy 66	Ho 67	Er 68	Tm 69	Yb 70	Lu 71

Transuranium Elements	Np 93	Pu 94	Am 95	Cm 96	Bk 97	Cf 98	Es 99	Fm 100	Md 101	No 102

Atomic masses are quoted where their order reverses the order of atomic number.

The anomalies have disappeared with the recent recognition of atomic *number* instead of atomic *mass* as the true basis of the Periodic Table. The atomic number of an element, being the number of protons on the atomic nucleus, determines the number of electrons in the atom, hence their arrangement and hence the properties of the element. The **Periodic Law** is now restated to read:

The properties of the elements are periodic functions of their atomic numbers.

The three anomalies mentioned on page 59 on the basis of atomic mass arose out of isotopy. The case of argon and potassium will illustrate the position. The principal isotopes of these elements are:

Argon				*Potassium*			
Nucleus	Protons	18	18	Nucleus	Protons	19	19
	Neutrons	18	22		Neutrons	20	22
		Ar = 36	Ar = 40			K = 39	K = 41
Atomic number 18 for both				Atomic number 19 for both			
Electrons 2, 8, 8 for both				Electrons 2, 8, 8, 1 for both			

In the case of argon, the *heavier* isotope predominates, giving an average atomic mass of 39.9; with potassium, the *lighter* isotope predominates, giving an average atomic mass of 39.1. But the order of atomic numbers clearly puts argon *before* potassium and there is no anomaly on this basis. The explanation is similar in the case of *cobalt* (no isotopy, mass number 59) and *nickel* (five isotopes, 58–64); and *tellurium* (eight isotopes, 120–130) and *iodine* (no isotopy, mass number 127).

Isotopes of a given element do not, in general, exceed a range of eight units of atomic mass. This range keeps the atomic mass order the same as the atomic number order, except for the three cases mentioned. This enabled Mendeléeff to recognize the Periodic Law and Table on the atomic mass basis though its real basis is atomic number.

Lothar Meyer's Atomic Volume Curve

The work of Mendeléeff was mainly concerned with periodicity of chemical properties, whereas, during the same year, Lothar Meyer noted a periodicity in a physical property, namely **atomic volume.** The atomic volume of an element is given by the relationship

$$\text{atomic volume} = \frac{\text{relative atomic mass}}{\text{density}}$$

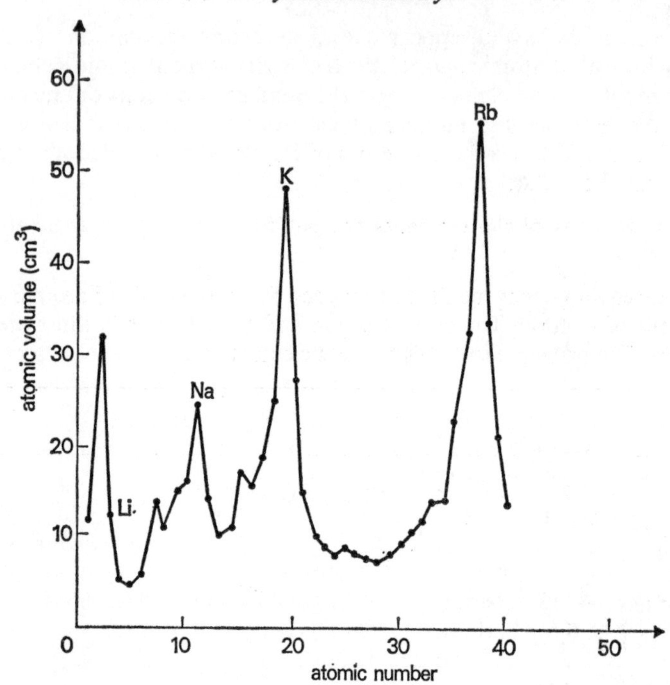

Figure 6.1. Variation of atomic volume with atomic number

The graph of atomic volume against atomic number (Figure 6.1) or, as in the case of Lothar Meyer, atomic mass, consists of a series of maxima and minima showing that the atomic volumes of elements are periodic functions of their atomic numbers. There is a close resemblance between this periodicity and that noted by Mendeleéff. For example, the elements which appear at the maxima of the curve all occur in the same group of Mendeléeff's Periodic Table.

Periodic Table and Electron Arrangements

With the development of theories of atomic structure and, in particular, the energy levels occupied by electrons, it became apparent that periodicity in physical and chemical properties is related to the electronic configurations of the elements. Considering the first twenty elements and their principal quantum energy shells, it can be seen that elements in the same group have similar outer electron arrangements:

	H 1		He 2				
Li 2,1	Be 2,2	B 2,3	C 2,4	N 2,5	O 2,6	F 2,7	Ne 2,8
Na 2,8,1	Mg 2,8,2	Al 2,8,3	Si 2,8,4	P 2,8,5	S 2,8,6	Cl 2,8,7	Ar 2,8,8
K 2,8,8,1	Ca 2,8,8,2						

When the elements beyond the first twenty are considered and the distribution of electrons into subshells is taken into account, it becomes desirable to modify the layout of the Periodic Table. The elements are better arranged in what is known as the *long form* of the Periodic Table (Figure 6.2).

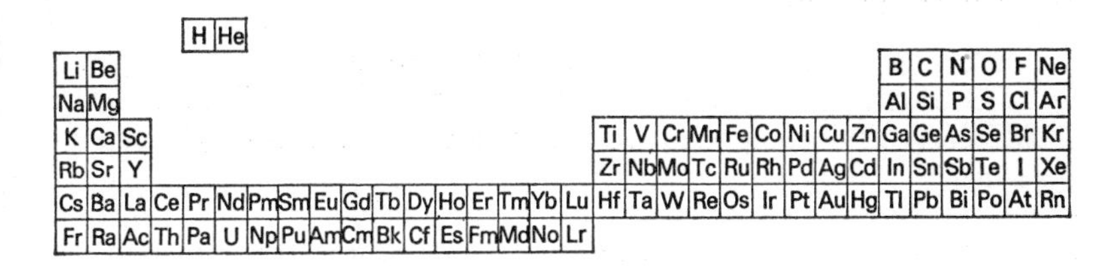

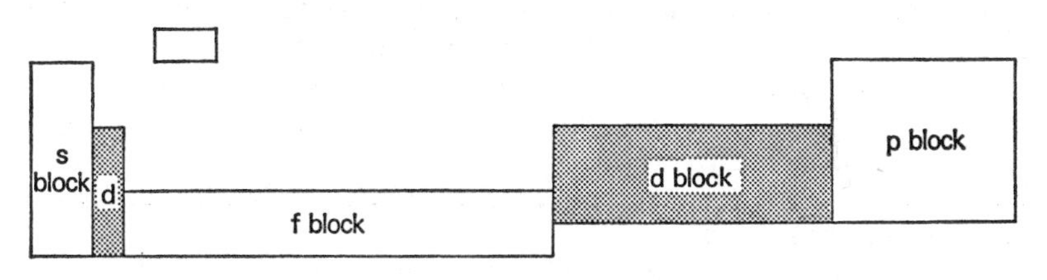

s, p, d, and f blocks of elements

Figure 6.2. ***s, p, d, and f blocks of elements in the long form of the Periodic Table***

On referring to the table on pages 286–7, it can be seen that the elements in Period 4 (atomic numbers 19–36) consist of two elements, K and Ca, with their last electrons added to an s subshell, ten elements, Sc to Zn, with their last electrons added to a d subshell, and six elements, Ga to Kr, with their last electrons added to a p subshell. Hence this part of the Periodic Table can be divided into s, d and p blocks of elements. Starting at element 58 it is also necessary to include an f block of elements.

A more detailed discussion of the variation in properties of elements and compounds in relation to the Periodic Table is given in *Inorganic Chemistry*, the companion volume to this book.

7. Bonding I: Ionic and Covalent

All pure solid substances consist of definite arrangements of particles. These particles may be atoms, charged atoms, or groups of atoms (called **ions**), or collections of atoms (called **molecules**). The physical properties of a substance give an indication of the type of particle it contains and the nature of the bonding forces involved.

1. **Conduction** If a compound, when it is converted to its liquid form, conducts electricity then the compound contains ions. Whereas, if the liquid form of a compound does not conduct it probably exists as molecules.

2. **Melting point** A high melting point indicates that the forces being overcome during melting are strong. This usually indicates that the solid exists in the form of a giant lattice. This is a large continuous arrangement of ions or atoms which can only be broken down (i.e. melted) by overcoming the main bonding forces between the atoms or ions.

A low melting point usually indicates that the substance exists as a molecular lattice. This consists of molecules held in a particular arrangement by weak attractive forces. When the substance melts, the weak attractive forces between the molecules are overcome but the molecules themselves remain intact.

3. **X-ray diffraction** When light is passed through a series of very fine slits (diffraction grating) each slit acts as a secondary wave source. This diffraction of the light only occurs when the width of the slits is of the same order as the wavelength of the light. The wave fronts from these secondary sources will meet and interference will occur. Where the waves are in phase they will reinforce each other and produce a brighter light, but where they are out of phase they will tend to cancel each other out and leave a dark area. Thus, on viewing the light through the grating, a series of maxima and minima, called a **diffraction pattern**, is observed. In a similar manner, when X-rays, which have a much smaller wavelength than visible light, are reflected from the electron clouds of successive layers of a crystal, reinforcement will occur when the reflected rays are in phase. The angles at which the reinforcements occur will depend on the dimensions of the crystal. The relationship between this angle, the wavelength of the

X-rays, and the distance between successive layers in the crystal was originally determined by Bragg in 1912.

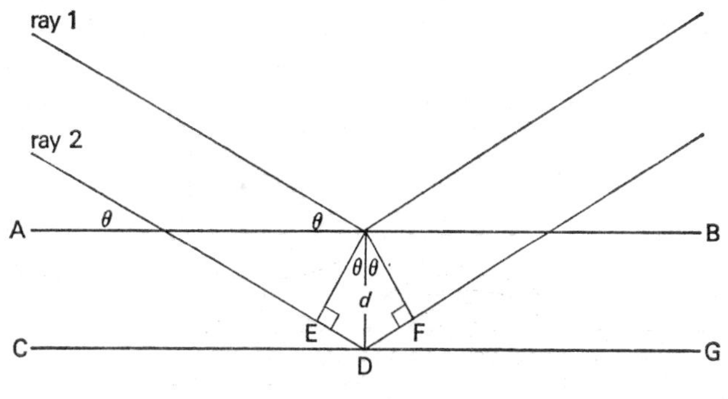

Figure 7.1.

In Figure 7.1. AB and CG represent successive layers in a crystal and d is the distance between the layers. A parallel beam of X-rays of wavelength λ is incident at an angle θ. Ray 2 travels a distance ED + DF further than ray 1. If this distance is equal to a whole number of wavelengths, the reflected rays will be in phase and reinforcement will occur.

$$ED = d \sin \theta$$
$$DF = d \sin \theta$$
$$ED + DF = 2d \sin \theta$$

For reinforcement
$$n\lambda = 2d \sin \theta$$

where n is a whole number.

It is therefore possible to use the results of X-ray diffraction to calculate a value for d and, if the crystal is investigated from various angles, the dimensions of the crystal lattice can be determined. This technique is called X-ray crystallography and a more detailed discussion of structures which have been elucidated by this method is given on page 86.

The wavelength of electrons (page 22) is such that electron diffraction patterns may be obtained in much the same way as X-ray diffraction patterns. Electron diffraction is of particular value in organic chemistry as it reveals the positions of hydrogen atoms more readily than the X-ray method.

In addition, X-ray and electron diffraction provide useful information concerning the way in which the atoms are bonded together. The X-rays are reflected by the electrons and the intensities of the reflections are related to the electron densities around the nuclei. The distribution of electrons can be represented by an electron density map which is drawn so that it is consistent with the results of the X-ray diffraction. Such a map consists of a series of contours where each contour joins all the points with identical electron densities. The electron density map of a substance containing ions is very different from a molecular substance (see Figure 7.2).

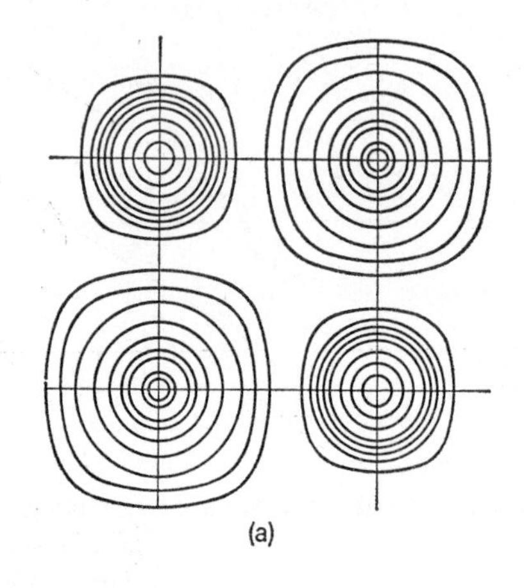
(a)

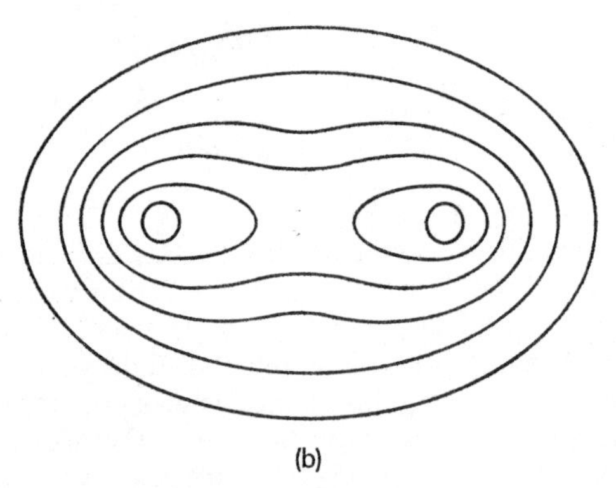
(b)

Figure 7.2. ***Electron density maps of (a) an ionic compound, sodium chloride, and (b) a molecule, hydrogen***

The electron density map of the ionic compound indicates that the substance consists of an arrangement of almost spherical ions with very low electron density between them, suggesting that the bonding forces which hold the ions together are due entirely to the electrostatic attraction between the oppositely charged ions.

The map of a molecular substance shows that a molecule consists of an arrangement of atoms with a relatively high electron density between the atoms. It therefore appears that the bonding forces in this case are due to the sharing of electrons. Any electrons shared between two nuclei will be attracting both nuclei, which will result in a net attractive force holding the nuclei together (see Figure 7.3).

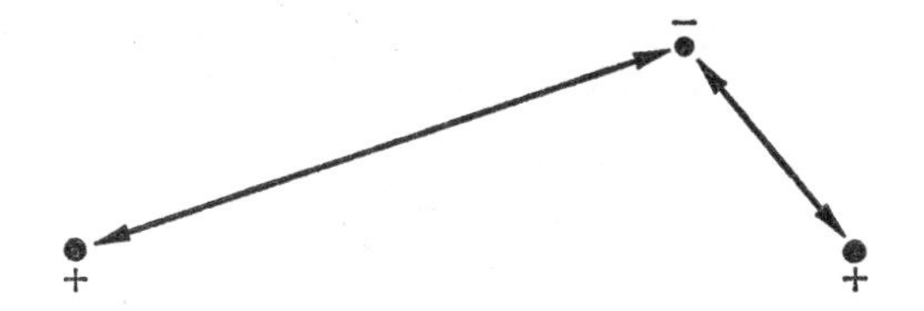

Figure 7.3. A shared electron holding two nuclei together

The bonding which involves ions is called **electrovalent** or **ionic bonding** and that which involves sharing of electrons is called **covalent bonding.** In order to see how the two types of bonding arise it is necessary to consider the electronic configurations of the elements involved.

Ionic Bonding

We shall first consider a specific case of electrovalent combination, that between sodium and chlorine. A sodium atom contains 11 nuclear protons and has the electron structure 2, 8, 1. It differs from the nearest noble gas electron structure, that of neon, 2, 8, by the presence of one extra electron in the third principal shell. A free chlorine atom contains 17 nuclear protons and has the electron structure 2, 8, 7. It differs from the argon electron structure, 2, 8, 8, by a deficiency of one electron in the third principal shell.

During chemical combination of the sodium and chlorine atoms, the single electron from the outermost orbit of the sodium atom passes over to the outermost orbit of the chlorine atom. In this way, two *ions* are produced. The sodium ion is positively charged, as Na^+, by the nuclear proton left in excess after the departure of the electron, and the electron structure is now 2, 8. The chlorine ion is negatively charged, as Cl^-, by the acquired electron, and its electron structure is now 2, 8, 8. In both cases, the ions have now the electron structure of a rare-gas (neon and argon respectively) with the external electron *octet.* They differ, however, from the neutral atoms of these noble gases by carrying their respective ionic charges. *No molecules* are formed in electrovalent combination. Any experimental quantity of sodium chloride contains an enormous number of chloride and sodium ions in equal proportions. The electrical attraction resulting from their opposite charges constitutes such chemical 'bond' as exists at all. The ions arrange themselves into a crystal lattice (see Figure 7.4) —in this case, a face-centred cube—but there is no specific pairing of ions

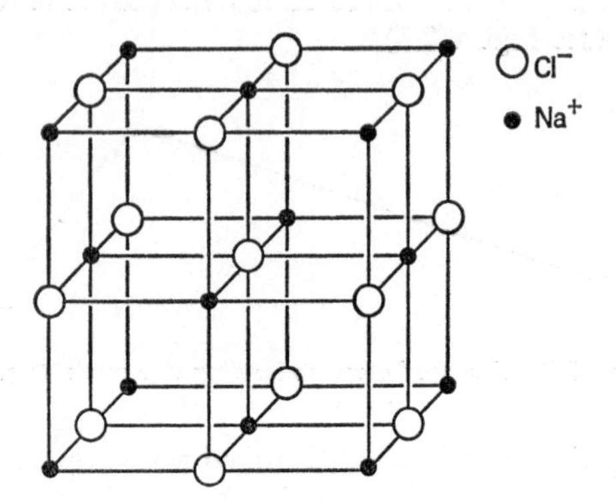

Figure 7.4. Crystal lattice of sodium chloride

which could be considered as a molecule of sodium chloride. The change during combination is expressed diagrammatically below.

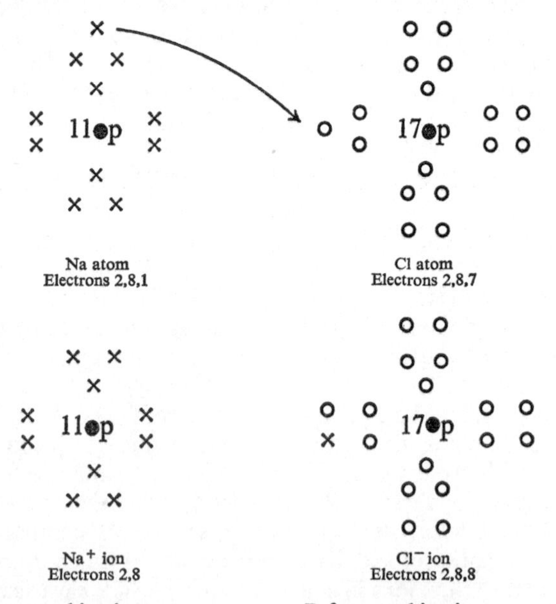

Before combination
Neutral Na atom, electrons, 2,8,1
After combination
Na^+ ion, electrons, 2,8

Before combination
Neutral Cl atom, electrons, 2,8,7
After combination
Cl^- ion, electrons, 2,8,8

In terms of subshells, each sodium atom loses one electron from a 3s subshell and each chlorine gains an electron into a 3p subshell.

$$\underset{1s^2\,2s^2\,2p^6\,3s^1}{Na} + \underset{1s^2\,2s^2\,2p^6\,3s^2\,3p^5}{Cl} \rightarrow \underset{1s^2\,2s^2\,2p^6}{Na^+} + \underset{1s^2\,2s^2\,2p^6\,3s^2\,3p^6}{Cl^-}$$

An ionic compound will be formed if the total energy change involved is favourable. In the case of sodium chloride, the energy required to remove an electron from sodium is more than made up for by the energy released when the electron is gained by chlorine and the oppositely charged ions come together to form the ionic lattice. A more detailed discussion of the energy changes involved in the formation of ionic compounds is given on page 189.

Other examples of ionic compounds are shown in the following table.

Potassium sulphide

Before combination:	*Two K atoms*	*One S atom*
Nuclear protons	19 each	16
Electrons	{2,8,8,1; 2,8,8,1}	2,8,6
After combination:		
Nuclear protons	19 each	16
Electrons	{2,8,8; 2,8,8}	2,8,8
	Two K^+ ions	One S^{2-} ion
	Potassium sulphide, $2K^+ \cdot S^{2-}$	

Magnesium oxide

Before combination:	*One Mg atom*	*One O atom*
Nuclear protons	12	8
Electrons	2,8,2	2,6
After combination:		
Nuclear protons	12	8
Electrons	2,8	2,8
	One Mg^{2+} ion	One O^{2-} ion
	Magnesium oxide, $Mg^{2+} \cdot O^{2-}$	

The three examples discussed so far exhibit certain features in their bonding which are common to a large number of ionic compounds, i.e.

1. A metal loses one or more electrons to form a positive ion (*cation*).
2. A non-metal gains one or more electrons to form a negative ion (*anion*).
3. By obtaining an external octet of electrons (complete s and p subshells), both elements attain an electron structure which is identical to that of the nearest noble gas.
4. The ratio of the numbers of ions combining to form the lattice is such that the compound is neutral over all.

The octet rule does not apply to d-block elements. For example, zinc has an electron structure of 2, 8, 18, 2 and zinc ions have a relative charge of 2 − and therefore an electron structure of 2, 8, 18. Also, more complex ions, such as NH_4^+ and SO_4^{2-}, which will be discussed in detail later, achieve octets by electron sharing within the ions.

Characteristic properties of ionic compounds

1. Ionic compounds contain no molecules. They are aggregates of ions. If the ions are made mobile by dissolving the compound in water or by melting it, the resulting solution or melt will conduct an electric current with decomposition, i.e. ionic compounds are *electrolytes*.

2. The electrical forces between the oppositely charged ions are powerful. Consequently, ionic compounds form comparatively rigid crystal lattices and are *solids*. They melt at high temperatures (compared with simple covalent compounds) and are non-volatile.

3. Ionic compounds are rarely soluble in organic liquids, but many of them dissolve in water.

Covalency

We shall consider, as typical of covalency, the formation of a chlorine molecule from two chlorine atoms. These two atoms each have an electron structure of 2, 8, 7. The seven electrons in the third principal quantum shell are distributed as

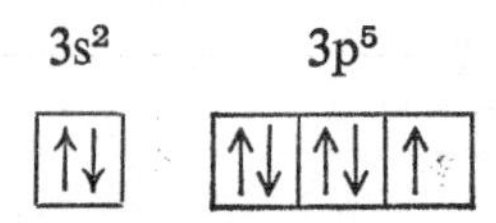

Each chlorine atom has one unpaired electron and is one electron short of the nearest noble gas structure. No other element is available from which electrons may be obtained; therefore the stable pairing of electrons can only be achieved by the unpaired electron from each chlorine atom being shared between the two atoms. This means that neither atom actually achieves a noble gas structure but, by forming the shared pair, they each have a share of eight electrons. The theory is expressed diagrammatically below.

```
  × ×   × ×
× Cl × Cl ×
×    ×    ×
  × ×   × ×
```

A shared pair of electrons is said to constitute one covalent bond and is conventionally represented by a stroke, —:

Cl—Cl

It is important that you are not misled, by either of these diagrams, into thinking that the sharing of a pair of electrons is a static affair. Electrons are in a continuous state of motion and, as discussed on page 55, it is not possible to define both the path and velocity of an electron. It is therefore best to regard the shared pair as a pair of electrons which spend most of their time between the two atoms. (See page 80 for a discussion of covalency in terms of atomic and molecular orbitals.) Other simple examples of covalency are shown below.

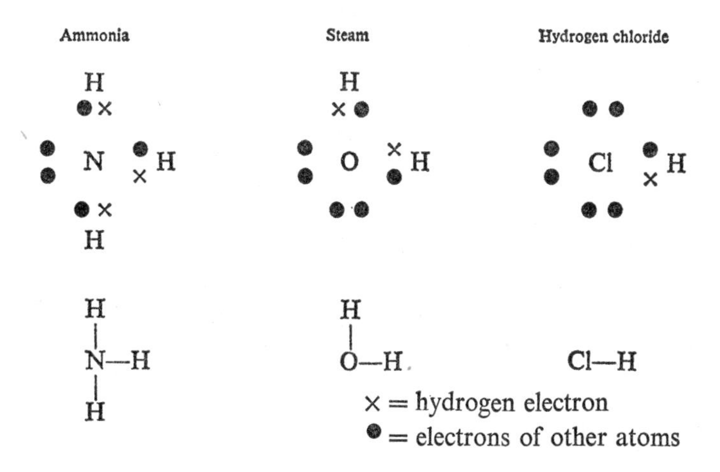

The following are rather more complicated examples of covalent molecules. Note that some of these molecules include atoms which share more than one pair of electrons and are thus being held together by multiple bonds. In all these examples, no attempt is made to indicate the three-dimensional shape of the molecules. This will be dealt with later, page 78.

Properties of covalent compounds

1. Covalency produces genuine molecules, never ions. Consequently, covalent compounds are non-electrolytes, except for a few examples which form ions when they dissolve in water (e.g. hydrogen chloride).

Ethane, C_2H_6

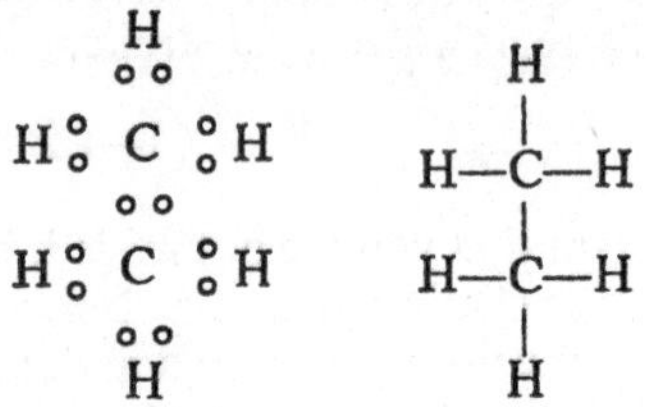

```
     H
     oo                H
  H: C :H              |
     oo             H—C—H
  H: C :H              |
     oo             H—C—H
     H                 |
                       H
```

Tetrachloromethane, CCl_4

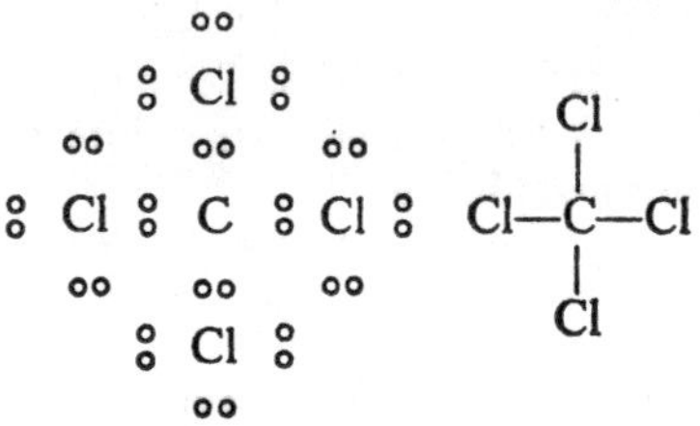

```
          oo
        : Cl :              Cl
  oo      oo      oo        |
: Cl  :   C   : Cl :    Cl—C—Cl
  oo      oo      oo        |
        : Cl :              Cl
          oo
```

Ethanol, C_2H_5OH

```
    H    H                 H  H
    oo   oo   oo           |  |
H : C :  C :  O : H     H—C—C—O—H
    oo   oo   oo           |  |
    H    H                 H  H
```

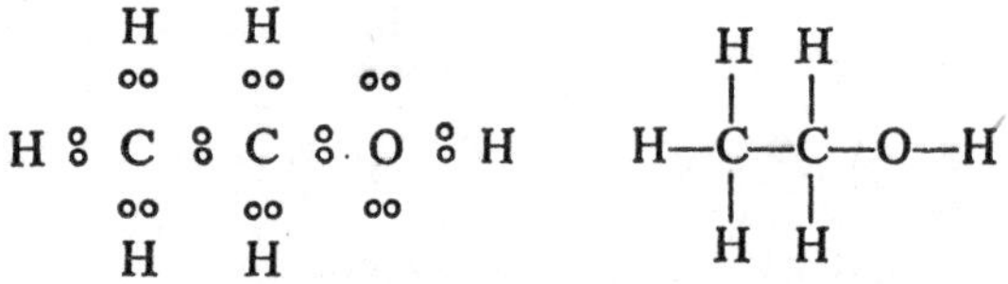

Ethanal (acetaldehyde), CH_3CHO

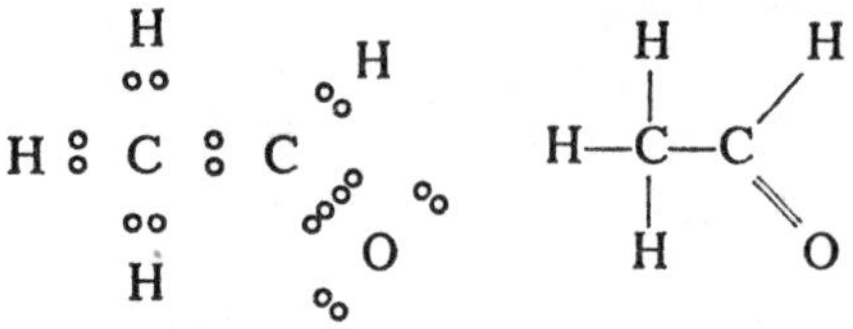

```
     H                     H     H
     oo      H             |    /
H :  C :  C                H—C—C
     oo        O           |    \\
     H                     H     O
```

Ethene, C_2H_4

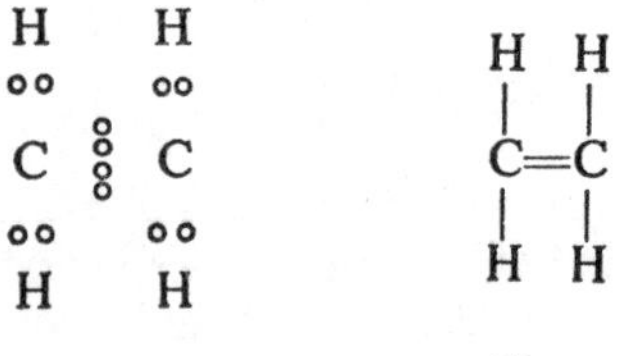

```
  H    H          H  H
  oo   oo         |  |
  C  ::  C        C=C
  oo   oo         |  |
  H    H          H  H
```

Methyl cyanide, CH_3CN

```
    H                        H
    oo                       |
H : C : C ::: N :         H—C—C≡N
    oo                       |
    H                        H
```

Carbon dioxide, CO_2

```
 oo           oo
 O  ::  C  ::  O          O=C=O
 oo           oo
```

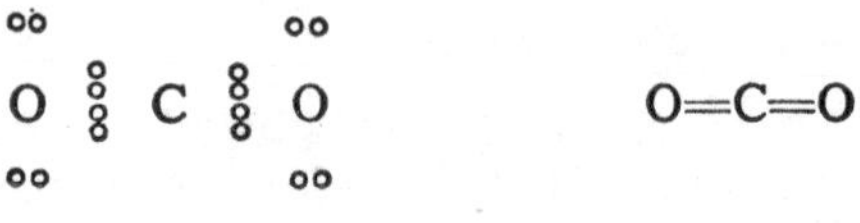

2. Unlike a solid ionic compound, which consists of a continuous giant lattice of oppositely charged ions, the covalent compound consists of an arrangement of molecules called a molecular lattice. Whereas the forces holding the atoms together within the molecules may be of comparable strength to the forces holding ions together, the forces which hold the molecules together are weak. The molecules which make up the molecular lattice are easily separated and simple covalent compounds are always volatile. Many of them are gases (e.g. chlorine, Cl_2; oxygen, O_2; methane, CH_4; ammonia, NH_3; and carbon dioxide, CO_2) or volatile liquids (e.g. trichloromethane, $CHCl_3$; ethanol, CH_3CH_2OH; and carbon disulphide, CS_2). More complex covalent compounds may be solids, e.g. the higher alkanes, C_nH_{2n+2}, where n is 18 or more. The weak forces which hold the molecules together in a molecular lattice are known as van der Waals' forces (page 103).

3. Covalent compounds are often insoluble in water, unless they contain such groups as the hydroxyl (—OH), but they usually dissolve in other covalent, organic solvents such as benzene or ether (ethoxyethane).

It is important to realize that the two types of bonding described so far, ionic and covalent, represent two extreme cases. In fact, covalent bonds between atoms of different elements can rarely be said to be entirely covalent as they usually have some ionic character (page 97) and only certain metal salts can be thought of as 100 per cent ionic (page 191).

Dative or Co-ordinate Covalency

This is a type of covalency. A shared pair of electrons is formed between the combining atoms in both these bonding types. The difference is that in covalency each atom contributes one electron to the shared pair, while in a dative bond one atom provides both electrons. The other atom only accepts the shared pair.

An example of this type of covalency is the formation of the stable molecular compound between ammonia, NH_3, and boron trichloride, BCl_3. The latter compound is purely covalent but boron, having only three electrons in its outer energy shell, makes up only a sextet of electrons by covalent sharing. Ammonia is a covalent compound with three shared pairs of electrons and an unshared pair (lone-pair). The lone-pair is donated from the nitrogen to be shared between the nitrogen atom and the boron atom, thus making up the boron octet of electrons.

```
         ××
        ×Cl×
        ×  ×                    H
  ××     o×                     o×
 ×Cl o   B                    o N o H
 ×   ×                        o   ×
  ××     o×                     ×o
        ×Cl×                    H
        ×  ×
         ××
```

```
         ××
        ×Cl×                    o = electron of B or N
        ×  ×  H.                × = electron of Cl or H
  ××     ×o   o×
 ×Cl o   B o  N o H
 ×   ×     o    ×
  ××     o×   o×
        ×Cl×  H
        ×  ×
         ××
```

A co-ordinate or dative bond may be represented in several ways: either as a stroke, —, as in normal covalent bonding; e.g.

```
     Cl  H
     |   |
 Cl—B—N—H
     |   |
     Cl  H
```

or as an arrow, →, which indicates the direction in which the electrons are donated, e.g.

```
     Cl   H
     |    |
 Cl—B ← N—H
     |    |
     Cl   H
```

or by a stroke and, because one atom has in effect lost an electron and the other has gained one, a positive and negative charge.

```
     Cl  H
     |- +|
 Cl—B—N—H
     |   |
     Cl  H
```

The arrow is the most common and preferred way of representing the bond. Two more important examples of co-ordinate or dative covalency are the ammonium, NH_4^+, and the hydroxonium, H_3O^+, ions.

The ammonium ion, NH_4^+*, and ammonium chloride*

The molecule of ammonia, NH_3, is covalent and contains a lone-pair of electrons. (See page 73). It is electrically neutral. If ammonia comes into contact with an acid, e.g. HCl in ionized condition, H^+Cl^-, the hydrogen ion (or proton) accepts a share in the lone-pair of electrons, combines in this way with the NH_3 molecule, and produces the ion, NH_4^+, positively charged by the accepted proton. This particle is associated by electrovalency with the Cl^-, which has accepted an electron from a hydrogen atom to supply the ion, H^+. Ammonium chloride thus shows all three of the common valency types. The electrovalency is predominant in determining the properties of the salt, which is, consequently, a crystalline solid, soluble in water and an electrolyte when in solution.

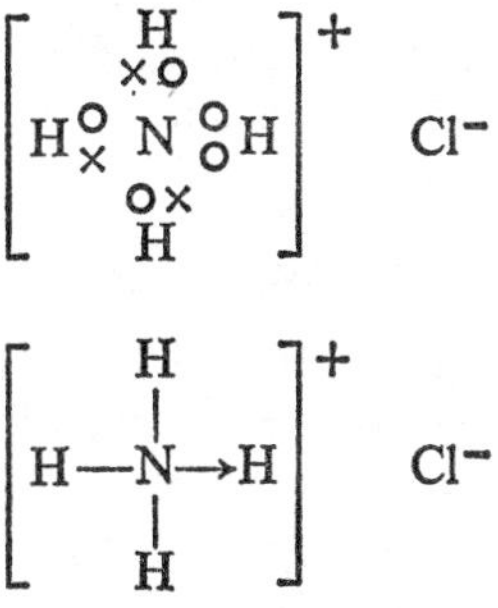

```
⌈     H     ⌉+
|    ×o     |
| H°× N °°H |      Cl⁻
|    o×     |
⌊     H     ⌋

⌈     H     ⌉+
|     |     |
| H — N → H |      Cl⁻
|     |     |
⌊     H     ⌋
```

Formula of ammonium chloride

The arrow, as a symbol for a co-ordinate bond, is a useful means of indicating the source of the electrons which constitute the bond, but in cases such as the ammonium ion it can be misleading as it suggests that one bond is different from the other four. In fact this is not the case as the nitrogen is bonded to four atoms of the same element and all four bonds, once they have been formed, are identical.

The hydroxonium ion, H_3O^+

The characteristic behaviour of an acid is the production of 'hydrogen ions' in water. In their simplest form, H^+, these are protons. The observed electrical conductivities of acids are, however, too low to admit of the presence of such a light, highly charged, mobile particle and it is certain that the proton is at once hydrated to form the hydroxonium ion.

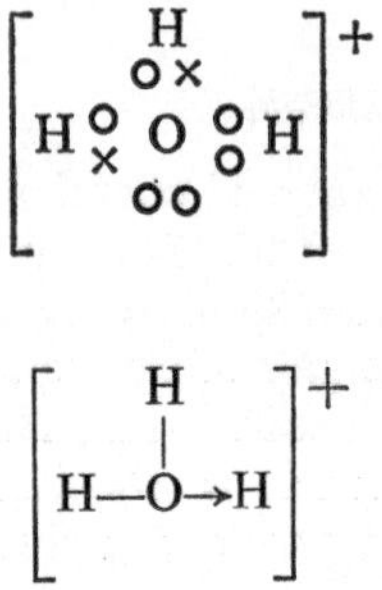

```
[     H     ]+
     o×
[ H°× O °°H ]
     oo
```

o = electron from O
× = electron from H

$$\left[\begin{array}{c} H \\ | \\ H—O\rightarrow H \end{array}\right]^{+}$$

Complex Ions

The so-called complex ions involve co-ordinate or dative covalency, and are formed by ions or molecules which possess lone-pairs of electrons, donating their lone-pairs to be shared between themselves and a metal ion. The group which donates the lone-pair is called a *ligand* and the metal ion is usually an ion of a transitional metal.

The hexacyanoferrate(II) ion, $[Fe(CN)_6]^{4-}$

This complex ion is formed by the addition of an excess of a solution of cyanide ions to a solution containing iron(II) ions. Each CN^- has a lone-pair of electrons which it is capable of donating. A neutral atom of iron has the following electron arrangement:

Fe 2, 8, 14, 2 or $1s^2, 2s^2, 2p^6, 3s^2, 3p^6, 3d^6, 4s^2$

An iron(II) ion has the structure

Fe^{2+} 2, 8, 14 or $1s^2, 2s^2, 2p^6, 3s^2, 3p^6, 3d^6$

Each iron(II) ion is capable of co-ordinating with six cyanide ions, thus giving it a share in another twelve electrons. This means that the iron atom becomes associated with the same number of electrons as the nearest noble gas, krypton, 2, 8, 18, 8.

The charge on the complex ion produced is 4−, which is the algebraic sum of the charges on the Fe^{2+} and the six CN^-.

$$Fe^{2+} + 6(CN^-) \rightarrow [Fe(CN)_6]^{4-}$$

Each iron combines with six ligands and is said to have a co-ordination number of six.

This particular ion is a complex anion, hence the use of the *-ate* ending in the name.

Tetraamminecopper(II) ion, $[Cu(NH_3)_4]^{2+}$

This ion is formed when an excess of ammonia solution is added to a solution containing copper(II) ions. The ammonia molecule is electrically neutral and it has a lone-pair of electrons. A neutral atom of copper has the electron structure

Cu 2, 8, 18, 1 or $1s^2, 2s^2, 2p^6, 3s^2, 3p^6, 3d^{10}, 4s^1$

A copper(II) ion has the structure

Cu^{2+} 2, 8, 17 or $1s^2, 2s^2, 2p^6, 3s^2, 3p^6, 3d^9$

Each copper(II) ion can co-ordinate with four ammonia molecules (the resulting number of electrons associated with the copper atom being one less than the krypton electron structure). The complex ion formed has a net charge of 2^+.

$$Cu^{2+} + 4NH_3 \rightarrow [Cu(NH_3)_4]^{2+}$$

Strictly, the formula of the ion is $[Cu(NH_3)_4(H_2O)_2]^{2+}$ and should be called tetraamminediaquocopper(II). The two water ligands are more distant from the copper ion and are held less strongly than the ammonia ligands. The structures of this ion and other complex ions are discussed in *Inorganic Chemistry*, the companion volume to this book.

The Importance of Electron Pairing

Some elements in the third period and above form compounds in which they do not adhere to the octet rule, thus indicating that the pairing off of unpaired electrons is of more fundamental importance than achieving a noble gas structure.

Phosphorus forms two chlorides: the lower one, PCl_3, conforms to the octet rule, but the higher one, PCl_5, does not. Phosphorus has five electrons in the outer shell and they would normally be distributed as

$3s^2$	$3p^3$		
↑↓	↑	↑	↑

Chlorine, with its electron structure of 2, 8, 7, has one unpaired electron. When PCl_3 is formed all the unpaired electrons are paired off and both atoms obtain a share in a complete octet of electrons.

The formation of PCl_5 can be accounted for by the electrons of phosphorus rearranging to a state of slightly higher energy in which there are five unpaired electrons.

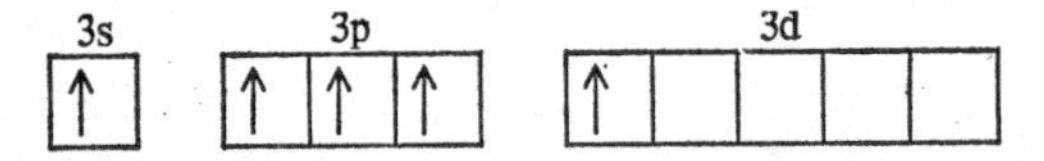

Each of these unpaired electrons forms a shared pair with an electron from a chlorine atom.

This results in phosphorus having a share in ten electrons in its outer shell. Another example of expansion of an octet is in the formation of sulphur hexafluoride in which the sulphur gains a share in twelve electrons. The expansion above the octet does not occur in the second period (Li→Ar) as there is not a corresponding d energy level available.

The Shape of Covalent Molecules

The covalent bonding force arises from the electrostatic attraction that a shared pair of electrons exerts on two nuclei and, as such, it is specific to the two atoms between which the two electrons are shared. On the other hand, the bonding forces within an ionic lattice are due to the electrostatic attraction between oppositely charged ions. Such forces will be non-directional as a positive ion will be able to attract negative ions from all directions.

Thus, one very important distinction between covalent and ionic bonding is that the covalent bond is directional and the ionic bond is not.

The directional nature of the covalent bond and the repulsive forces between the pairs of electrons involved account for the observed shapes of covalent molecules. (The shapes of covalent molecules can be deduced from X-ray and electron diffraction studies (page 64) and they are usually defined by stating the angles between the bonds.)

The four covalent bonds of carbon are arranged with an angle of 109.5° between any pair of bonds. This is explained in terms of the repulsions between the bonding pairs of electrons, resulting in them being as widely spaced as possible and forming a tetrahedral shape.

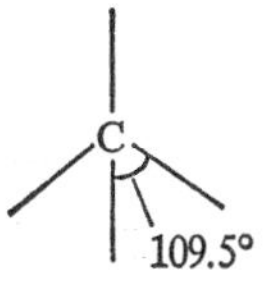

C

109.5°

The angles between the N — H bonds in ammonia are found to be 107.3°. This angle is slightly less than the tetrahedral angle and is explained by the repulsions between three bonding pairs and the slightly greater repulsion from a lone-pair (non-bonding pair). The reason for the greater repulsion of the non-bonding pair is thought to be that its electron cloud is not stretched out by the presence of another atom, which results in there being a greater concentration of negative charge nearer the nucleus and hence nearer to the other bonds.

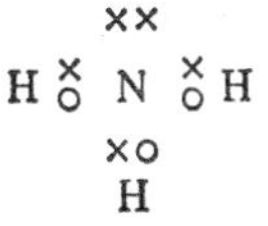

××
 H ×○ N ×○ H
 ×○
 H

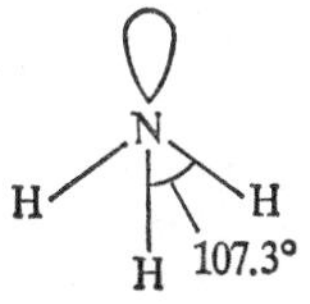

N
 H H
 H 107.3°

Water has two bonding pairs and two non-bonding pairs. Therefore, as would be expected, the angle between the OH bonds, 104.5°, is slightly less than that between the NH bonds in ammonia.

××
 H ×○ O ××
 ×○
 H

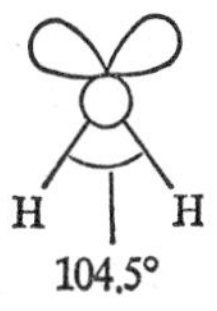

H H
 104.5°

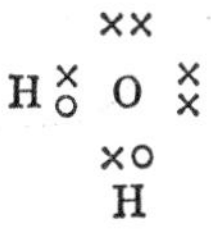

The shapes of some other molecules and ions are given below.

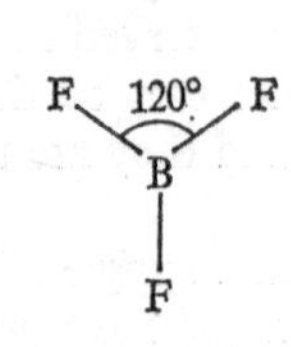

BF_3 is planar as boron does not have a lone-pair.

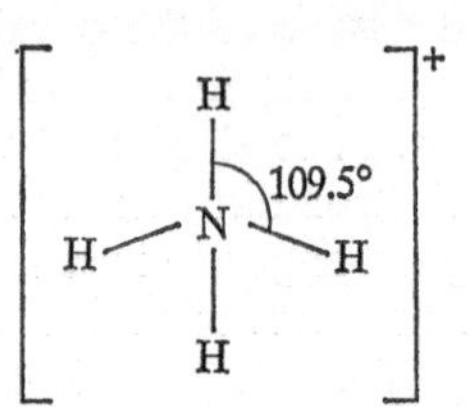

NH_4^+ is tetrahedral as there are four bonding pairs.

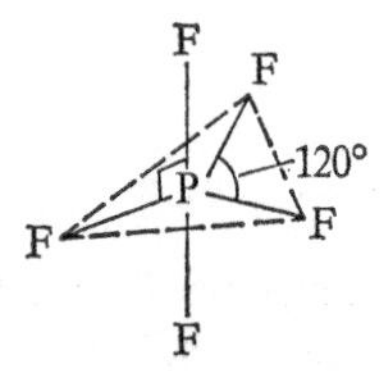

A single molecule of PF_5 exists in the form of a trigonal bipyramid.

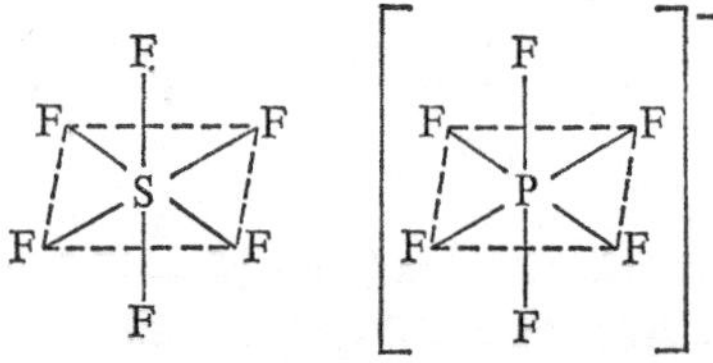

SF_6 and PF_6^- are both octahedral.

The Orbital Approach to Covalent Bonding

As previously stated (page 55), it is not possible to determine both the path and velocity of an electron but it is possible to determine the rough outline of a shape in which the electron is most likely to be found. This shape is called an **atomic orbital,**

The formation of a covalent bond can be considered as the overlapping of two atomic orbitals, each containing one electron, to form a new shape, called a **molecular orbital**, in which there is a high probability of finding the pair of electrons forming the bond.

When a hydrogen molecule is formed, the 1s orbital of one hydrogen atom overlaps the 1s orbital from another to form a molecular orbital as shown below

H + H → H_2

The bond formed by the overlapping of two s orbitals is called a σ(sigma; bond. When a covalent bond is formed by two p orbitals (as shown below), each containing one electron of opposite spin, it is called a π(pi) bond.

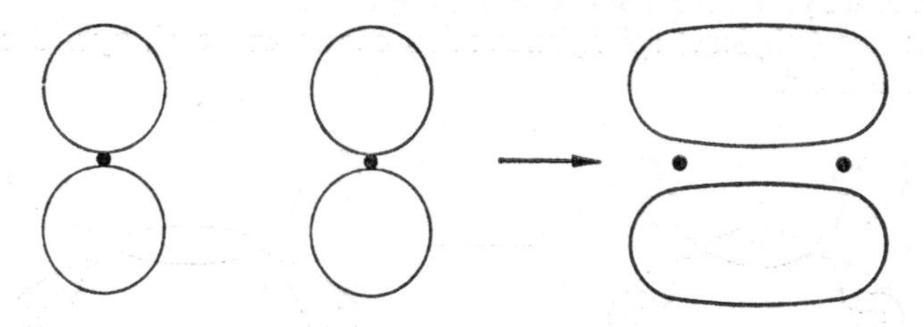

The π bond, unlike the σ bond, has a very low electron density along the axis which joins the two nuclei.

For carbon to form four covalent bonds it must have available four unpaired electrons. These are obtained by the electrons of carbon re-arranging to form a state of slightly higher energy than the ground state of carbon.

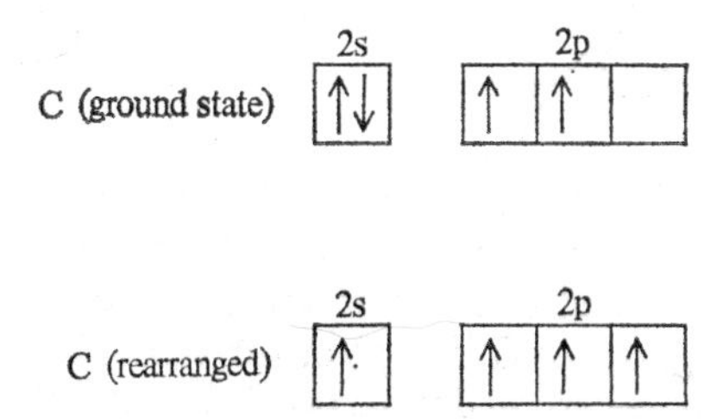

In order to form four identical bonds, as in methane, it is considered that the one electron from the s orbital and the three electrons from the p orbitals are amalgamated to form four bonds of equal energy which point towards the corners of a tetrahedron.

The changing of a number of orbitals of more than one type to form the same number of orbitals of equal energy is called **hybridization.** The hybridization involved in the tetrahedral covalency of carbon is called sp^3 hybridization which indicates the number and origin of the electrons involved. In methane each of the sp^3 hybrid orbitals overlaps with an s orbital of a hydrogen atom and forms a σ bond.

In ethene (ethylene) sp^2 hybridization occurs and each hybrid orbital forms a σ bond.

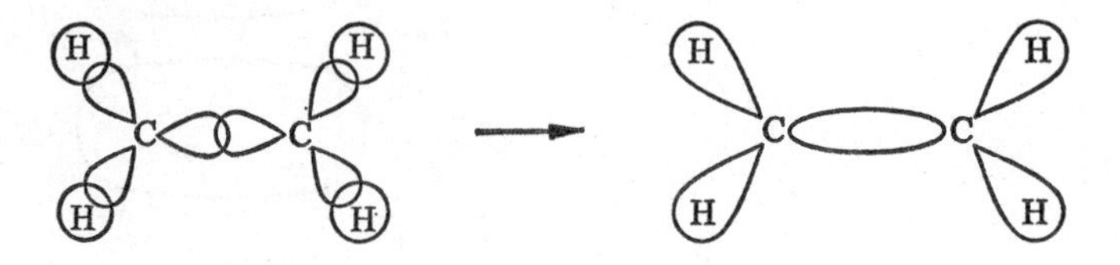

The fourth, unpaired, electron of each carbon atom is in a p orbital. The two p orbitals overlap to form a π bond.

It is the π bond of ethene which is particularly susceptible to attack by an electron seeking group (electrophilic) during the characteristic addition reactions. The π bond also accounts for the geometric isomerism (see below) exhibited by alkenes. This type of isomerism is due to the restricted rotation of the carbon–carbon double bond. The rotation of such a bond is difficult as it would destroy the overlap between the two p orbitals forming the π bond.

cis-butenedioic acid
(maleic acid)

trans-butenedioic acid
(fumaric acid)

Delocalization

Benzene

Using the theory that a covalent bond results from a shared pair of electrons, benzene requires an arrangement of alternate double and single bonds.

```
          H
          |
          C
        //  \
   H—C        C—H
     |        ||
   H—C        C—H
       \\   /
          C
          |
          H
```

There is a lot of evidence to suggest that this does not represent the true structure.

1. X-ray and electron diffraction experiments show that the carbon–carbon internuclear distance is identical for all six bonds. The bond length is 0.139 nm which is intermediate between that of a single carbon–carbon bond, 0.154 nm, and a double carbon–carbon bond, 0.134 nm.

2. The experimentally determined standard enthalpy of hydrogenation of benzene to form cyclohexane, i.e.

   $$C_6H_6(l) + 3H_2(g) \rightarrow C_6H_{12}(l) \qquad \Delta H^{\ominus} = -208 \text{ kJ mol}^{-1}$$

   is less than three times the standard enthalpy of hydrogenation of a normal carbon–carbon double bond as found in alkanes.

   $$C_2H_4(g) + H_2(g) \rightarrow C_2H_6(g) \qquad \Delta H^{\ominus} = -120 \text{ kJ mol}^{-1}$$

   For three double bonds the predicted enthalpy change is -360 kJ. The difference between these values, 152 kJ mol^{-1}, represents the margin by which the actual structure of benzene is more stable than the theoretical alternate double and single bond structure.

From such evidence it is clear that all six bonds in benzene are identical and involve more electron sharing than a normal carbon–carbon single bond but less sharing than a normal double bond. It is considered that such a structure is achieved by the six electrons, which might be expected to form the second bond of each double, being shared over all six carbon atoms. The sharing of electrons over more than two atoms is referred to as delocalization and energy associated with the resulting extra stability is known as the delocalization energy.

Using the molecular orbital approach, each carbon atom forms three normal covalent bonds by sp^2 hybridization. The fourth electron from each carbon atom is in a p orbital at right angles to the plane of the benzene ring. Each of the six p orbitals overlaps with both adjacent orbitals producing delocalized π orbitals as follows:

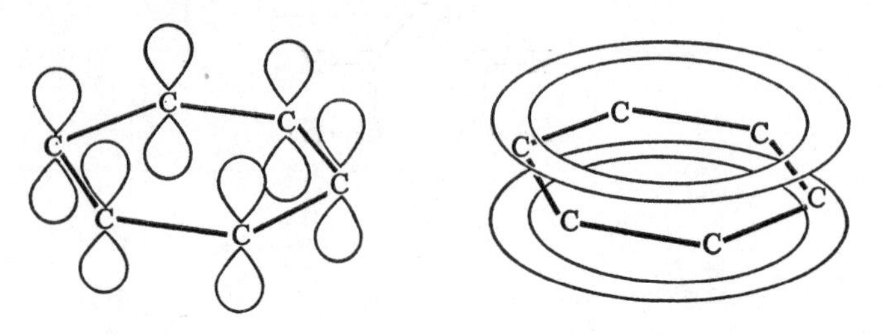

This view of the bonding in benzene leads to the coventional symbol for benzene being

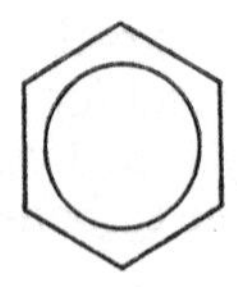

Carboxylic acid anions

Using normal covalent bonds the ethanoate (acetate) ion could be written in two ways:

$$CH_3{-}C({=}O){-}O^- \quad \text{or} \quad CH_3{-}C({-}O^-){=}O$$

It is not suggested that these two structures are isomers but that, if one particular ion is being considered, there are two possible ways in which the bonds could be arranged.

The carbon–oxygen bond lengths are found to be identical, which indicates that neither of the above formulae are correct. The true structure is intermediate between those given above and is formed by electrons being delocalized over all three atoms. The true structure could be represented as follows:

$$\left. CH_3{-}C\begin{matrix} \cdots O \\ \cdots O \end{matrix} \right\}^-$$

Nitrate ion

The bonds in a particular nitrate ion could be arranged in several ways; three are

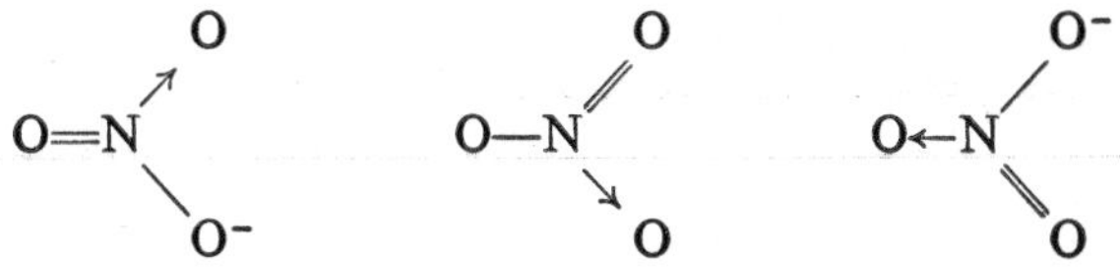

O   O O^-

O=N   O—N   O←N

O^- O   O

However, delocalization occurs and all three nitrogen–oxygen bonds are identical. The bond length is intermediate between the length of a N—O bond and a N=O bond.

Sulphate ion

Using the available electrons, there are several different ways of representing the bonds in a sulphate ion, but it is found that the ion contains four identical bonds arranged tetrahedrally around the central sulphur atom. Thus, once again, delocalization must have occurred.

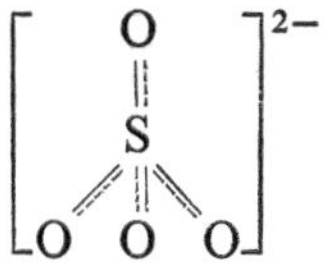

$$\left[\begin{array}{ccc} & O & \\ & \| & \\ & S & \\ \diagup & \| & \diagdown \\ O & O & O \end{array}\right]^{2-}$$

8. Crystal Structure

It is important to make the general distinction between **giant lattices** and **molecular lattices**. The term giant lattice is used to describe the large continuous arrangements of particles (atoms or ions), which make up certain solids. The bonding forces in a giant lattice may be directional (covalent), as in diamond, or non-directional, as in metals and ionic compounds. A molecular lattice, unlike the above, consists of an arrangement of molecules in which the forces holding the molecules together are very much weaker than the forces which hold the atoms together within the molecules.

As previously stated, the geometrical arrangement of the particles (atoms, ions, or molecules) which make up a lattice is determined by X-ray and electron diffraction techniques (page 64).

Giant Lattices

1. *Metal lattices*

The bonding forces within a metal are considered to arise from the delocalization of the 'valency' electrons of the metal atoms. The picture is that of an arrangement of metal ions with the valency electrons free to move within the metal. It is the mobility of these electrons which accounts for the electrical conductivity of the metal.

In a metal, unlike the ionic lattice of a salt, there is only one type of ion. Therefore, from the geometrical point of view, the packing of ions in a metal can be compared to the packing of spheres of identical size.

When a single layer of identical atoms or spheres is packed as closely as possible, it is found that each atom is surrounded by six near neighbours, as shown in Figure 8.1. If a second layer of spheres is added so that they rest in the hollows of the first, the plan view would look like Figure 8.2.

A careful examination of Figure 8.2 reveals that if yet another layer of atoms is added, they can be placed in one of two possible positions.

(a) The third layer can be placed vertically above the first layer (position 1). If the atoms in the first layer are denoted by A and those in the second layer by B, then, because the third layer is a kind of repetition

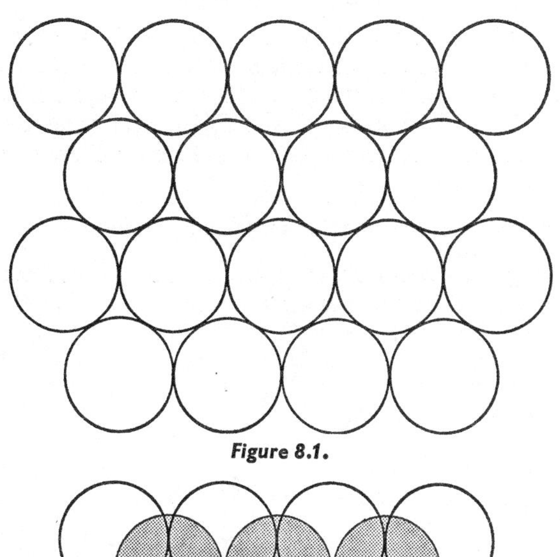

Figure 8.1.

Figure 8.2.

of the first, and the fourth of the second, and so on, the structure is called an ABA type.

(b) Alternatively the third layer can be placed in position 2 where they would not be vertically above either of the two previous layers. This structure is called an ABC type.

It is worthwhile looking more closely at each of these alternative structures.

ABA (called hexagonal close-packing)

Consider atom X (Figure 8.2) which is in the second layer; it has twelve near neighbours, six in its own layer, three in the first layer, and three in the third layer (the positions of which are indicated by the three 1s around atom X). The atom is said to have a co-ordination number of 12.

If the structure is rotated about an axis at right angles to the layers of atoms, six identical positions are possible. This axis is called a six-fold axis of symmetry and the structure is known as hexagonal close-packing.

ABC (called cubic close-packing or face-centred cubic)

As in the hexagonal close-packing, each atom has a co-ordination number of 12, having six near neighbours in its own layer, three in the layer below, and three in the layer above.

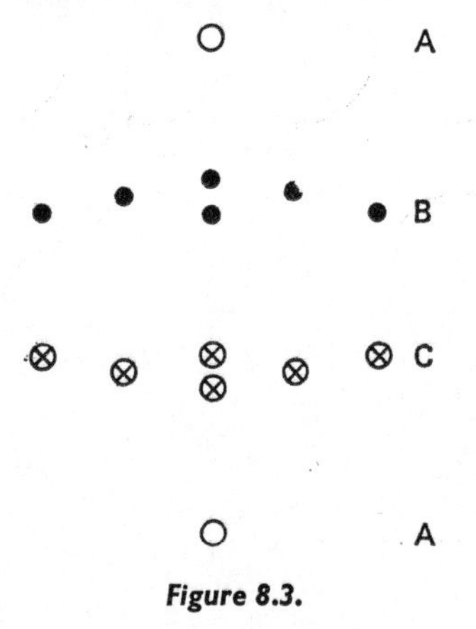

Figure 8.3.

Figure 8.3 represents the side elevation of parts of four layers (i.e. ABCA) looked at from slightly above the plane of the layers. One atom in the uppermost layer (A) is shown, six atoms of the next (B), six atoms of the next (C), and one atom of the lowest (A).

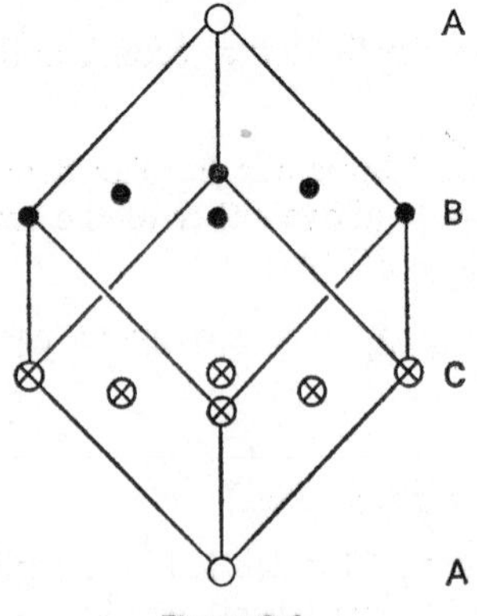

Figure 8.4.

Figure 8.4 is the same as Figure 8.3 except that construction lines have been added to show that the atoms form a cube, standing on one corner with an atom at the centre of each face. The structure is called cubic close-packing or face-centered cubic. In this case, a vertical line through the planes of atoms is a three-fold axis of symmetry.

Body-centred cubic

This third type of structure which is adopted by some metals cannot be described as close-packed. Each atom has a co-ordination number of 8. The structure is described as body-centred as each atom is at the centre of a cube formed by eight others (Figure 8.5).

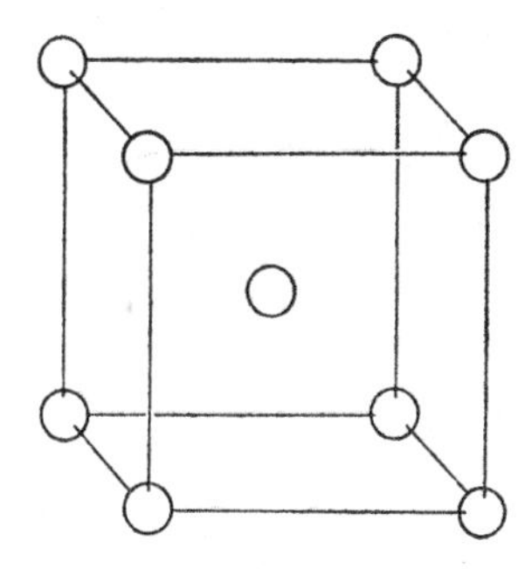

Figure 8.5.

Examples of metals which adopt each type of structure

Hexagonal close-packed	magnesium and zinc,
Cubic close-packed	copper and aluminium,
Body-centred cubic	alkali metals.

A considerable number of metals can adopt two structures. For example, iron is known to occur in both cubic close-packed and hexagonal close-packed forms.

2. *Ionic lattices*

When a sphere is placed in a hollow in a previous layer, it comes into contact with three other spheres and the overall shape of this unit is tetrahedral. In between the four spheres there is a small space which is called a **tetrahedral site** (see Figure 8.6).

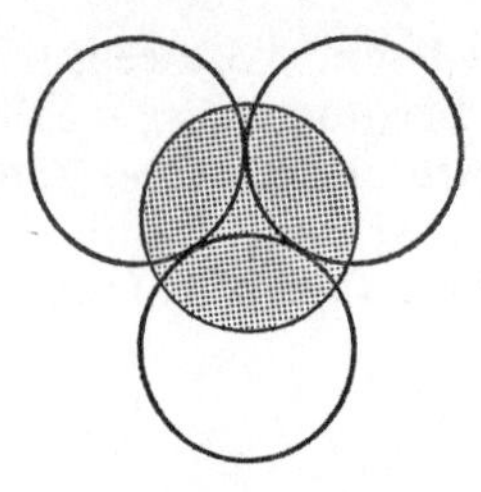

Figure 8.6.

When a triangle of spheres is placed on the top of another triangle of spheres, in such a manner that the centre of one triangle is immediately above the centre of the other, the overall shape of the spheres is octahedral. The space which is left at the centre of these spheres is called an **octahedral site** (see Figure 8.7).

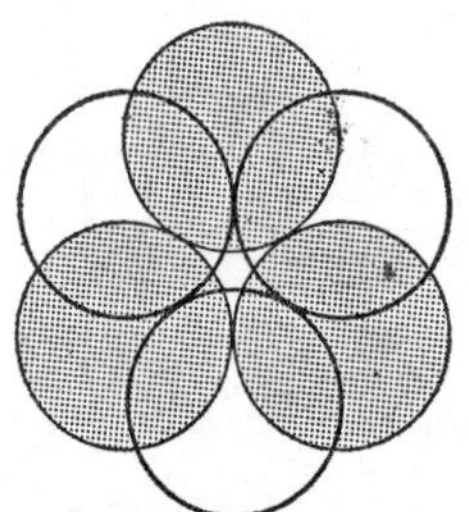

Figure 8.7.

Figure 8.8 shows all the tetrahedral (T) and octahedral (O) interstitial sites for two layers of close-packed atoms. It is very useful to relate the structures of ionic compounds to the various metal structures and the interstitial sites.

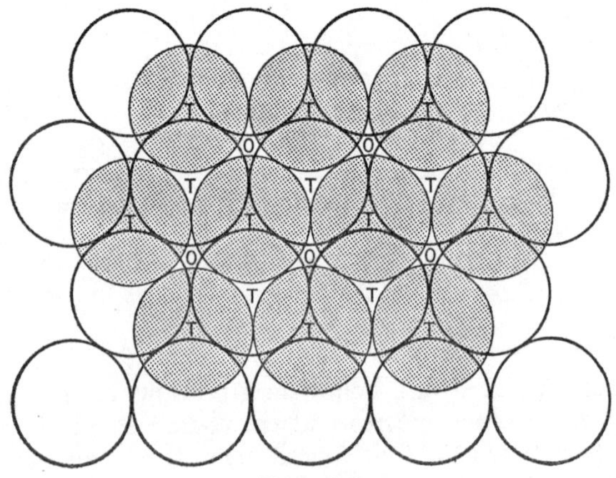

Figure 8.8.

The main factors which determine the type of lattice adopted by an ionic compound are:

1. The relative charges on the ions which dictates the relative numbers of ions present.
2. The relative sizes of the ions.

The size of an ion is denoted by its ionic radius which is determined by X-ray diffraction methods. The X-ray analysis gives only the inter-nuclear distance which is the sum of the two ionic radii. It is necessary to decide on a value for one particular ion and then by investigating a whole range of salts it is possible to build up a table of ionic radii.

	Atomic radius (nm)		*Ionic radius* (nm)
Na	0.19	Na^+	0.095
K	0.235	K^+	0.133
Cl	0.099	Cl^-	0.181
O	0.074	O^{2-}	0.140

It can be seen from the table that for a positive ion the ionic radius is much smaller than the atomic radius (e.g. K^+ and K), and in the case of a negative ion the reverse is true (e.g. Cl^- and Cl). This difference is consistent with the protons being in excess of the electrons in a positive ion and *vice versa* for a negative ion.

Some of the more common types of ionic lattice are described below.

Sodium chloride (*ionic radii*, $Na^+ = 0.095$ nm, $Cl^- = 0.181$ nm)

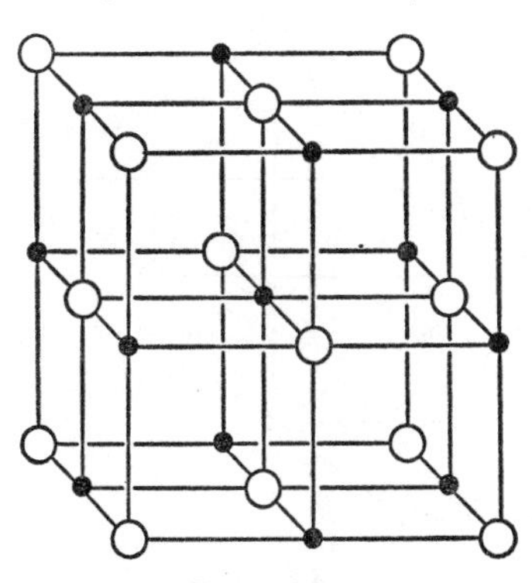

Figure 8.9.

The geometry of this structure may be described in the following manner.

1. Each chloride ion is in close contact with six sodium ions and each sodium ion with six chloride ions. The crystal is said to have 6:6 co-ordination.
2. Considering the chlorides only, it can be seen that although the ions are not actually in contact, they can be regarded as being in a face-centred cubic arrangement (i.e. ABC type which is cubic close-packing). The sodium ions are also in a face-centred cubic arrangement. The whole lattice consists of two interlocking face-centred cubes with the sodium ions appearing in the octahedral sites of the chloride arrangement and *vice versa* for the chlorides (Figure 8.10).

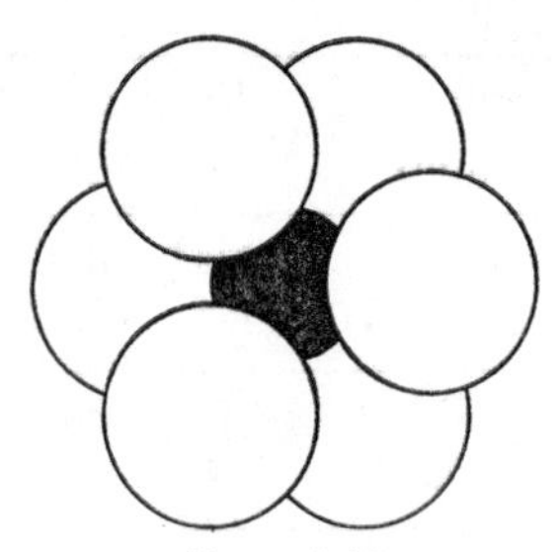

Figure 8.10.

Caesium chloride (*ionic radii*, $Cs^+ = 0.167$ nm, $Cl^- = 0.181$ nm)

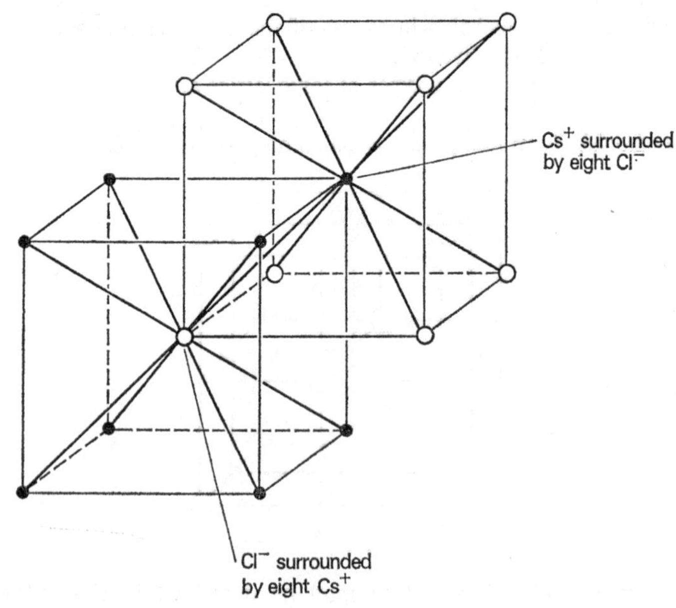

Figure 8.11.

This structure resembles sodium chloride in that it is a cubic structure, but it differs in the following respects:

1. Each caesium ion has eight chloride ions as near neighbours and each chloride has eight ceasiums as near neighbours. Hence the co-ordination is 8:8.

2. Each type of ion is arranged in a simple cubic form. The whole structure may be regarded as two interlocking simple cubic forms with each caesium ion appearing at the centre of a cube of chlorides and *vice versa* for chloride ions.

Note that, although in some ways Figure 8.11 resembles a body-centred cubic arrangement, it is not correct to label it as such, because the ion at the centre is of a different type from the ions at the corners of the cube.

Zinc blende (*ionic radii*, Zn^{2+} = 0.074 nm, S^{2-} = 0.190 nm)

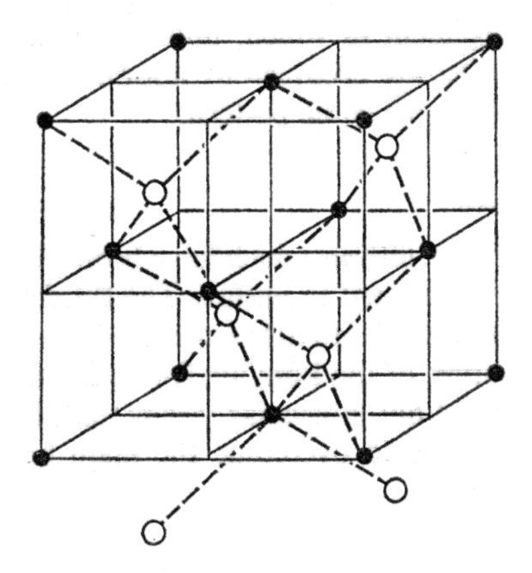

Figure 8.12.

1. There is 4:4 co-ordination.

2. The sulphide ions are arranged in a face-centred cubic pattern.

The zinc ions can be regarded as being in a simple cubic pattern with each zinc ion appearing in a tetrahedral site of the sulphide cubic close-packed system.

Note that for every atom in a close-packed system there are two tetrahedral sites. In the zinc blende type of structure half of the sites are occupied.

Fluorite (ionic radii, Ca^{2+} = 0.094 nm, F^- = 0.133 nm)

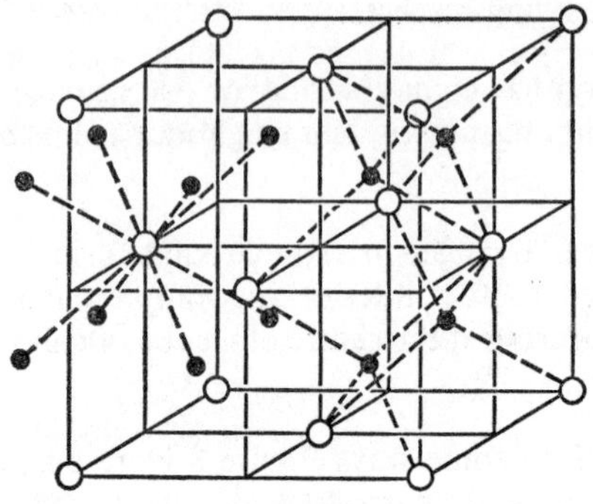

Figure 8.13.

1. Each calcium ion has 8 co-ordination around it and each fluoride has 4 co-ordination.
2. The calcium ions can be considered as being in a face-centred or cubic close-packed arrangement in which, unlike the zinc blende structure, all the tetrahedral sites are occupied by fluoride ions. The fluoride ions themselves are in a simple cubic pattern.

3. *Giant covalent lattices*

In the two types of giant lattices which have been discussed, the bonding forces have not been directional in nature. The third type of giant lattice involves covalent bonds which are directional.

Even though the nature of the bonding forces is different, the fact that they are giant lattices and not molecular means that, like metals and ionic compounds, it is necessary to overcome the main bonding forces when melting these substances.

Diamond

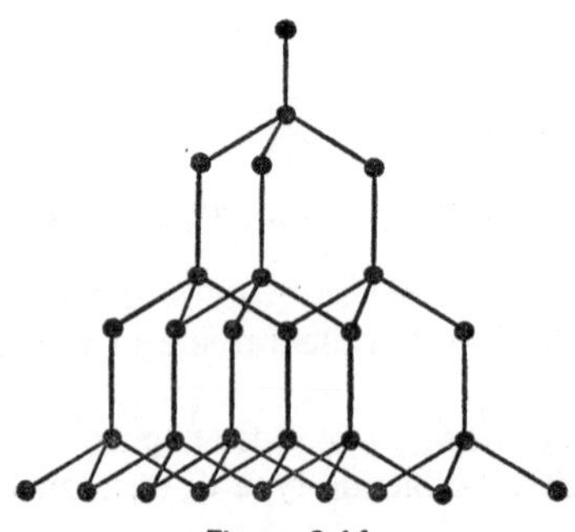

Figure 8.14.

Each carbon atom is covalently bonded to four other carbon atoms giving the structure a co-ordination of 4. The system is cubic and the atom at the centre of a tetrahedron can be compared to a zinc ion in a tetrahedral site in the zinc blende structure. Remember, in the case of diamond it is the number of unpaired electrons possessed by each carbon atom which determines the 4 co-ordination.

Graphite

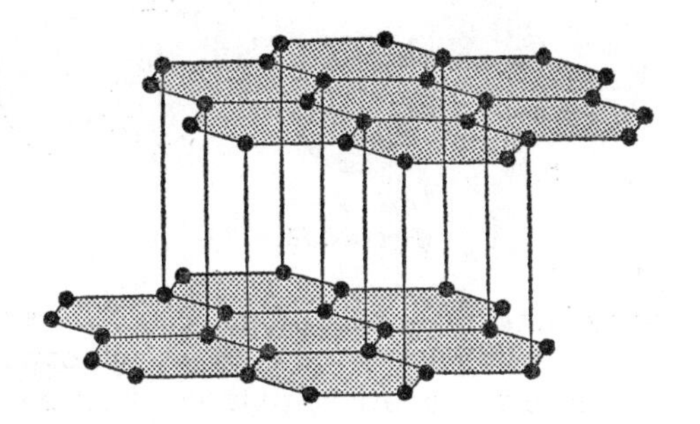

Figure 8.15.

Graphite consists of a layer structure. Within each layer, each carbon atom is covalently bonded in a planar arrangement to three others. The C—C bond length in diamond is 0.154 nm which is the normal length for a single covalent bond between carbon atoms. In graphite the bond length is 0.142 nm which suggests that there is some delocalization of the fourth electron which results in extra bonding. However, the extra bonding is not as powerful as that in benzene in which the bond length between the carbon atoms is 0.139 nm. The fact that graphite conducts electricity suggests that the delocalization occurs throughout the layer.

The distance between the layers in graphite is quite large at 0.34 nm and indicates weaker bonding forces. These are van der Waals' forces (page 103) and it is their weakness which allows the layers to move over each other and so result in graphite being soft and suitable for use as a lubricant.

Molecular Lattices

In particular, these lattices are characterized by their low melting and boiling points. These properties arise because the bonding forces between the molecules are weak. The important point to remember is that when these substances melt the covalent bonds within the molecules are unaffected and the molecules remain intact.

Iodine

Figure 8.16.

This lattice consists of an arrangement of iodine molecules each of which contains two iodine atoms bonded together by a single covalent bond. To melt iodine it is only necessary to overcome the weak attractive forces (van der Waals' forces) between the molecules and, as expected, its melting point and boiling point are comparatively low at 387 K and 456 K respectively.

Naphthalene

The molecular formula of naphthalene is $C_{10}H_8$ and its structural formula is

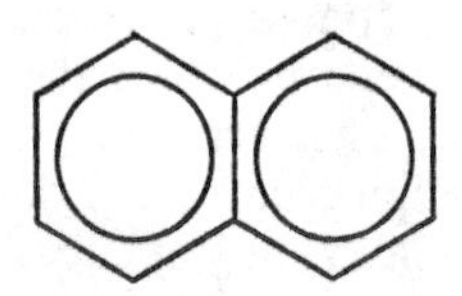

Solid naphthalene consists of an arrangment of these molecules held together by weak van der Waals' forces which result in a relatively low melting point 353 K, and boiling point, 491 K.

9. Bonding II: Intermolecular

The wide range of energies associated with intermolecular bonding suggests that not all of the bonds are formed in the same way.

Bond Polarity

The covalent bond which holds two atoms together within a molecule is envisaged as a pair of electrons shared between the atoms. Unless the atoms are identical, the electrons will not be equally shared. When unequal sharing does occur it will result in a slight separation of charge. The atom with the greater share of the electrons will be slightly negative and the other atom will be slightly positive. This phenomenon is referred to as **bond polarity** and the bond is said to be of *intermediate* type, i.e. partly covalent and partly ionic.

The relative tendencies of elements to attract electrons are expressed numerically on an arbitrary scale, as their **electronegativities.** The larger the difference between the electronegativities of the atoms forming the bond, the more unequal the sharing will be. The comparative electronegativities of some elements are as follows:

F	O	N	Cl	C	H
4.1	3.5	3.1	3.0	2.5	2.1

The values quoted above show that there will be unequal sharing in the C—Cl bond and in the C—O bond, but that the electrons forming the C—H bond will be more equally shared.

Polar Compounds

Bond polarity within a diatomic molecule results in a separation of charge over the whole molecule. The molecule is said to be polar and to possess a dipole moment which can be expressed numerically by taking into account the magnitude of the charge and the distance separating the charges.

A molecule made up of more than two atoms and exhibiting some bond polarity does not necessarily have a dipole moment. If the molecule is symmetrical, the effects of the individual dipoles of the bonds, as far as the molecule as a whole is concerned, cancel each other out. Thus, trichloromethane has a dipole whereas tetrachloromethane does not.

```
        H                          Cl
        |                          |
        C                          C
      / | \                      / | \
    Cl  |  Cl                  Cl  |  Cl
        Cl                         Cl
```

Trichloromethane   Tetrachloromethane

The lack of a dipole where one might be expected is an indication of symmetry within a molecule. For example, carbon dioxide is made up of two elements with differing electronegativities but because the compound is linear, there is no resultant dipole.

O=C=O

Carbon dioxide

Dipole–Dipole Interactions

Electrostatic attractions between the oppositely charged ends of dipoles account for the intermolecular bonding in such compounds as propanone and the energy associated with this type of bonding is about 1/100 of that associated with a covalent bond.

```
   CH3              CH3
     \    δ−          \    δ−
   δ+C=O · · · · · δ+C=O
     /                /
   CH3              CH3
```

Hydrogen Bonds

When electrons are withdrawn from a hydrogen atom, either by it being bonded to an electronegative atom (as in water) or by the presence of electronegative atoms in the rest of the molecule (as in $CHCl_3$), it acquires a slight positive charge. Due to the smallness of the hydrogen atom, the positive charge is more concentrated and is therefore more effective in forming links with other molecules. Hydrogen atoms in this type of situation are found to make particularly strong intermolecular bonds with electronegative atoms which possess lone-pairs (non-bonding pairs) of electrons. Fluorine, oxygen, and nitrogen are all good examples of such elements.

An intermolecular bond of this type is called a hydrogen bond and the energies associated with such bonds are usually of the order of 20 to 40 $kJ\,mol^{-1}$ as compared with energies of the order of 300 to 400 $kJ\,mol^{-1}$ for normal covalent bonds.

Certain compounds, HF, H_2O, and NH_3, possess both of the essential requirements for hydrogen bond formation, i.e. one or more hydrogen atoms in a suitable environment and an electronegative element with a lone-pair of electrons.

The effect of the unusually powerful intermolecular forces which exist between molecules of these compounds is made obvious if their melting points and boiling points are compared to those of hydrides of the other elements in their respective groups (see Figure 9.1).

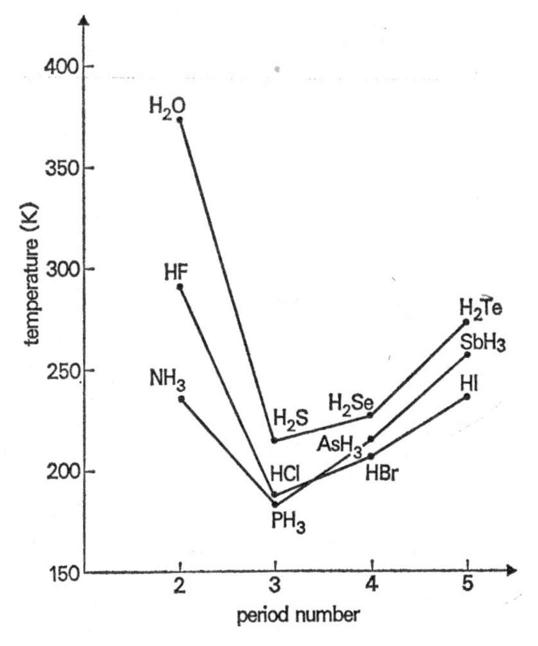

Figure 9.1. Boiling points of the hydrides of Groups 5, 6, and 7

The melting points and boiling points of HF, H_2O, and NH_3 do not fit in with the general trends within their groups showing that they possess unusually powerful intermolecular forces.

Methane, CH_4, which is in the corresponding position in Group IV, fits in perfectly with the general trend for the hydrides of its group as it possesses neither of the requirements for hydrogen bond formation (see Figure 9.2, on page 100).

The hydrogen bond is a very specific case of intermolecular bonding and a few important examples of its occurrence will be discussed in more detail.

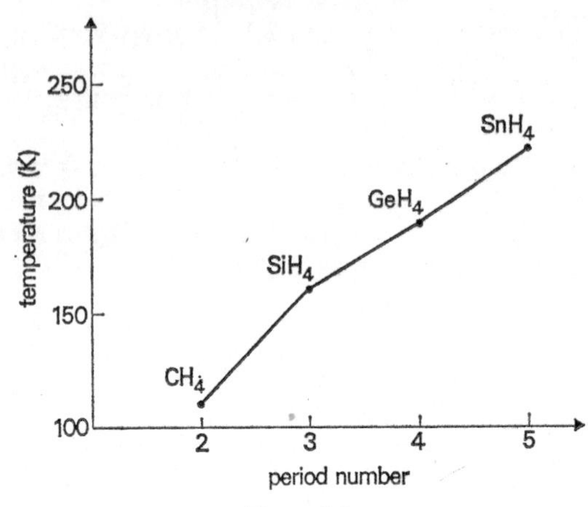

Figure 9.2.

1. *Ice*

In the crystal lattice of ice each H_2O molecule forms:

(a) two hydrogen bonds by virtue of its non-bonding pairs being attracted to the slight positive charges on hydrogens of other H_2O molecules,

(b) two hydrogen bonds by virtue of the slight positive charges on its hydrogen atoms being attracted to non-bonding pairs of electrons associated with oxygens of other H_2O molecules.

The four bonds form an approximately tetrahedral arrangement around each oxygen atom. The resulting structure is cubic and wurtzite in type. Compared to the more random arrangement of H_2O molecules in water, the ice structure is more open, which results in ice having a lower density than water. Thus when water is cooled its density increases to a maximum at 4 °C, after which the more open structure begins to form and the density decreases.

Hydrogen bonds to lone-pairs of other H_2O molecules

H

O

H

Hydrogen bonds to hydrogen atoms of other H_2O molecules

Hydrogen bonding in ice

2. *Gypsum*

Calcium sulphate occurs naturally in two forms, $CaSO_4,2H_2O$ which is gypsum, and the anhydrous form which is called anhydrite. The physical characteristics of these two forms are quite different. Anhydrite exhibits no obvious cleavage and is fairly hard (Mohs' scale 3.5), whereas gypsum is much softer (Mohs' scale 2) and has a very obvious cleavage. These observations, together with evidence from X-ray crystallography, are interpreted as:

(a) Anhydrite consists of a giant lattice of Ca^{2+} and SO_4^{2-} ions. This lattice is not easily cleaved along any plane.

(b) Gypsum is a layer structure, with each layer consisting of a lattice of Ca^{2+} and SO_4^{2-} ions. The layers are held together by hydrogen bonds which act between the sulphate ions in the layers and the water molecules which are situated between the layers. Due to the comparative weakness of the hydrogen bonds, very obvious cleavage planes exist between the layers.

3. *Carboxylic acids*

In non-polar solvents, such as benzene, a number of organic acids are found to have relative molecular masses which are approximately twice the expected values. The double molecules, or dimers, are formed by two molecules being held together by hydrogen bonds.

```
          O—H·········O
         /             \
CH3—C                   C—CH3
         \\             /
          O·········H—O
```

A dimer of ethanoic acid

Ethanoic (acetic) acid, CH_3CO_2H, and benzenecarboxylic acid (benzoic acid), $C_6H_5CO_2H$, both form dimers in this way.

4. *Biochemistry*

Proteins are essential constituents of all living material and the structure adopted by a protein depends to a large extent on hydrogen bond formation. Amino acids combine in various sequences to form compounds of large relative molecular mass which are known as proteins. In general terms;

```
H    R    O              H    R'   O
 \   |   //               \   |   //
  N--C--C        +         N--C--C
 /   |   \                /   |   \
H    H    O--H           H    H    O--H

          H  R     O  R'    O
          |  |    //  |    //
  =       N--C--C     C--C
         /   |   \   /|   \
      etc    H    N  H     etc
                  |
                  H
```

The amino acid chain has the necessary requirements for hydrogen bond formation, namely, hydrogen atoms (H) bonded to an electronegative atom (N) and electronegative atoms with lone-pairs of electrons (O). It is hydrogen bonds between these sites which keep the protein chain in a coiled form, as shown in Figure 9.3.

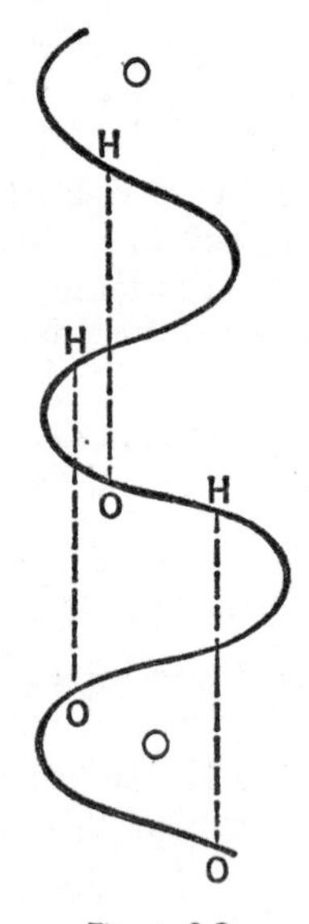

Figure 9.3.

Another striking illustration of the occurrence of hydrogen bonding is in the structure of deoxyribonucleic acid (DNA), which is the substance responsible for the process of heredity. It consists of two nucleic acid helices joined together by hydrogen bonds between the hydrogen atoms of N—H groups and other nitrogen atoms or the oxygen atoms of C=O groups. When the hydrogen bonds break, each helix can then, by reforming the specific hydrogen bonds, produce a replica of the original DNA.

van der Waals' Forces

Certain compounds and elements, for example alkanes, iodine, and the noble gases, possess none of the characteristics required for hydrogen bond formation or dipole–dipole interaction. However, these substances can be obtained in liquid and solid form, and so intermolecular forces of some description must exist. The forces are called van der Waals' forces and are thought to be electrostatic in nature and due to slight molecular dipoles which are temporarily induced when molecules come into close contact. In support of this explanation, it is found that 'straight' chain alkanes are more easily liquefied than ones with branched chains. The more linear shape of the straight chain, as opposed to the roughly overall spherical shape of the branched chain, provides more possibility of interactions between the electron clouds of neighbouring molecules.

The energy associated with van der Waals' forces is usually about ten times less than that associated with hydrogen bonds.

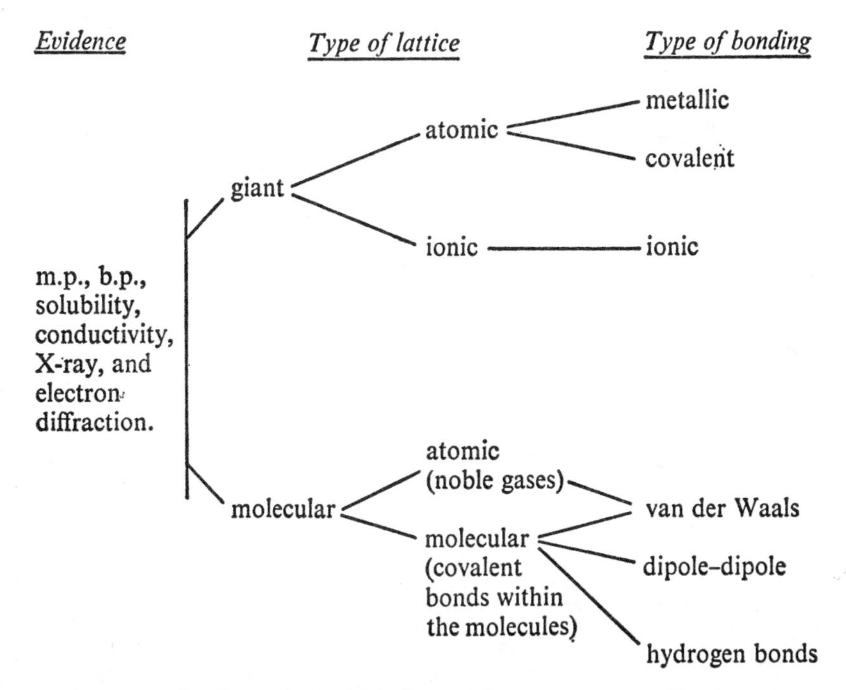

In the case of a diatomic hydride (e.g. HF) there are contributions from both dipole–dipole interactions and the more directional hydrogen bonds.

Covalent and van der Waals' radii

As mentioned on page 96, solid iodine is a molecular lattice in which each molecule consists of two atoms covalently bonded together and the intermolecular forces are of the van der Waals' type. This arrangement results in two different internuclear distances for iodine atoms. Iodine atoms within a molecule are relatively close together and half the internuclear distance is called the covalent radius of iodine. Adjacent atoms from neighbouring molecules are further apart and half the internuclear distance is known as the van der Waals' radius for iodine.

The types of structure and bonding forces which have been discussed in this text are shown summarized on page 103.

10. Colloids and the Colloidal State

The founder of the science of colloids is generally taken to be Graham because of work done by him in this field between 1851 and 1861.

Colloids were once regarded as a separate class of materials, such as starch, gum, gelatine, and glue, which had the characteristics of *not* diffusing through a colloidal membrane, such as parchment paper, and not being generally known in crystalline form. The word **colloid** is derived from the Greek *kolla*, meaning *glue*. The contrast was with **crystalloids,** such as common salt, copper sulphate, or oxalic acid, which diffused readily through the membrane and were readily obtained as crystals. It is now known, however, that most elements and compounds can be obtained in colloidal solution in a medium in which they are not soluble in the ordinary sense of that term, so that colloids are regarded collectively as being in a particular state of matter rather than a separate class of materials.

In 'true' solution, the dissolved substance is (with very few exceptions) in the state of single molecules or ions which are distributed in the solvent. The whole is regarded as a homogeneous mixture. The solute is in its ultimate state of division for chemical purposes. At the other end of the scale, there are insoluble particles so large that they settle out of the liquid. Such a mixture is definitely heterogeneous and is called a **suspension.** A colloidal solution occupies the intermediate region between true solution and suspension. At both ends, the line of division is indefinite and the size of colloidal particles passes gradually into that of suspended particles at one end of the scale and that of the molecules and ions of true solution at the other. In general, colloidal particles have a diameter of the order of 1–10^2 nm. Above this range, suspended particles begin to appear; below it, the particles cannot be distinguished from those of true solution.

Colloidal solutions are generally referred to as **sols**; if the dispersion medium is water, the term **hydrosol** (or **aquasol**) is employed. In some cases, fairly concentrated colloidal solutions are obtainable, and they may set to a jelly-like material. In that case, they are referred to as **gels** (or **hydrogels** if aqueous). The colloidal particles are called the **disperse phase** and the liquid in which they are carried (usually water) is called the **dispersion medium.**

Preparation of Sols

Since the disperse phase is, on the average, intermediate in particle size between suspension and true solution, sols can be made either by aggregating the particles of true solution or further dispersing those of suspension. The following are some typical examples of both types. In general, each sol will need purification by **dialysis**. This process is described after the preparations of sols.

1. *Dispersion methods*

(a) *Mechanical grinding* If sulphur is finely ground using an agate pestle and mortar, with urea (acting as an abrasive) and the mixture is stirred with water, some of the sulphur will be fine enough to form a sol with the water. The rest can be filtered out and the urea removed by dialysis.

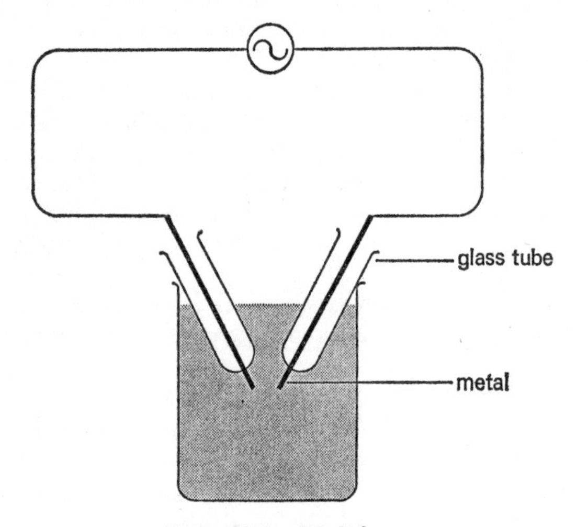

Figure 10.1. Bredig's arc

(b) *Bredig's Arc method for metal sols* In this process an arc is struck, usually under conductivity water, between two wires of the metal required in the sol (Figure 10.1.) The containing vessel may be cooled in ice. For the preparation of gold sols, a trace of alkali in the water is advantageous. The repeated striking of the arc detaches minute particles of the metal and they pass into colloidal solution. Larger particles may be filtered out. In aqueous sols, contamination by traces of metallic oxide or hydroxide is common.

(c) *By peptization* Peptization is the dispersal of a precipitated material into colloidal solution by the action of an electrolyte in solution. The following is an example of it. The gradual mixing of solutions of potassium hexacyanoferrate(II) and iron(III) chloride (both 3 per cent) precipitates Prussian blue. After a few minutes, the precipitate is filtered off and washed well. Oxalic acid solution (5 per cent), acting as the peptizing electrolyte, is poured through the filter-paper and the Prussian blue passes into a sol. It should be dialysed to remove most of the oxalic acid.

2. *Aggregation methods*

In these cases, aggregation of molecular or ionic materials occurs to the size of colloidal particles. Among the chemical methods used for this purpose are the following.

(a) *Reduction* This method is commonly used to produce metal sols. For example, the preparation of a silver sol can be carried out as follows. To 5 cm^3 of a solution of silver nitrate (1 per cent), add very dilute ammonia, drop by drop, with shaking, till the precipitated silver oxide is just redissolved. Dilute with 95 cm^3 of water. Then add 0.5 cm^3 of tannin solution (0.5 per cent) as reducing agent. A silver sol will be obtained, brown in colour.

(b) *Hydrolysis* This process is usually applied to the production of oxide or hydroxide sols. An example of this is the production of a sol of hydrated iron(III) hydroxide by pouring a few cm^3 of a concentrated solution of iron(III) chloride into a large volume (say 1.5 dm^3) of boiling distilled water. The sol is ruby-red and can be dialysed to remove most of the electrolyte, HCl, though over-dialysis will cause precipitation of the colloidal material

$$2FeCl_3(aq) + 3H_2O(l) \rightleftharpoons Fe(OH)_3(sol) + 3HCl(aq)$$

(c) *Double decomposition* An arsenic(III) sulphide sol can be made by utilizing the reaction

$$As_2O_3(aq) + 3H_2S(g) \rightarrow As_2S_3(sol) + 3H_2O(l)$$

To 100 cm^3 of hot distilled water add 1 g of arsenic(III) oxide and let it dissolve as fully as possible. Cool and filter the solution. Saturate 200 cm^3 of distilled water with hydrogen sulphide. While still passing this gas, gradually add the arsenical solution till a bright yellow sol is formed. Remove the electrolyte, H_2S, by passing a stream of hydrogen.

Dialysis

This is the process of removing crystalloids (usually electrolytes) from sols by the use of a colloidal membrane. The crystalloid can pass through the membrane; the colloid cannot. The membrane most used has been parchment paper, but cellophane is now available and collodion membranes are also suitable. Though it is not absolutely necessary, the membrane is usually used with a **dialyser**; a U-shaped hollow tube of parchment paper is also quite efficient (Figure 10.2).

A slow, continuous flow of fresh water is maintained in the outer vessel. As a result, the concentration of crystalloid in the outer vessel continually tends to zero and any crystalloid in the sol passes gradually through the membrane and is washed away. Eventually the sol is left pure, though in fact most sols require a trace of electrolyte present to stabilize them and over-dialysis may precipitate the colloid.

Lyophobic and Lyophilic Sols

Sols are classified into **lyophobic** and **lyophilic** sols, these terms meaning **solvent-hating** and **solvent-loving** respectively. (When water is the dispersion medium, the terms **hydrophobic** and **hydrophilic** are used.) These sols are distinguished by the following features:

Lyophobic sols	*Lyophilic sols*
1. On evaporation or cooling give *solids* which do not readily form sols again by the action of water.	1. With similar treatment give *gels*, which easily form sols again by the action of water.
2. Are precipitated by relatively small amounts of electrolyte.	2. Are not precipitated by relatively small amounts of electrolyte. (May be salted out by large amounts of electrolyte.)
3. Have surface tension and viscosity similar to those of the dispersion medium.	3. Generally have higher viscosity and lower surface tension than the dispersion medium.
4. The colloidal particles all migrate in one direction under the influence of an electrical P.D.	4. In similar conditions, the particles may not migrate at all, or may move in either direction.
5. The colloidal particles are easily detected in the ultra-microscope.	5. The particles are not easily detected in the ultra-microscope.
Examples colloidal metals, metallic sulphides and hydroxides; other inorganic colloids.	*Examples* gums, starch, proteins.

Properties of Sols

1. *Optical properties*

The smallest particles visible in the ordinary optical microscope are about 0.2 μm in diameter. This is greater than the diameter of the largest

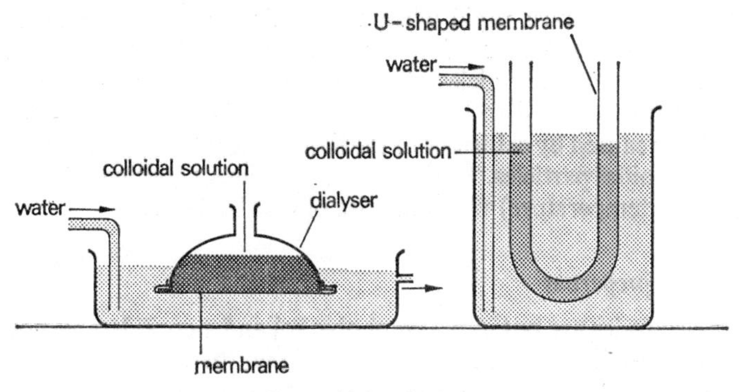

Figure 10.2. Dialysis

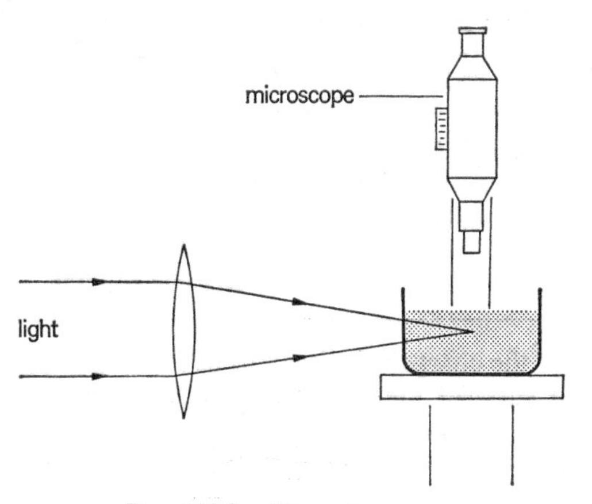

Figure 10.3. Ultra-microscope

colloidal particles, consequently colloidal particles cannot be directly viewed. They can, however, be detected in the ultra-microscope. In this instrument (Figure 10.3), the sol is brightly illuminated laterally and is observed by a microscope in a direction at right angles to the direction of illumination. The colloidal particles scatter the light and can be recognized as points of light against a dark background. The particles in hydrophobic sols are easily detected, those of hydrophilic sols much less easily and those of true solutions not at all.

2. *The Brownian movement*

The Brownian movement is so called from its discovery by the botanist Brown (1827) as a property of pollen particles in water. The colloidal particles of a sol can be observed by the ultra-microscope, as described in the above paragraph, and are found to be generally in a state of random motion, which is most rapid when the particles are very small and the liquid least viscous. This motion is called the **Brownian movement**. In the case of large colloidal particles, the movement may be only an oscillation. At first, it was ascribed to convection currents in the liquid, but is now known to be caused by collisions of the molecules of the dispersion medium with the colloidal particles. The Brownian movement continues indefinitely (at constant temperature) with particles of colloidal size.

3. *Electrical properties*

If inert electrodes are put into a lyophobic sol and a considerable potential difference is applied across them, it is found that the colloidal particles will migrate towards one of the electrodes, showing that the particles are electrically charged. This phenomenon is called **electrophoresis**. For particles of colloidal silver, a typical rate of flow is 3×10^{-4} cm s^{-1} for a potential gradient of 1V cm^{-1}. A summary of the situation in water is:

Colloids positively charged	*Colloids negatively charged*
Metallic hydroxides	Metallic sulphides Metals Organic colloids

The source of the charge is believed to be ionic. The colloidal particles adsorb (at the surface) ions derived from electrolyte dissolved in the

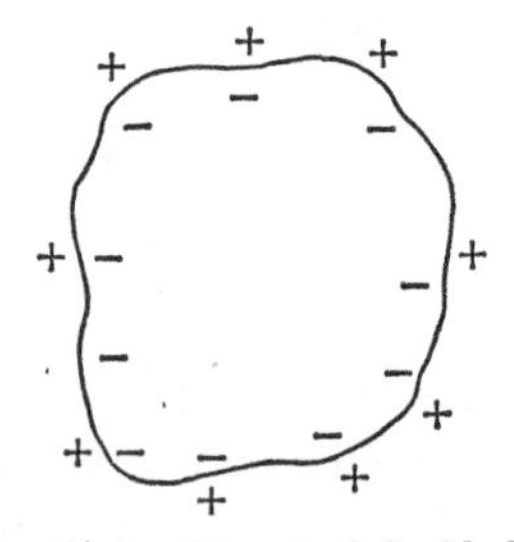

Figure 10.4. Electrical double layer

dispersion medium and acquire the corresponding charge, which, in the majority of cases, is negative by adsorption of anions. Cations are dispersed in the neighbouring liquid to form an *electrical double layer* (Figure 10.4) and this system normally remains comparatively undisturbed, moving as a unit. Under the influence of a potential gradient, the charged colloidal particle moves in one direction towards the oppositely charged pole and the ionic layer moves in the other direction. For a positively charged colloidal particle, the position is the reverse of that shown in Figure 10.4.

Inorganic lyophobic sols undergo precipitation if the particles lose their electric charge. This can happen if dialysis is carried so far that the electrolyte concentration in the sol becomes virtually nil and none is available for adsorption. An over-dialysed iron(III) hydroxide sol, for example, precipitates gelatinous iron(III) hydroxide. Precipitation can also be caused if colloidal particles are made to adsorb an ion opposite in charge to their usual condition. For example, the particles of a silver sol, normally negative, will adsorb Al^{3+} ions from an added aluminium salt until they reach neutrality (the *isoelectric point*) and then precipitate out. If the isoelectric point is passed *rapidly* so that the colloid has no time to coagulate, it may be stabilized with a charge opposite to its usual charge.

It is obvious, also, that colloidal particles of opposite electrical charge will bring about each other's precipitation when mixed, by neutralizing each other's electrical charges. For example, iron(III) hydroxide sol (positive) and arsenic(III) sulphide sol (negative) precipitate each other when mixed and, in electrical equivalency, precipitation is complete.

4. *Coagulation of colloids*

Lyophobic sols can be forced to precipitate the colloid by addition of electrolyte. It is found (as would be expected) that the colloid is precipitated by an ion of opposite charge to its own. The ion of opposite charge is known to be adsorbed by the colloid because it can be detected in the precipitated material.

For precipitation of a given colloid, a certain minimum concentration of a given electrolyte is needed. The values of these minimum concentrations have been closely studied. Figures for an arsenic(III) sulphide sol (1.85 g dm^{-3}) are given on page 107. It is a negatively charged colloid, precipitated by cations. The cations were present as metallic chlorides in all cases except that of Ce^{3+}, which was added as nitrate.

Cation	*Minimum conc. for coagulation in 10^{-3} mol dm^{-3} (as chlorides)*
Li^+	58
Na^+	51
K^+	50
Mg^{2+}	0.72
Ca^{2+}	0.65
Sr^{2+}	0.63
Zn^{2+}	0.68
Al^{3+}	0.093
Ce^{3+}	0.080

It will be observed that all the univalent cations require about the same minimum concentrations to precipitate the colloid; so do the divalent cations, but the concentration is much smaller; so also do the trivalent cations, but the concentration is smaller still. That is, the effectiveness of an ion in precipitating an oppositely charged colloid increases rapidly with increase of the charge of the ion. A rough average shows that the minimum concentrations required for tri-, di-, and univalent cations are in the proportion of 1:100:700 respectively. (The situation is similar for anions in the precipitation of positively charged colloids.) This is the reason why alum is so effective as a styptic, i.e. in checking minor bleeding by coagulation of colloids in blood. The effective agent is the trivalent ion, Al^{3+}. This property of the aluminium ion is also made use of during the treatment of sewage.

Another example of the coagulation of sols by electrolytes is the formation of deltas when material carried in colloidal form by rivers is precipitated on contact with sea water.

Lyophilic sols are much more stable towards added electrolyte. They can be obtained in an electrically neutral (*isoelectric*) state in some cases. They are usually little affected by small concentrations of electrolytes. For this reason, a hydrophilic sol, such as gelatine or gum-arabic, will often prevent or delay precipitation of a hydrophobic sol by added electrolyte. This effect is called **protection.**

The phenomenon called **thixotropy** is interesting. It occurs when the addition of electrolyte to a sol (e.g. a gelatine sol) causes the production of a pasty gel, which will liquefy when shaken but recover its pasty character on standing. This phenomenon has been exploited in the recent introduction of thixotropic paints of a pasty consistency, which spread readily when brushed but do not spill in use.

Adsorption Indicators

Adsorption indicators are used for titrations with silver salts in which precipitation of silver halide occurs. The use of *eosin* in the titration of potassium bromide by silver nitrate solution is typical.

Silver bromide is precipitated and can, in the appropriate conditions, adsorb Br^-ion, Ag^+ ion or the indicator dye, eosin. So long as potassium bromide remains in excess (i.e. the titration is incomplete), bromide ions are adsorbed on to the precipitate, which remains pinkish-yellow. As soon as the titration is completed and Ag^+ ion has gone into excess, Ag^+ ion and eosin are adsorbed on to the precipitate, which turns red in colour. This adsorption indicator is very effective with dilute solutions, say about 0.01 M. With adsoption indicators, the colour change takes place *on the precipitate*, not in the solution.

11. Physical Behaviour of Gases: The Kinetic Theory of Gases

The following laws were formulated from the results of the experimental observation of the behaviour of gases.

Boyle's Law

Boyle's Law states the relation between gaseous pressure and volume in the following way:

> The volume of a given mass of gas is inversely proportional to its pressure at constant temperature.

This is expressed mathematically in the form:

$$pV = k_1 \quad \text{(when } T \text{ is constant)}$$

Charles' Law

Charles' Law states the relation between the volume of a gas and its temperature in the following way:

> The volume of a given mass of gas is directly proportional to its absolute temperature at constant pressure.

This is expressed mathematically in the form:

$$\frac{V}{T} = k_2 \quad \text{(when } p \text{ is constant)}$$

In the above expressions, p, V and T denote the pressure, volume, and absolute temperature respectively of a given mass of gas and k_1 and k_2 are constants.

The two expressions given above as the mathematical expressions of Boyle's Law and Charles' Law can be combined into a single expression, which is:

$$\frac{pV}{T} = k$$

This equation can also be stated in the form:

$$pV = kT$$

If the equation refers to *one mole* of a gas, the constant k has a value denoted by R and the equation is written

$$pV = RT$$

Expressed in terms of joules, R has the value of 8.314 J K^{-1} mol^{-1}.

If a given mass of a particular gas has pressure, volume, and absolute temperature, p_1, V_1, and T_1 in one set of conditions, and p_2, V_2, and T_2 in another set,

$$\frac{p_1V_1}{T_1} = \frac{p_2V_2}{T_2}$$

both fractions being equal to k, which is always constant for the same mass of the same gas. This is a very valuable equation, which enables a gaseous volume (relating to a fixed mass of a particular gas) to be 'corrected' from one set of temperature and pressure conditions to another. For reference purposes, the internationally agreed set of *standard temperature and pressure conditions* (s.t.p.) is 0 °C or 273 K and 760 mm Hg (101 325 N m^{-2}). These figures should be memorized and are in continual use in physical science.

Students are reminded that the *Kelvin scale* of temperature uses the Celsius degree, but starts from a zero which is, for approximate purposes, 273° *below* the Celsius zero, which is the equilibrium temperature between ice and water at 760 mm Hg (101 300 Nm^{-2}). Consequently, the temperature on the absolute scale corresponding to t °C is $(t + 273)$. For further details of the absolute scale, the reader should consult a text-book of physics.

Dalton's Law of Partial Pressures

For a mixture of gases which do not chemically react, the total pressure is equal to the sum of their partial pressures, where the partial pressure of a gas is the pressure the gas would exert if it occupied the volume alone. Thus the partial pressure of a gas depends on the number of moles of that component present and is related to the total pressure by the equation

$$\text{partial pressure of a component} = \text{total pressure} \times \text{mole fraction},$$

where

$$\text{mole fraction} = \frac{\text{number of moles of that component present}}{\text{total number of moles of gas present}}$$

Graham's Law

It is a common observation that gases, unlike solids and liquids, tend to fill any container into which they are placed. Also, a low density gas will pass more rapidly through a porous material than a high density gas. Thus, in the apparatus represented by Figure 11.1 the levels of liquid in the U-tubes will move in the directions indicated. The movement of gases is called diffusion and it is obvious from experiments such as those illustrated in Figure 11.1 that the rate of diffusion of a gas is inversely related to its

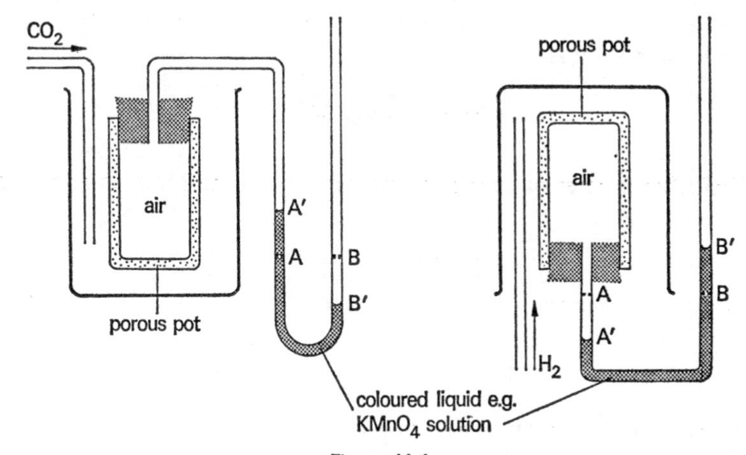

Figure 11.1.

density. In fact, Graham found from experimental results that the rate of diffusion of a gas is inversely proportional to the square root of its density.

Kinetic Theory of Gases

The above experimental laws can be accounted for qualitatively by considering a gas as being made up of widely spaced molecules which are free to move in any direction. If this picture of a gas is treated quantitatively, it is possible to derive theoretically the same gas laws. This agreement between theory and experimental results is good evidence for the validity of the theory.

In the mathematical treatment of the kinetic theory it is necessary to make several assumptions, in particular:

1. the individual molecules of a gas have negligible volume,
2. the attractive forces between the molecules are negligible,
3. the average kinetic energy of the molecules of a gas is proportional to the absolute temperature on the Kelvin scale.

Consider n molecules, each of mass m, in a cubical box of side l. The velocity, c_1, of any one molecule can be resolved into three components, x_1, y_1, and z_1, along the directions of the edges of the cube.

$$c_1^2 = x_1^2 + y_1^2 + z_1^2$$

Considering the x component only, a molecule travelling with this velocity will experience a change of momentum of $2mx_1$ on colliding with one wall of the box.

The molecule would collide x_1/l times per unit time. Therefore the change of momentum per unit time, or the rate of change of momentum, for this component is

$$2mx_1 \times \frac{x_1}{l} = \frac{2mx_1^2}{l}$$

Similarly the rates of change of momentum for the other components are

$$2my_1^2/l \text{ and } 2mz_1^2/l$$

Thus the total change of momentum per unit time for one molecule is

$$\frac{2mx_1^2}{l} + \frac{2my_1^2}{l} + \frac{2mz_1^2}{l} = \frac{2mc_1^2}{l}$$

For n molecules the total change in momentum is

$$\frac{2m}{l}(c_1^2 + c_2^2 + c_3^2 \ldots + c_n^2)$$

This can be simplified to $2mn\bar{c}^2/l$, where $\bar{c}$ is the root mean square velocity. The rate of change of momentum is equal to the force exerted by the collisions. This force is acting on a total area of $6l^2$ and, as pressure is equal to force per unit area,

$$p = \frac{2mn\bar{c}^2}{6l^3} = \frac{mn\bar{c}^2}{3l^3}$$

But l^3 is equal to the volume of the cube, and therefore

$$pV = \tfrac{1}{3}mn\bar{c}^2$$

This fundamental gas equation can be used to derive the gas laws.

Boyle's Law

As stated on page 115, the kinetic energy of a gas is proportional to the temperature on the Kelvin scale and so, for a fixed mass of gas at a constant temperature, the kinetic energy, $\frac{1}{2}mn\bar{c}^2$, is a constant. Rearranging the fundamental gas equation,

$$pV = \tfrac{2}{3} \times \tfrac{1}{2}\, mn\bar{c}^2$$

therefore

$$pV = \text{constant}$$

which is Boyle's Law.

Charles' Law

Re-arranging the fundamental gas equation,

$$V = \frac{1}{p} \times \tfrac{2}{3} \times \tfrac{1}{2}\, mn\bar{c}^2$$

At constant pressure,

$$V = k \times \tfrac{1}{2}mn\bar{c}^2$$

and the kinetic energy is proportional to the absolute temperature, T, therefore

$$V = kT$$

which is Charles' Law.

Dalton's Law

Consider a mixture of two gases in the same volume, V. Then for each gas,

$$p_1V = \tfrac{2}{3} \times \tfrac{1}{2}m_1n_1\bar{c}_1^2$$

$$p_2V = \tfrac{2}{3} \times \tfrac{1}{2}m_2n_2\bar{c}_2^2$$

In the mixture of the gases the total kinetic energy is equal to the sum of the individual kinetic energies. Then, where p is the total pressure,

$$pV = \tfrac{2}{3}(\tfrac{1}{2}m_1n_1\bar{c}_1^2 + \tfrac{1}{2}m_2n_2\bar{c}_2^2)$$

$$pV = \tfrac{2}{3}(\tfrac{3}{2}p_1V + \tfrac{3}{2}p_2V)$$

$$p = p_1 + p_2$$

which is a statement of Dalton's Law.

Graham's Law

Rearranging the fundamental gas equation,

$$\bar{c}^2 = \frac{3pV}{mn}$$

but mn/V is equal to the density of the gas. Therefore

$$\bar{c}^2 = \frac{3p}{d}$$

$$\bar{c} = \sqrt{\frac{3p}{d}}$$

At a constant pressure,

$$\bar{c} \propto \frac{1}{\sqrt{d}}$$

and as the rate of diffusion is directly proportional to the velocity of the molecules,

$$\text{rate of diffusion} \propto \frac{1}{\sqrt{d}}$$

which is Graham's Law.

Molecular Velocities

By substituting values for the density and pressure, using appropriate units, into the equation

$$\bar{c} = \frac{3p}{d}$$

it is possible to calculate the root mean square velocity of the molecules of a gas.

Distribution of molecular velocities

Although the total kinetic energy of a gas is constant for a constant temperature, not all the molecules of the gas travel with the same velocity.

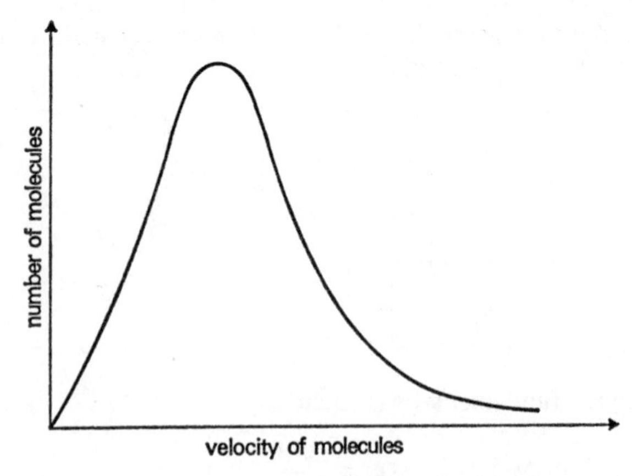

Figure 11.2. Distribution of molecular velocities

The distribution of molecular velocities was determined theoretically by Maxwell and Boltzman and it is represented graphically in Figure 11.2.

At a higher temperature, T_2, the shape of the distribution curve will be similar, i.e. the majority of the molecules will have velocities near to the mean, but the mean velocity will be higher. The importance of the variation of the distribution with temperature is discussed in the chapter on reaction kinetics, page 261.

Variation of the Behaviour of Gases from Boyle's Law

It has been shown on page 117 that an ideal gas, which obeys Boyle's Law exactly, would fulfil the requirement

$$pV = k \quad \text{(when } T \text{ is constant)}$$

In fact, no actual gas does so. If a graph of the product pV at 273 K is drawn against p (as in Figure 11.3), it is found that, for most gases pV is

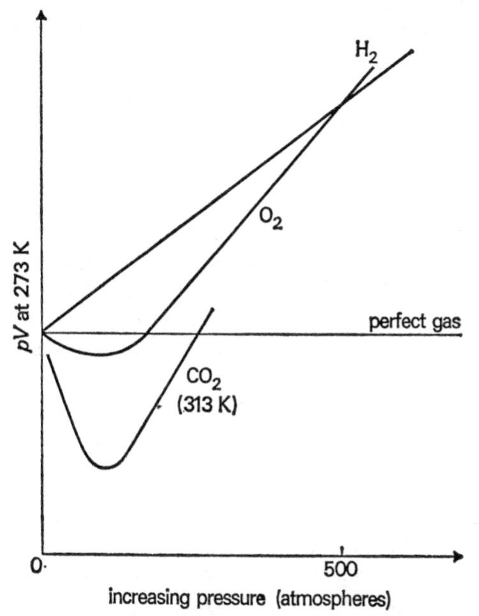

Figure 11.3. ***Variation of value of pV with pressure***

lower in value than Boyle's Law requires until very high pressures are reached (of the order of 250 atmospheres or more). Hydrogen and the noble gases, however, show values of pV which are too high for the requirements of Boyle's Law throughout the whole range of pressures from one atmosphere upwards. The variations from Boyle's Law are greatest in gases which, like carbon dioxide, are most easily liquefied.

These variations arise from two assumptions which are inherent in the formulation of Boyle's Law. These assumptions are:

1. That the gaseous molecules occupy a volume which is completely negligible compared with the space occupied by the gas. This assumption is correct only at very low pressures.
2. That there are no attractive forces between the gaseous molecules.

The best-known attempt to deal with these variations from ideal gaseous behaviour, and to express them mathematically, is that of *van der Waals*.

To allow for the reality of the volume occupied by the gaseous molecules, van der Waals introduced a correction to the volume, V, which became $(V - b)$, where b is a constant appropriate to each gas and related to the volume occupied by the molecules.

The attractive forces normally exercised between the molecules of gases cancel each other when exerted on a molecule in the interior of the gas, because they are exercised at random. In the case of a molecule near the containing wall of the vessel, however, the resultant attractive force must be *inwards*, since there are no gaseous molecules on the outside. Consequently, the pressure exerted on the containing wall by such a molecule is lessened by this attractive force of the other molecules from inside. The attractive force is proportional to the number of molecules in the gas, i.e. to its density. Further, the number of molecules of gas striking the containing wall at a given moment is also proportional to the density of the gas, so that the total inward attracting force, representing loss of pressure, is proportional to the *square* of the density of the gas. This is inversely proportional to the square of the volume of the gas so that the *loss* of pressure resulting from molecular attraction can be represented by a quantity a/V^2, where a is a constant. This quantity must be added to the observed pressure of the gas to represent its 'true' pressure.

Allowing for these two corrections, van der Waals' equation takes the form

$$\left(p + \frac{a}{V^2}\right)(V - b) = RT$$

and, in this form, relates to one mole of the gas.

Each gas has a particular value of its own for constant, a, and another value for constant, b. The constant, a, was at first thought to be independent of temperature but, in fact, is not so.

Since gases do not obey Boyle's Law exactly, they also show small variations from other laws and generalizations relating to gases.

The Work of Andrews on Carbon Dioxide

Some experiments of Andrews (1861) on carbon dioxide have come to be regarded as classical. They are closely related to the theoretical work of van der Waals.

Andrews investigated the effect of changes of temperature and pressure on the volume of carbon dioxide.

The experimental results are expressed by isothermals as in the approximate graph (Figure 11.4). Each isothermal refers to the same mass of carbon dioxide and shows the variation of the volume of carbon dioxide with pressure for a particular temperature. The isotherm of 321 K, MN, approximates to a rectangular hyperbola, which is the correct plot of the

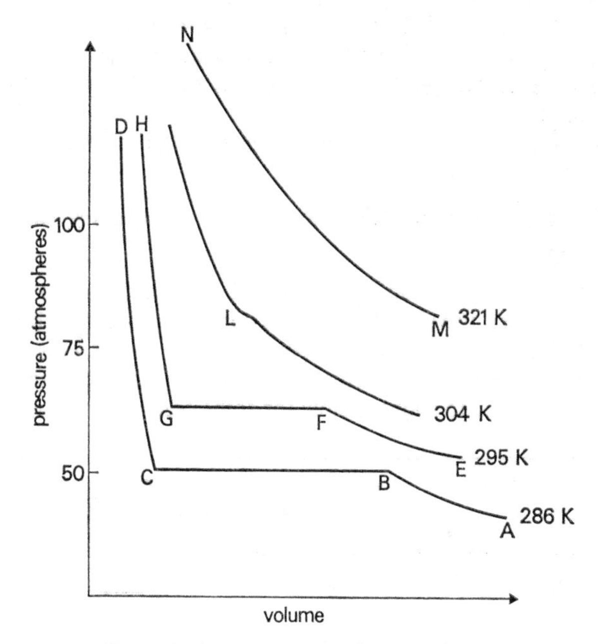

Figure 11.4. Isothermals of carbon dioxide

equation, $pV = k$, which is the expression of Boyle's Law. In this isotherm, carbon dioxide is showing the 'normal' behaviour of a gas.

The isotherm of 286 K, ABCD, is markedly different. Its course is interpreted in the following way. The portion AB represents the usual contraction of the volume of a gas as pressure increases, and approximates to a hyperbola. The portion BC represents a large contraction in volume for almost no pressure change. The portion CD represents almost no volume change for a very large pressure change. Thus at B the pressure has become large enough to start the *liquefaction* of carbon dioxide at 286 K, and from B to C this liquefaction continues and completes itself with only slight increase of pressure. At C liquefaction is completed and, liquids being almost incompressible, further increase of pressure produces no appreciable volume change. The isotherm of 295 K shows similar effects, with EF corresponding to AB, FG (a shorter stretch at the higher temperature) representing the liquefaction stage, and GH the behaviour of liquid carbon dioxide as almost incompressible.

The isotherm of 304.1 K is very significant. It shows only a slight inflection (at L). This corresponds to the point at which the liquefaction stage of CB and GF in the lower isothermals has decreased so that, in effect, the points C,B or G,F have come to coincide. This means that the liquefaction stage just barely occurs, and above this temperature (as in MN) does not occur at all; that is, 304.1 K represents a temperature above which carbon dioxide cannot be liquefied no matter what pressure may be applied to it. This temperature is known as the **critical temperature** of carbon dioxide.

This situation is quite general and every gas or vapour has its own **critical temperature,** which is defined as

> that temperature above which no amount of pressure will serve to liquefy the material.

The least pressure which suffices to liquefy the material at its critical temperature is called its **critical pressure.** The volume occupied by one mole of the material at the critical temperature and pressure is called the **critical volume.** The values of critical temperature (to the nearest degree) and critical pressure are shown below for some common materials:

	Critical temperature (K)	*Critical pressure (atmospheres)*
Hydrogen	38	20.0
Nitrogen	127	33.0
Oxygen	155	50.0
Carbon dioxide	304	73.0
Chlorine	417	76.1
Sulphur dioxide	430	77.6

It will be noticed that the last three gases tabulated have critical temperatures *above* room temperature and so can be liquefied at room temperature by the exercise of pressure alone. These gases were among the earliest to be liquefied. The first three gases, however, have very low critical temperatures. These could not be attained in the early days of experimentation on liquefaction; consequently these gases (and others such as carbon monoxide and methane) were regarded as *permanent gases.* There is, in fact, no truly permanent gas. All known gases have now been liquefied by cooling each below its critical temperature and applying the necessary pressure.

Liquefaction of Gases

The work of Andrews on carbon dioxide showed clearly that to liquefy any gas it is necessary to cool it below its critical temperature and then to apply enough pressure to induce liquefaction. The amount of pressure required is at its maximum at the critical temperature and falls as the temperature of the gas is reduced progressively below the critical temperature. This gives the following general methods of liquefaction of gases.

1. *By cooling only*

If room temperature is well below the critical temperature of a gas, a moderate amount more of cooling may reduce the temperature to the point at which the gas will liquefy under one atmosphere pressure, i.e. the gas can be liquefied merely by cooling. For example, nitrogen dioxide and sulphur dioxide can be liquefied by passage through a tube immersed in a freezing-mixture of ice-salt, giving about 261 K. The gas must be *dry* to prevent formation of ice.

2. *By the joint application of cooling and pressure*

Though others also worked on this method of liquefaction, the classical work in this field is associated with Faraday. His work on the liquefaction of chlorine is typical. He used an inverted, V-shaped glass tube, which was sealed, with one end containing chlorine hydrate, $Cl_2{\cdot}8H_2O$. The empty end was cooled in a freezing-mixture and the chlorine hydrate was warmed. It liberated a relatively large volume of chlorine, which generated high pressure. Under the influence of this high pressure and the cooling effect of the freezing-mixture, drops of liquid chlorine were obtained. Chlorine is now liquefied by pressure alone and stored dry in steel containers.

3. *By using the Joule-Thomson effect. Liquefaction of air*

If a highly compressed gas is allowed to expand into a region of low pressure, a cooling effect is observed. This is called the Joule-Thomson effect. It arises because as the molecules separate during expansion, internal work is done in overcoming the attractive forces between them. A perfect gas, with no attractive forces between the molecules, would show no Joule-Thomson effect.

The Joule-Thomson effect is more marked at lower temperatures and was used in the Linde process for the liquefaction of air. In this process, air is purified by passage over soda-lime to remove carbon dioxide and is then dried. It is pumped to about 200 atmospheres pressure, cooled back to ordinary temperature, and passed through closely spiralled thin copper tubing surrounded by thicker copper tubing (Figure 11.5). At the valve, A,

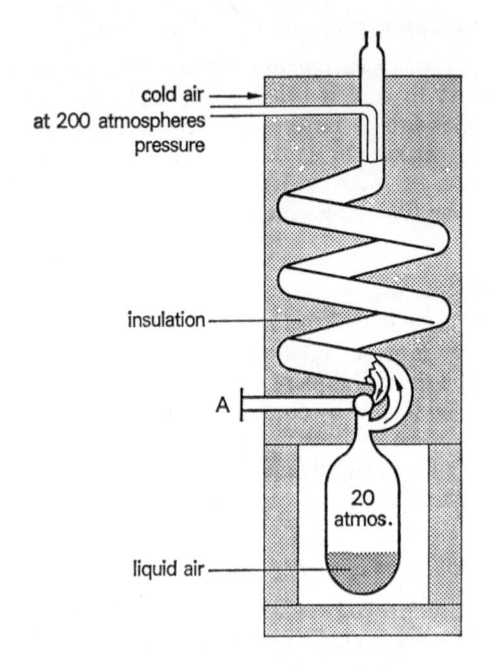

Figure 11.5. Liquefaction of air

the air expands to about 20 atmospheres pressure and is cooled by the Joule-Thomson effect. The cooled air passes out to the pumps in the wider copper tube, cooling the incoming, highly compressed air in the inner tube. This, in turn, is further cooled by the Joule-Thomson effect and still further cools incoming air as it passes out. Eventually, this cascade effect so reduces the temperature of the air that, at 20 atmospheres pressure, it liquefies. All the tubing is protected from heat loss by insulating material. The process was later improved by Claude by making the expanding gas do work, and so lose heat energy, in an expansion engine. The engine drives a dynamo, which recovers some of the energy used in compressing the gas and also makes the liquefaction process more rapid.

Hydrogen is anomalous with regard to the Joule-Thomson effect, showing a *rise* of temperature on expansion at room temperature. This continues down to 193 K; then, at this point (the *inversion temperature*), hydrogen becomes normal and *cools* under expansion. Hydrogen was liquefied (1898) by Dewar, who cooled the gas below the inversion temperature by liquid air and then used the principle of the Linde process as described above. Helium is anomalous like hydrogen, with a very low inversion temperature (33 K).

Densities of Gases and Relative Molecular Masses

Cannizzaro's use of the densities of gases relative to the density of hydrogen for the determination of relative molecular masses and ultimately relative atomic masses was discussed on page 10. It is possible to use the ideal gas equation for the calculation of the relative molecular mass of a gas from its density.

For example, it was found that 100 cm^3 of butane at 298 K and a pressure of 102 700 N m^{-2} had a mass of 0.233 g.

The ideal gas equation is

$$pV = RT$$

where p is the pressure in N m^{-2},
V is the volume in m^3 of 1 mole of gas,
T is the temperature on the Kelvin scale,
and R the gas constant, is equal to 8.31 J K^{-1} mol^{-1}.

Re-arranging the equation, the volume occupied by one mole is given by

$$V = \frac{RT}{p}$$

Under the stated conditions of temperature and pressure,

$$V = \frac{8.31 \times 298}{102\,700}$$

$$= 2.41 \times 10^{-2} \text{ m}^3$$

But 100 cm^3 of butane has a mass of 0.233 g. Therefore 1 m^3 has a mass of

$$\frac{0.233 \times 10^6}{10^2} = 2.33 \times 10^3 \text{ g}$$

Therefore 2.41×10^{-2} m^3 will have a mass of

$$2.33 \times 10^3 \times 2.41 \times 10^{-2} = 56.15 \text{ g}$$

Thus this density measurement gives a value of 56 for the relative molecular mass of butane.

Experimental Methods for Determining the Densities of Gases

1. *Direct weighing*

The density of a substance which is a gas at room temperature can be found with sufficient accuracy by direct weighing. The mass of a container of known volume, such as a glass syringe or an air-tight plastic bottle, is determined. Then, using a predetermined value for the density of air, the mass of air in the container is calculated and subtracted from the first weighing to give the mass of the container. The container is then filled with the gas under investigation and reweighed. It is thus possible to find the mass of a known volume of gas.

The densities of the vapours of volatile liquids can also be used for calculating their relative molecular masses. The densities of such compounds can be determined by the traditional nineteenth century methods or their modern equivalents.

2. *Victor Meyer's method for finding the vapour density of a volatile liquid*

This method is used chiefly in connection with liquids such as diethyl ether (ethoxyethane), trichloromethane, ethanol, and carbon disulphide,

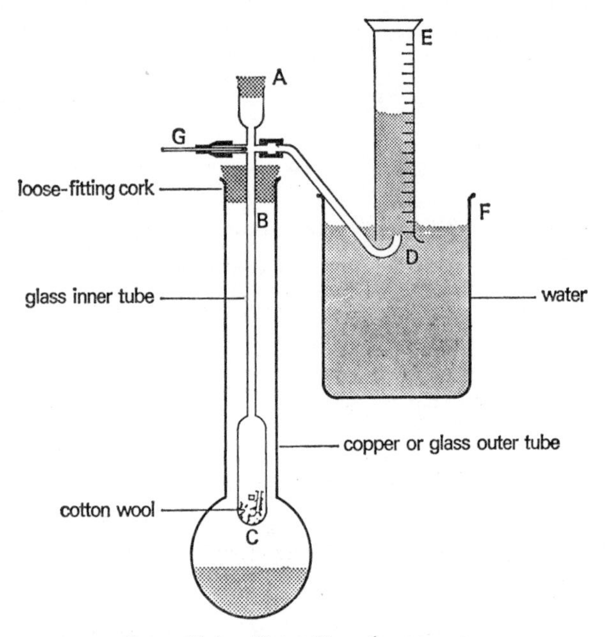

Figure 11.6. *Victor Meyer's apparatus*

all of which have relatively low boiling points. The apparatus is shown in Figure 11.6. The experiment will be described for the special case of *ether*. A small stoppered tube is weighed empty, and then weighed again (checking immediately before use in the apparatus) full of ether. The mass of ether contained is usually about 0.1 g.

The water below C in the outer tube is heated, with the tube ED not yet in position. Air bubbles will be seen escaping from the narrow tube at D as the inner tube warms up, but eventually steam will begin to escape from the loose cork below B. The flame heating the water is adjusted to maintain a slight but steady escape of steam near B. Shortly after this, escape of bubbles at D will cease, showing that the inner tube from B to C is in a *steady state* with most of its length at about 100 °C. The *graduated* tube ED is placed in position as shown, but full of water.

The small weighed tube of ether is then inserted at A and held at B by the glass rod, G, until stopper A is replaced. The rod G is pulled gently aside so that the tube of ether falls to C where cotton-wool prevents breakage. The ether vaporizes, forcing out the stopper of the small tube, and air is expelled rapidly at D and collected (as shown) in the graduated tube. When no more air is expelled, the tube ED is moved clear of the narrow tube at D and lowered into the water till E and F are at the same level, i.e. the air in the tube is at atmospheric pressure. The volume of the air is read off; the temperature of the water in the collecting vessel is noted and the barometer is read. The vapour pressure of water at the relevant temperature is also required (from tables).

The following points are important:

(a) The temperature in the tube BC need not be known, but should remain steady during the determination and should be high enough to vaporize the experimental liquid easily, i.e. should be at least 20 K above the boiling point of that liquid.

(b) The vapour of the experimental liquid should remain near C and, in particular, must not reach B where it could liquefy. The wide bulb near C, and the length of narrower tubing below B, through which diffusion is slow, are designed to secure this. The air expelled from the tube by the experimental liquid then represents the volume which that liquid would occupy if it could exist *as vapour* under room conditions.

(c) The advantages of Meyer's method are its rapidity and the small amount of liquid required. It is not very accurate, but this is a minor matter. For a consideration of this point, see after the calculation below.

Calculation

Mass of stoppered tube	= 0.672 g
Mass of stoppered tube and ether	= 0.783 g
Mass of ether	= 0.111 g
Volume of air displaced	= 36.0 cm^3
Temperature	= 285 K
Pressure	= 100 800 N m^{-2} (756 mmHg)
Vapour pressure of water at 285 K	= 1467 N m^{-2} (11 mmHg)
True air pressure = 100 800 − 1467	= 99 333 N m^{-2}

Volume of air corrected to s.t.p. $= 36.0 \times \frac{99\,333}{101\,300} \times \frac{273}{285}$

$= 33.8\ \text{cm}^3$

Mass of this volume of hydrogen $= 33.8 \times \frac{0.09}{1000} = 0.003\,04\ \text{g}$

(since 1000 cm^3 of hydrogen at s.t.p. has a mass of 0.09 g).

Relative vapour density of ether $= \frac{\text{Mass of ether}}{\text{Mass of hydrogen}}$

$= \frac{0.111}{0.003\,04} = 36.5$

Relative molecular mass of ether $= 36.5 \times 2 = 73.0$

Alternatively the calculation may be based on the ideal gas equation in a similar manner to the example on page 125.

The true molecular mass of ether is 74. The method of Meyer is not very accurate but this hardly matters, for the following reason. If ether were a new material, analysis would show its molecular formula to be $C_4H_{10}O$ or some multiple of this, i.e. its molecular mass must be 74 or 2 × 74 or 3 × 74, etc. The method of Meyer is needed merely to determine which of these numbers, 74, 148, 222, etc., is the correct multiple. Moderate inaccuracy hardly matters in such a case.

3. *Dumas' method for finding the relative molecular mass of a volatile liquid*

This method will be described for the special case of ether. A Dumas bulb (Figure 11.7) is weighed full of air at room temperature and pressure, which must be known. The bulb is then warmed and allowed to cool with its jet, A, under ether. As the bulb cools, ether will be drawn into it and should fill about one-quarter of the bulb. The bulb is then placed in position as shown in Figure 11.7, and held so that the greatest possible fraction of the bulb is immersed in the water. The water is then heated. In a short time, the ether begins to vaporize and escape from A, where it can be burnt. As it does so, air is swept from the bulb. Eventually the water boils, by which time the flame at A will be very small. The water is kept boiling for five minutes or so to secure constant temperature and the jet A is sealed off by heating with a burner while the water is still boiling. In a successful experiment this gives a bulb full of ether vapour at the boiling point of water, which is read from the thermometer, and barometric pressure, which is also read. The bulb is dried and weighed when cool.

A file scratch is gently made near the sealed end, A, of the bulb and the end is broken off under the surface of cold, *boiled-out* water. This water should rush in and fill the bulb. Boiled-out water is used because it contains no air, bubbles of which could prevent the bulb from filling properly with

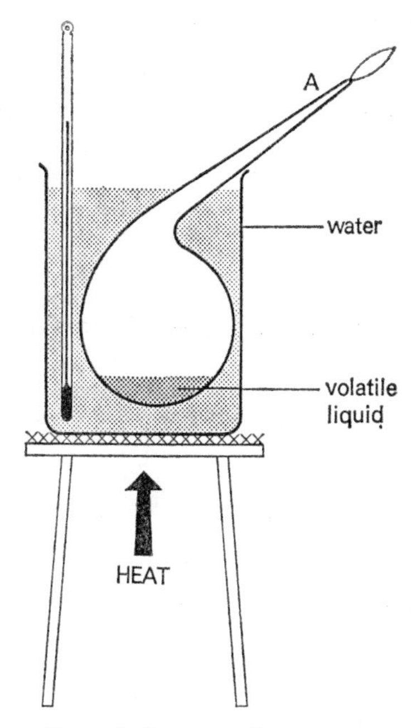

Figure 11.7. Dumas' apparatus

water. The bulb is dried and weighed full of water, including the broken end of the jet.

Calculation

Mass of bulb full of air	= 78.385 g
Temperature of room	= 287 K
Atmospheric pressure	= 102 900 N m^{-2} (772 mmHg)
Temperature at which bulb was sealed	= 373 K
Mass of bulb and ether	= 78.765 g
Mass of bulb full of water	= 355.7 g
Capacity of bulb = (355.7 − 78.4)	= 277.3 cm^3

This capacity referred to s.t.p. is

$$277.3 \times \frac{102\,900}{101\,300} \times \frac{273}{287} = 268 \text{ cm}^3$$

Mass of air in bulb

$$= 1.293 \times \frac{268}{1000}$$

$$= 0.347 \text{ g}$$

(since 1000 cm^3 of air at s.t.p. has a mass of 1.293 g).

Mass of evacuated bulb $= (78.385 - 0.347)$
$= 78.038$ g

Mass of ether vapour sealed in $= (78.765 - 78.038)$
$= 0.727$ g

Volume of ether vapour at s.t.p. $= 277.3 \times \frac{273}{373} \times \frac{102\,900}{101\,300}$
$= 206 \text{ cm}^3$

Mass of this volume of hydrogen $= 0.09 \times \frac{206}{1000}$
$= 0.0185$ g

(since 1000 cm^3 of hydrogen at s.t.p. has a mass of 0.09 g).

Relative vapour density of ether $= \frac{0.727}{0.0185}$
$= 39.3$

Relative molecular mass of ether $= 78.6$

A more rigorous method of calculation would regard the first weighing as the mass of the bulb alone and would consider that the mass of the bulb full of vapour was too small by an amount equal to the mass of the air displaced by the sealed bulb. The final answer will be the same whichever method is employed.

A slight error is involved in taking the capacity of the bulb to be the same at 373 K as at room temperature. The expansion of the glass is ignored.

Dumas' method is capable of greater accuracy than Meyer's, but, in a general way, the value obtained need only be accurate enough to distinguish between multiples of an empirical molecular mass as was indicated at the end of the account of Meyer's method. The principal disadvantages of Dumas' method are:

(a) It needs a large quantity of material.
(b) If the material contains any impurity of higher boiling point than itself, this impurity accumulates in the final vapour and makes the vapour density inaccurate.
(c) It is difficult to expel the *whole* of the air. A small bubble is usually left after filling with water.

In the modern equivalent of the last two methods a known mass of the volatile liquid is injected through a serum cap into a glass syringe enclosed in a transparent steam bath. The volume of vapour produced can be read directly from the syringe. Thus the volume of a known mass of vapour at the temperature of the steam bath and at atmospheric pressure is known.

12. Phase Equilibria

A **phase** is a physically distinct part of a system. Thus examples of two-phase systems are those containing:

a gas and a solid,
a gas and a liquid,
a solid and a liquid,
a solid and a solid,
two immiscible liquids.

Examples of three-phase systems are:

a solid, a liquid and a gas,
two immiscible liquids and a gas.

Phase equilibria involves a study of the conditions (temperature, pressure, and concentration) under which different phases are in equilibrium.

The number of **components** in a system is the minimum number of chemical entities required to make up the system. A liquid and its vapour constitute a two-phase, one-component system, whereas a solution of a solute and the solvent vapour is a two-phase, two-component system.

One-component Systems

Concentration is not a variable when there is only one component in a system and the phase equilibria are only affected by changes in temperature and pressure.

A liquid in a closed container will reach a state of equilibrium with its vapour. This state is realized when the rate of movement of molecules from liquid to vapour is exactly balanced by the rate of movement from vapour to liquid. At equilibrium the concentration of molecules in the vapour phase will be constant which will result in the liquid exerting a constant vapour pressure for that temperature (saturated vapour pressure). The saturated vapour pressure and the temperature represent conditions for equilibrium between the liquid phase and the vapour phase.

An increase in temperature causes an increase in the rate of movement of molecules from liquid to vapour phase and a new equilibrium is set up at a higher vapour pressure.

A graph representing the equilibrium conditions, a **phase diagram**, over a range of temperatures and pressures is given in Figure 12.1.

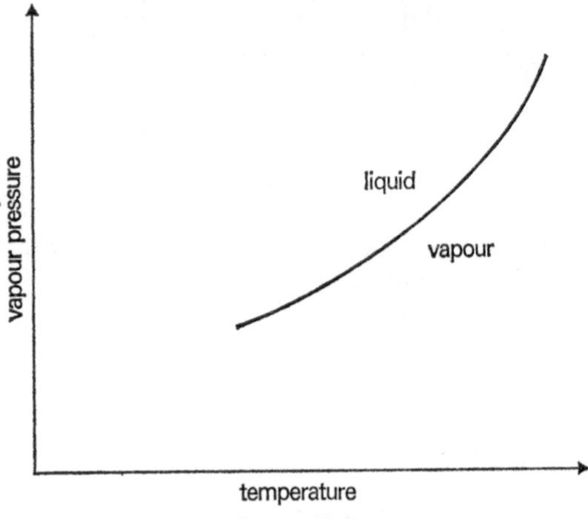

Figure 12.1.

Also for a one-component system it is possible to determine the conditions for equilibrium between solid and liquid and at low pressures between solid and vapour. The complete phase diagram for water is given in Figure 12.2.

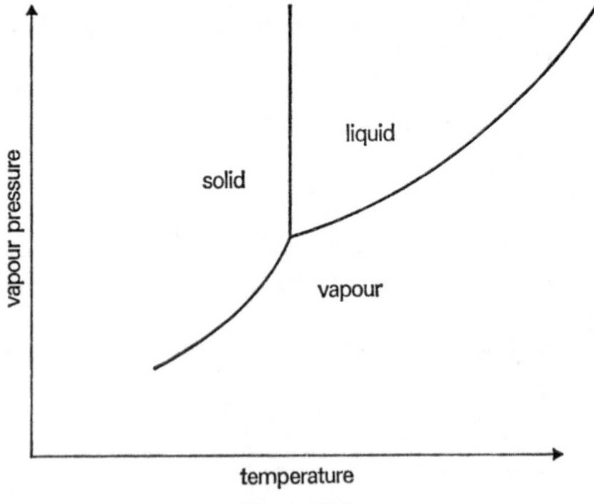

Figure 12.2.

Two-component Systems

In the case of two-component systems all three variables, temperature, pressure, and concentration, may affect the equilibria. It is customary to keep one of these conditions constant and to investigate the effects of variations in the other two.

Six general types of two-component systems are considered in this chapter.

1. *Dilute solutions of non-volatile solutes*

(a) *Lowering of vapour pressure*

The presence of a non-volatile solute reduces the rate of movement of molecules from the liquid phase to the vapour phase and so lowers the saturated vapour pressure. This phenomenon was investigated by Raoult, who, keeping the temperature constant, found that the lowering of vapour pressure was related to the concentration of the solution by the equation

$$\frac{p_1 - p_2}{p_1} = \frac{n}{N + n}$$

where p_1 = vapour pressure of pure solvent,
p_2 = vapour pressure of solution,
n = number of moles of solute,
N = number of moles of solvent.

The left-hand side of the equation is the relative lowering of the vapour pressure and the right-hand side is the mole fraction of solute. This relationship, which is one form of Raoult's law, only holds for dilute solutions of non-volatile non-electrolytes.

If the solution is very dilute, n is small relative to N and the law can then be used in the approximate form

$$\frac{p_1 - p_2}{p_1} = \frac{n}{N}$$

The following calculation illustrates the application of the law.

Calculation The vapour pressure of ether (molecular mass 74) is 58 930 $N m^{-2}$ (442 mmHg) at 293 K. If 3 g of a compound A are dissolved in 50 g of ether at this temperature, the vapour pressure falls to 56 800 $N m^{-2}$ (426 mmHg). Calculate the molecular mass of A.

Using the same symbols as above,

$$n = \frac{3}{M}$$

where M is the molecular mass of A.

$$N = \frac{50}{74};\ p_1 = 58\,930;\ p_2 = 56\,800$$

Using the complete formula:

$$\frac{58\,930 - 56\,800}{58\,930} = \frac{3/M}{50/74 + 3/M}$$

From this, by the usual algebraic methods, $M = 118$.

The approximate form of the equation gives:

$$\frac{58\,930 - 56\,800}{58\,930} = \frac{3/M}{50/74}$$

$$M = 123$$

Direct measurement of vapour pressure change is rather difficult. The following gravimetric method can be applied to water and its solutions. A current of *dried* air is passed slowly through several bulbs containing a solution of known concentration of the compound of which the molecular mass is required. The air becomes saturated to the vapour pressure, p_2. The air then passes on through a set of weighed bulbs containing pure water. Here the air is saturated to the vapour pressure, p_1. The bulbs are weighed after the experiment and the *loss* of mass is proportional to $(p_1 - p_2)$. The air then passes through weighed calcium chloride tubes, which are weighed again after the experiment. The *gain* in mass of these tubes is proportional to p_1.

If the *loss* of mass in the bulbs of water is m_1 and the *gain* in the calcium chloride tubes is m_2,

$$\frac{m_1}{m_2} = \frac{n}{N + n}$$

and the calculation is on the same lines as before.

When one particular solution is considered the concentration will remain constant and the variation of the vapour pressure of the solution with temperature can be investigated. The vapour pressure curve of a solution will be similar in shape to but lower than that of the pure solvent. Figure 12.3 shows vapour pressure curves of the pure solvent and two solutions. The concentration of solution 2 is greater than that of solution 1.

(b) *Elevation of boiling point*

The third variable, vapour pressure, is most easily kept constant by considering the temperatures required to raise the vapour pressures of solutions to the pressure of the atmosphere at which points the solutions will boil. It is clear from Figure 12.3 that the presence of a non-volatile

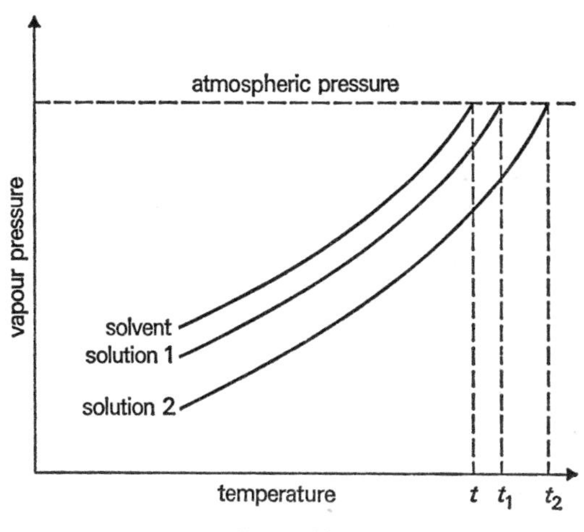

Figure 12.3.

solute raises the boiling point of a solvent. The boiling points of the pure solvent, solution 1, and solution 2 are t, t_1, and t_2, respectively. The magnitude of the elevation is related to the lowering of the vapour pressure which in turn, according to Raoult's Law, is related to the mole fraction of solute. The elevation of boiling point must therefore be dependent on the mole fraction of solute. In fact, it is found that a constant mole fraction of various solutes in one particular solvent always produces the same elevation. If the mole fraction used is 1 mole of solute in 1000 g of solvent the elevation produced is known as the **boiling-point constant,** K. The elevation of boiling point provides a method for the determination of relative masses of solutes, the limitations being that the solutes must be non-volatile and non-electrolytes and that only dilute solutions are used in the experimental work.

Determination of relative mass by boiling-point elevation

Several methods are available for this purpose. The one described below was introduced by Landsberger and modified by Walker and Lumsden. The apparatus is shown in Figure 12.4. A quantity of the pure solvent is placed in the inner tube and the apparatus is completed as shown. Some more of the pure solvent is boiled in a flask and the vapour is passed in as shown until the thermometer registers a *constant* temperature, which is read (with a lens) as the boiling point of the pure solvent. (Notice that the inlet tube for vapour ends in a small perforated bulb to break up the vapour into small bubbles.)

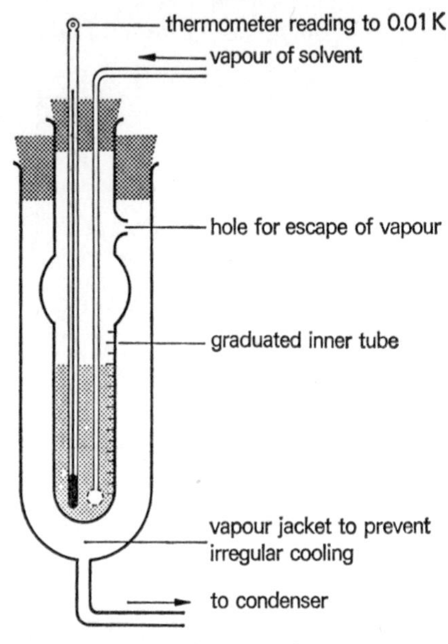

Figure 12.4. Elevation of boiling point

A *known* mass of the solute (of which the relative molecular mass is required) is then dissolved in the solvent and a second (higher) boiling point of the solution is observed as before. At once, to avoid further dilution, the supply of vapour is cut off. The fittings are removed from the inner tube and the volume of solution is read off from the graduations. If desired, more solvent vapour can then be passed, which will dilute the solution, and other similar observations of boiling point and volume of liquid can be made.

The following is a typical set of results:

Boiling point of trichloromethane	= 332.80 K
Mass of camphor added	= 0.400 g
Boiling point of the solution	= 333.10 K
Volume of solution	= 23.0 cm^3

(Density of trichloromethane is 1.50 g cm^{-3}. K is 3.9 K per 1000 g of trichloromethane.)

The density of the solution can be taken as the same as that of trichloromethane, so that the mass of trichloromethane used is (1.5 × 23.0 g) = 34.5 g. The boiling-point elevation is (333.10 − 332.80) K = 0.30 K. From the results:

0.30 K is the elevation in 34.5 g of trichloromethane by 0.40 g of camphor.

3.9 K is the elevation in 1000 g of trichloromethane by

$$0.40 \times \frac{3.9}{0.30} \times \frac{1000}{34.5} \text{ g} = 151 \text{ g of camphor}$$

That is, the relative molecular mass of camphor is 151.

If desired, this kind of calculation can be generalized into the following formula. Let x K be the elevation of boiling point produced by dissolving a g of a solute A in b g of a solvent B. Then, if y K is the boiling-point constant per 1000 g of B,

$$M = a \times \frac{y}{x} \times \frac{1000}{b}$$

where M is the relative molecular mass of the solute, A.

(c) *Depression of freezing point*

The vapour pressure curves of solutions 1 and 2 cross the line which represents the transition to solid (Figure 12.5) at lower and lower temperatures, indicating that a non-volatile solute depresses the freezing point of a solvent.

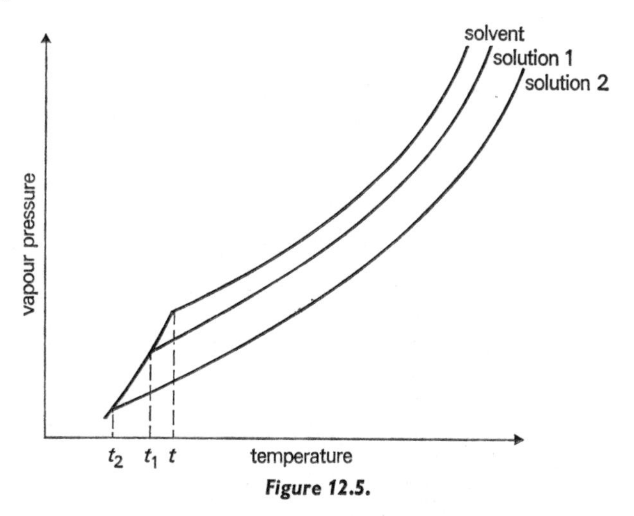

Figure 12.5.

In Figure 12.5, t, t_1 and t_2 are the freezing points of the pure solvent, solution 1, and solution 2 respectively. It is clear from the graph that the magnitude of the depression is related to the lowering of the vapour pressure, which, according to Raoult's law, is related to the mole fraction of solute. The laws which govern depression of freezing point are identical to those for elevation of boiling point, that is, a constant mole fraction of any non-volatile non-electrolyte, when dissolved in the same solvent, will produce identical depressions. When the mole fraction is 1 mole of solute dissolved in 1000 g of solvent, the depression is known as the freezing-point constant, for that solvent. Freezing-point depressions can also be used for relative molecular mass determinations.

Determination of molecular mass by freezing-point depression

Apparatus suitable for this determination is shown in Figure 12.6. The freezing mixture is not necessarily ice or ice-salt. It is a material which can

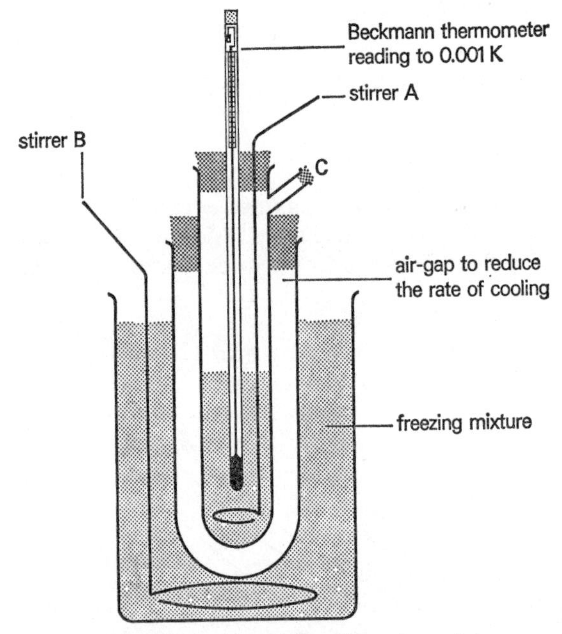

Figure 12.6. Depression of freezing point

maintain a temperature about 7 K below the freezing point of the solvent used in the inner tube. If this solvent is water, ice-salt is suitable, but, in some cases, the freezing mixture may be warm water. The choice is relative to the freezing point to be observed.

In making the determination, a known mass of the solvent is placed in the inner tube. Stirrer B is used occasionally to keep the freezing mixture reasonably uniform. The solvent is stirred vigorously by stirrer A and temperature observations are made every quarter-minute. From them a graph can be drawn which will usually take the form shown in Figure 12.7.

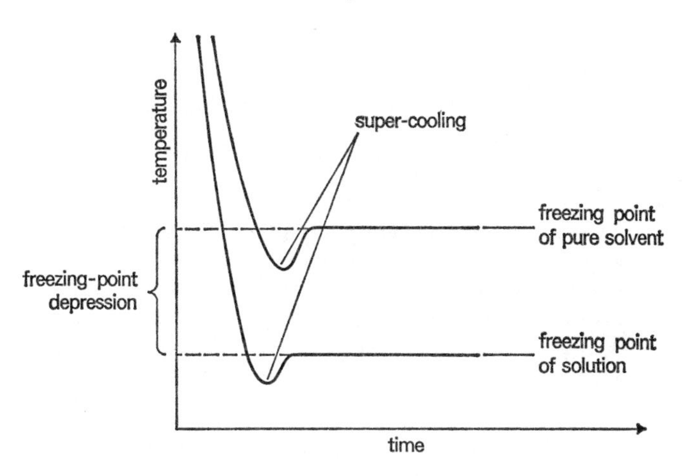

Figure 12.7. Cooling curves

There is usually some degree of super-cooling. If this occurs, the temperature will rise as the solvent starts to freeze and will become steady at the freezing point, as shown in Figure 12.7. The solvent is then allowed to warm up and melt, and a known mass of the solute (preferably in pellet form) is introduced by the side-tube, C. (If it is more convenient, a solution of known concentration can be substituted for the pure solvent.) When the solute has dissolved, a cooling curve is plotted from observations as before. Super-cooling will occur again (see Figure 12.7) and should not exceed 0.5 K. If it does, so much solid solvent will separate rapidly when freezing begins that the concentration of the solution will be appreciably altered. Super-cooling can usually be kept at a tolerable level by vigorous stirring, but in some cases 'seeding' by a small crystal of the solid solvent may be required to induce freezing. The solution freezes at a lower temperature than the pure solvent and the freezing-point depression is read off as shown on the graph. Other observations can be made by addition of further amounts of the solute.

The following are typical results:

Volume of benzene used $= 60.0\ cm^3$
Freezing point of benzene $= 278.495$ K
Mass of naphthalene dissolved $= 1.250$ g
Freezing point of the solution $= 277.515$ K
(Density of benzene $= 0.880\ g\ cm^{-3}$
$K = 5.1$ K per 1000 g of benzene.)
Mass of benzene used $= 60 \times 0.880$ g
$= 52.8$ g

From the results:

0.980 K is the depression in 52.8 g of benzene by 1.250 g of naphthalene. That is, 5.1 K is the depression in 1000 g of benzene by

$$1.250 \times \frac{5.1}{0.980} \times \frac{1000}{52.8} = 123 \text{ g of naphthalene}$$

From this, the relative molecular mass of naphthalene is 123.

This type of calculation can be generalized in the following way. Let x K be the depression of freezing point produced by dissolving a g of a solute A in b g of a solvent B. Then, if y is the freezing-point constant per 1000 g of B,

$$M = a \times \frac{y}{x} \times \frac{1000}{b}$$

where M is the relative molecular mass of the solute, A.

The freezing-point method for relative molecular mass determination is probably more accurate than the boiling-point method, but in practice neither really needs to show a high degree of accuracy for the following reason. Consider the above result for naphthalene. Analysis shows that the composition of this hydrocarbon corresponds to the simplest formula, C_5H_4. The molecular formula of naphthalene must then be $(C_5H_4)_n$ and the relative molecular mass 64 × 1 or 64 × 2 or 64 × 3, etc. The method for determination of relative molecular mass need only be accurate enough to distinguish among these multiples, and the occurrence of a few per cent of error is unimportant. The figure 123 in the calculation above clearly indicates the multiple 128 and the formula $C_{10}H_8$.

Freezing-point and boiling-point constants for some common solvents are given below:

Solvent	K per 1000 g of solvent Boiling point (K)	 Freezing point (K)
Water	0.52	1.86
Benzene	2.60	5.10
Camphor	—	40.0
Tribromomethane	—	14.3

Method of Rast for determination of relative molecular mass by freezing-point depression

This method utilizes the exceptionally high freezing-point constant of camphor, which results in a large depression for a small mole ratio of solute to solvent and hence there is no need to use a very accurate thermometer.

A known mass (probably a few mg) of the solute is melted with a known mass (10–12 times its own mass) of camphor. The product is cooled and ground to powder. Its melting point is determined in a small capillary tube attached to a thermometer bulb and heated in propane-1,2,3,-triol (glycerol) as in the usual method of finding melting points of organic compounds. The melting point of the pure camphor having been similarly determined, the depression is known and the calculation can be made as before.

The Beckmann thermometer

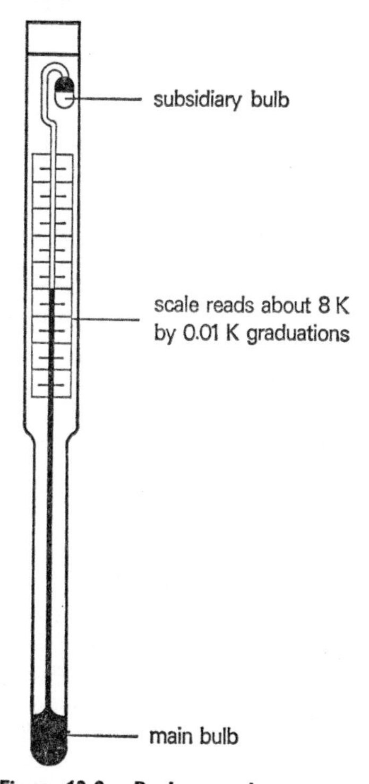

Figure 12.8. Beckmann thermometer

The Beckmann thermometer is a very accurate and sensitive form of centigrade thermometer. Its important features are:

(i) A large bulb of mercury and a very thin capillary tube in which the mercury moves. This combination makes the instrument very sensitive. It is usually graduated in hundredths of a degree.

(ii) The instrument has only a small range of temperature on its scale, usually about 8 K. It is used for measuring temperature *differences*.

(iii) It has a subsidiary bulb at the top of its capillary tube which can be used to vary the temperatures registered by the scale. The system works in the following way. Suppose the Beckmann thermometer (Figure 12.8) has been used for observations on the freezing points of aqueous solutions. It is then adjusted so that the scale reads about − 6 K to + 2K. If it is to be used for boiling-point observations on aqueous solutions, the scale should read about 371–379 K. This requires some mercury to be forced out of the main bulb into the upper, subsidiary bulb. To allow for the thread of mercury above the scale, the thermometer is gradually heated in a liquid boiling at about 383 K (say toluene) and this, at the boiling point, forces mercury up into the subsidiary bulb. When the thermometer is registering the temperature of 383 K it is tapped sharply so that the mercury expanded into the upper bulb is detached from the thread of mercury in the capillary tube. The top limit of the thread then registers 383 K at the entrance to the upper bulb. On cooling 3–4 K, the end of the thread will fall back on to the top of the scale which then represents about 379 K. The 8 K range of the scale will take it down to about 371 K as required.

To reverse this process, the detached mercury is tapped back to the top of the upper bulb and the main bulb is then gradually heated in a bath of suitable liquid till the mercury thread rejoins.

(d) *Osmosis and osmotic pressure*

It is found that certain membranes exist which allow water to pass through them, but do not allow dissolved solutes to pass. Such membranes are said to be *semi-permeable*. The best known and most efficient semi-permeable membrane is an artificially prepared one of copper hexacyanoferrate(II). The preparation is described below. Animal membranes, such as pigs' bladders, are often approximately semi-permeable but rarely perfectly so.

The different phases to be considered in this case are two aqueous solutions of the same solute, but of different concentrations, which are separated by a semi-permeable membrane. The rate of movement of solvent molecules through the membrane from the less concentrated solution to the more concentrated will be greater than in the reverse direction and so there is a net movement in that direction and the solutions tend to become equal in concentration. When a solution is separated from pure solvent by a semi-permeable membrane, the pressure which must be applied to the solution in order to balance and hence prevent the movement of solvent into the solution, is called the *osmotic pressure* of the solution. The passage of the solvent through the membrane is called *osmosis*. This can be illustrated by the apparatus of Figure 12.9.

A sugar solution inside the pot is separated from distilled water (acting here as an infinitely dilute sugar solution) in the outer vessel by the semi-permeable membrane. At the beginning, the inner and outer liquid levels are equal. With the passage of time, distilled water moves through the membrane into the pot and the level of liquid rises. As it does so, the hydrostatic pressure of the rising column of sugar solution from the pot opposes the effects of osmotic pressure. Eventually the column of sugar solution becomes so long that the hydrostatic pressure of the column of sugar solution just equals the osmotic pressure of the solution.

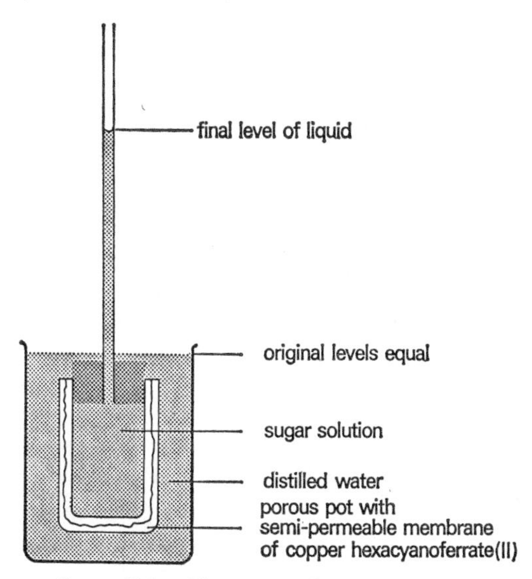

Figure 12.9. Illustration of osmotic pressure

The solution inside the pot at this stage is more dilute than the original and it is the osmotic pressure of this more dilute solution which is being balanced and not that of the original solution.

Preparation of a semi-permeable membrane

Natural semi-permeable membranes, such as animal bladders, are not much use for scientific purposes. They are variable in quality, not fully semi-permeable, and weak. A copper hexacyanoferrate(II) membrane is, for all practical purposes, completely semi-permeable, i.e. it allows water to pass quite freely but prevents the passage of dissolved solutes. It is, however, very weak. Pfeffer (1877) strengthened the membrane by depositing it in the walls of a porous pot so that the membrane acquired the mechanical strength of the walls of the pot while losing little in semi-permeability.

In Pfeffer's method, the porous pot was cleaned with alkali solution and then with potassium nitrate solution. It was thoroughly washed to remove these materials, after which it was filled with water and subjected to pressure so that water was forced into the pores. The pot was filled with a 3 per cent solution of potassium hexacyanoferrate(II) and placed in a similar solution of copper(II) sulphate. The two solutions diffused into the walls of the pot, depositing the membrane of copper hexacyanoferrate(II).

$$2Cu^{2+}(aq) + Fe(CN)_6^{4-}(aq) \rightarrow Cu_2Fe(CN)_6(s)$$

The pot was then thoroughly washed with distilled water. A later method of Morse and Frazer improved on Pfeffer's by assisting the production of the membrane by electrolysis as well as diffusion.

Measurement of osmotic pressure

Pfeffer's method

The apparatus used is illustrated diagrammatically in Figure 12.10. It must be kept at constant temperature. With the passage of time, water

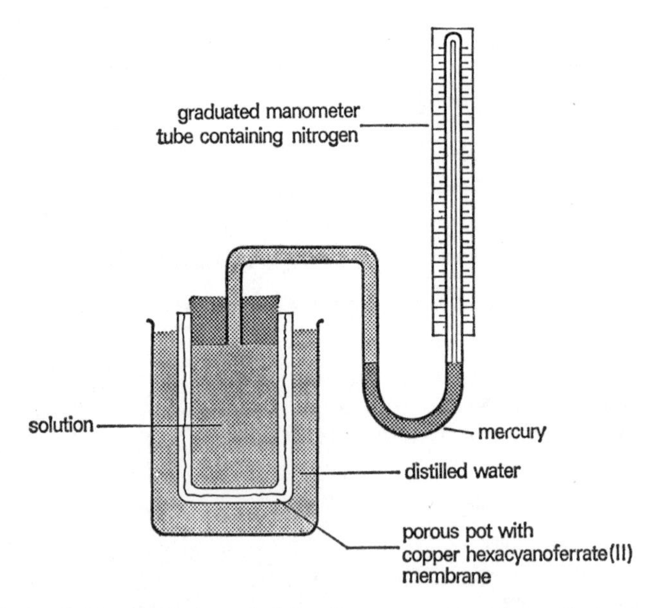

Figure 12.10. ***Pfeffer's method for the measurement of osmotic pressure***

enters the porous pot by the operation of osmotic pressure, forcing the mercury up the manometer tube and compressing the nitrogen. Eventually the reverse pressure of the nitrogen equals the osmotic pressure of the solution and equilibrium is reached. The pressure can then be read from the manometer scale. The weaknesses of this method are:

(a) Water enters the porous pot diluting the solution. The extent of the dilution is known by the movement of mercury in the manometer tube but it may not be equally distributed in the solution. Hence there is some slight doubt about the effective concentration of the solution.
(b) The pressure is exerted against the walls of the porous pot. The strength of the pot sets a limit to the osmotic pressures which can be measured.

In the modern equivalent of Pfeffer's method the membranes are usually made of cellulose or cellulose nitrate supported by perforated metal plates. The osmotic pressure is determined either by the rise in a capillary tube or by measuring the flow rates for certain applied pressures and interpolating to find the pressure for zero flow rate.

Laws of Osmotic Pressure

These laws apply to unionized solutes only. The effects of ionization on osmotic pressure are considered later.

1. *Concentration effect*

For dilute solutions of a given solute at a constant temperature, osmotic pressure, π, is directly proportional to concentration, C.

The mathematical expression of this law is

$$\frac{\pi}{C} = k \quad \text{(if } T \text{ is constant)}$$

Since concentration is inversely proportional to the volume of the solution, this law can also be written

$$\pi \times V = k \quad (T \text{ constant})$$

This is analogous to Boyle's Law for gases, expressed in the form

$$p \times V = k \quad (T \text{ constant})$$

2. *Temperature effect*

The osmotic pressure of a given solution is directly proportional to its absolute temperature.

The mathematical expression of this law is:

$$\frac{\pi}{T} = k \quad \text{(if concentration is constant)}$$

This is analogous to the law for gases, expressed in the form

$$\frac{p}{T} = k \qquad \text{(if } V \text{ is constant)}$$

3. *Molecular effect*

Equimolar dilute solutions of different solutes at the same temperature have the same osmotic pressure.

This law shows that osmotic pressure is determined by the *number* of molecules of solute present in unit volume of solution and not by their size or chemical nature. Thus, if 46 g of alcohol, 60 g of urea, 180 g of glucose, and 342 g of cane sugar are each dissolved separately to make, say, 10 dm^3 of solution, all the solutions will have the same osmotic pressure at a given temperature. The figures are the relative molecular masses of the solutes. Notice that, because of its lighter molecule, urea is, mass for mass, three times as effective as glucose and nearly six times as effective as cane sugar in setting up osmotic pressure.

By combining the mathematical statements of laws (1) and (2) above, we obtain the composite equation

$$\frac{\pi \times V}{T} = \text{a constant}$$

This is analogous to the expression of the gas laws, which takes the form

$$\frac{p \times V}{T} = \text{a constant}$$

This similarity is not merely qualitative. It is quantitative too, the two constants being identical in numerical value, as the following calculations show.

For gases

1 mole of any gas at 273 K in 22.4 dm^3 exerts a pressure of 1 atmosphere. Applying these figures,

$$\text{constant} = \frac{p \times V}{T} = \frac{1 \times 22.4}{273}$$

$$= 0.082 \text{ dm}^3 \text{ atmosphere K}^{-1} \text{ mol}^{-1}$$

For osmotic pressure

Experiment shows that 10 g of cane sugar (molecular mass 342) in 1 dm^3 of solution produce an osmotic pressure of 0.66 atmospheres at 273 K. From this,

$$\text{constant} = \frac{\pi \times V}{T} = \frac{0.66 \times 34.2}{273}$$

$$= 0.083 \text{ dm}^3 \text{ atmospheres k}^{-1} \text{ mol}^{-1}$$

(V is the number of dm^3 containing 342 g of cane sugar.)
The same calculations using SI units are given below.

For gases

1 mole of any gas at 273 K in 2.24×10^{-2} m^3 exerts a pressure of 1.013×10^5 N m^{-2}.

The gas equation for 1 mole is $pV = RT$. The gas constant

$$R = \frac{1.013 \times 10^5 \times 2.24 \times 10^{-2}}{273}$$

$$= 8.31 \text{ J mol}^{-1} \text{ K}^{-1}$$

For osmotic pressure

Experiment shows that 10 g of cane sugar (molecular mass 342) in 1×10^{-3} m^3 of solution produces an osmotic pressure of 6.68×10^4 N m^{-2} at 273 K.

$$\text{Volume of solution containing 1 mole} = \frac{1 \times 10^{-3} \times 342}{10}$$

$$= 3.42 \times 10^{-2} \text{ m}^3$$

$$\text{Using } \pi V = RT,$$

$$R = \frac{6.68 \times 10^4 \times 3.42 \times 10^{-2}}{273}$$

$$= 8.36 \text{ J mol}^{-1} \text{ K}^{-1}$$

The slight difference in the figures is caused by experimental error in the osmotic pressure data. This quantitative identity means that:

A solute produces in dilute solution the same osmotic pressure as it would produce gas pressure, if it existed as a gas in the same volume as the solution at the same temperature.

That is, suppose a dilute sugar solution to be contained in an exactly full, covered beaker. Then suppose the water to be suddenly whisked away, leaving the sugar molecules suspended in otherwise empty space. The sugar then functions as a gas, exerting a pressure equal to its former osmotic pressure in solution. From this it follows that, just as 1 mole of any gas in 22.4 dm^3 at 273 K exerts 1 atmosphere pressure, so 1 mole of any exerts 1 atmosphere osmotic pressure. This offers a means of finding the molecular mass of a soluble compound, for the relative molecular mass is that mass in grammes which, at 273 K in 22.4 dm^3 of solution, gives 1 atmosphere osmotic pressure. Any observation of the osmotic pressure of a suitable solution at a known temperature enables this calculation to be made. An example is given below.

Calculations on osmotic pressure

These calculations can be performed by formula or from first principles. Both methods are given below. The most convenient form of the osmotic pressure formula is

$$\frac{\pi}{C \times T} = \frac{\pi_1}{C_1 \times T_1}$$

where C is concentration and T is absolute temperature. This formula combines laws (1) and (2) given above.

Example 1 20 g of cane sugar in 2 dm^3 of solution at 283 K produce an osmotic pressure of 0.68 atmosphere. Calculate the relative molecular mass of cane sugar.

In using the above formula, one side of it (say the left-hand side) expresses the experimental data given; the other side expresses the fundamental principle that the relative molecular mass of cane sugar in grammes in 22.4 dm^3 of solution at 273 K gives 1 atmosphere osmotic pressure.

Let the relative molecular mass of cane sugar be M.

$$\frac{\pi}{C \times T} = \frac{\pi_1}{C_1 \times T_1}$$

$$\frac{0.68}{20/2 \times 283} = \frac{1}{M/22.4 \times 273}$$

From this, $$M = \frac{20}{2} \times 283 \times \frac{1}{0.68} \times \frac{1}{273} \times 22.4$$

i.e. relative molecular mass of cane sugar = 341

Working from first principles, the calculation takes the form:

0.68 atmospheres is the osmotic pressure developed in 2 dm^3 of solution at 283 K by 20 g cane sugar

1 atmosphere is the osmotic pressure developed in 22.4 dm^3 of solution at 273 K by

$$20 \times \frac{1}{0.68} \times \frac{22.4}{2} \times \frac{283}{273} = 341 \text{ g cane sugar.}$$

i.e. relative molecular mass of cane sugar = 341.

In the above fraction, the part 1/0.68 expresses the fact that the change from 0.68 atmosphere to 1 atmosphere of osmotic pressure requires an *increase* in the mass of cane sugar; the fraction 22.4/2 expresses the fact that an *increase* in the volume of solvent from 2 to 22.4 dm^3 requires a corresponding *increase* in the mass of cane sugar to maintain the same osmotic effect; the fraction 283/273 expresses the fact that the *fall* in temperature from 283 K to 273 K requires an *increase* in the mass of cane sugar to maintain the same osmotic effect.

The same calculation may be done by direct application of the gas equation. SI units will be used in this case.

20 g of cane sugar in 2×10^{-3} m^3 of solution at 283 K produce an osmotic pressure of 6.89×10^4 N m^{-2}. The gas constant is 8.31 J mol^{-1} K^{-1}. The volume of solution containing 1 mole is given by

$$V = \frac{RT}{\pi}$$

$$= \frac{8.31 \times 283}{6.89 \times 10^4}$$

$$= 3.414 \times 10^{-2} \text{ m}^3$$

But, 2×10^{-3} m^3 contains 20 g of sugar; therefore 3.414×10^{-2} m^3 contains

$$\frac{20 \times 3.414 \times 10^{-2}}{2 \times 10^{-3}} = 341.4 \text{ g}$$

The relative molecular mass of the sugar is 341.

Osmotic pressure effects for comparable mole fractions are much higher than the effects on freezing and boiling points. Thus when the relative molecular mass of a polymer, from which it is only possible to prepare a solution with a small mole fraction, is determined the osmotic pressure method is preferred.

Abnormal relative molecular masses

When the relative molecular masses of certain compounds are determined by the foregoing methods, abnormalities appear in the results. Some compounds show higher relative molecular masses than would be expected from their accepted chemical formulae; others show lower values.

Abnormally high relative molecular masses

These are comparatively rare and arise from *polymerization* of the compound, i.e. from the formation of complex molecules by the combination of two or more normal molecules. The normal relative molecular mass is multiplied by the factor n.

$$nA \rightarrow A_n$$

Examples are benzoic acid and ethanoic acid (acetic acid) in benzene solution, where dimerization occurs to $(C_6H_5CO_2H)_2$ and $(CH_3CO_2H)_2$. This phenomenon probably involves hydrogen-bonding page 101. Ethanoic acid is also dimerized in the vapour just above its boiling point. Methanoic acid (formic acid) dimerizes in the vapour state, but only partially.

Abnormally low relative molecular masses

These arise when the observed effect (osmotic pressure, elevation of boiling point, etc.) is greater than expected. These high observed values are due to the solute dissociating into ions.

Consider an electrolyte such as ethanoic acid (acetic acid) in solution. Let 1 mole of the acid be dissolved to make a definite volume of solution. Suppose that a fraction α of this mole is ionized when equilibrium is reached at constant temperature (say 298 K.). The position is then:

	$CH_3CO_2H \rightleftharpoons$	$CH_3CO_2^- +$	H^+
Originally	1 mole	—	—
At equilibrium	$(1-\alpha)$ mole	α	α mole

α is called the *degree of ionization or dissociation* of the acid.

The 'normal' osmotic pressure developed in the solution would be related to 1 mole of unionized acid. But each of the ions exercises the same influence in solution as the molecule from which it was derived; therefore the total number of effective particles produced per molecule of the original acid is $(1-\alpha) + 2\alpha$, and the abnormally high osmotic pressure (and freezing point and boiling point changes) is related to this figure. From this

$$\frac{(1-\alpha)+2\alpha}{1} = \frac{\text{osmotic pressure observed in practice}}{\text{osmotic pressure calculated for no ionization}}$$

The numerator of the fraction on the left-hand side of the equation is always greater than unity. This explains the abnormally high osmotic pressure produced by ethanoic acid in solution and, if the actual osmotic pressure is experimentally determined, provides a means of finding α, the degree of ionization of ethanoic acid. Parallel conclusions follow for freezing-point depression and boiling-point elevation.

In general, if an electrolyte produces n ions per molecule and has, under the relevant conditions of concentration and temperature, a degree of ionization α,

$$\frac{(1-\alpha)+n\alpha}{1} = \frac{\text{osmotic pressure observed in practice}}{\text{osmotic pressure calculated for no ionization}}$$

and correspondingly for freezing-point depression and boiling-point elevation.

A dilute solution of a strong binary electrolyte such as sodium chloride always produces an effect which is twice that calculated from the assumption that it exists as an ion pair of relative formula mass 58.5. This indicates that sodium chloride is completely dissociated in dilute aqueous solutions.

2. *Solubility of a solid in a liquid*

In this system there are two components, the solute and the solvent, and two phases, the solution and the undissolved solid.

Definition of solubility

If a fixed amount of water is shaken with, say, common salt (preferably in a state of fine division) at a fixed temperature, say 288 K, it will be found that the first small quantities of common salt added will disappear into the water, leaving a colourless mixture. This mixture is called a **solution** (more strictly an **aqueous solution**) of common salt; it consists of two parts: (a) the liquid which is said to **dissolve** the common salt and is called the **solvent**, (b) the solid, which undergoes **dissolution**, and is called the **solute**.

If more and more common salt is added (with no temperature change), it will finally be found that some of it remains undissolved and settles to the bottom of the container. The solution then contains as much common salt in solution as it can for that quantity of water and at that temperature; it is said to be a *saturated solution*, and can be defined as follows:

A **saturated solution** is one which, at the temperature concerned, is in equilibrium with undissolved solute.

The solubility of a substance must be defined with reference to a saturated solution of it at a stated temperature in a standard amount of solvent, and the following definition has been adopted:

The **solubility** of a substance in water at a given temperature is the mass of the substance in grams which will saturate 100 grams of water at that temperature.

Supersaturation

If a boiling-tube is filled with water to a depth of about 2 cm, then filled up with sodium thiosulphate crystals and heated, the sodium thiosulphate will dissolve to give a colourless solution. If the liquid is stirred (to make it uniform) and then cooled under the tap *while being held quite still*, it will be found that no crystals separate. If then the liquid is *seeded* (or inoculated) with a very small sodium thiosulphate crystal, crystals will at once begin to separate, starting from the seeding crystal as centre and growing steadily downwards into the solution. An almost solid mass of crystals will be formed (and the temperature will rise considerably as they separate, see page 191. After the crystals have separated, the solution left must still be saturated because it is in equilibrium with crystals of the solute. Before the crystals separated it must have been more-than-saturated or *supersaturated*; this may be defined as follows:

A **supersaturated** solution is one that contains more of the solute in solution than it can hold, at that temperature, in the presence of crystals of the solute.

It is important to notice that supersaturated solutions are unstable. They usually require the following conditions:

1. If aqueous they are given (to any marked extent) by relatively few compounds, e.g. sodium sulphate $Na_2SO_4.10H_2O$, and sodium thiosulphate $Na_2S_2O_3.5H_2O$. Supersaturation is more common with organic solvents.
2. They generally require exclusion of dust particles, which might act as centres of crystallization.
3. They usually require *slow* cooling. Sodium thiosulphate is exceptional in giving supersaturation with rapid cooling.
4. They require avoidance of stirring or shaking.

Determination of the solubility of a solid, e.g. sodium chloride, in water at room temperature

A suitable amount of distilled water (say 150 cm^3) is placed in a beaker and warmed to a temperature appreciably (but not greatly) above room temperature, say to 300 K. Finely divided sodium chloride (pure) is added, with stirring, until a little is left undissolved, i.e. the solution is saturated at about 300 K. The liquid, protected from dust by a cover-glass, is then cooled to room temperature. As it cools, more crystals of common salt will separate. The mixture is allowed to stand, with occasional stirring, till it has had time to reach room temperature throughout and attain equilibrium between solution and crystals. It is then filtered with *completely dry* apparatus. The filtrate should be an exactly saturated solution of common salt at room temperature, which should be noted.

In principle, the easiest analysis of the solution consists in adding some of it to a weighed evaporating dish and reweighing, then evaporating the liquid to expel all the water and weighing the (cooled) dish and solid. The calculation is as follows:

Mass of evaporating dish $= a$ g
Mass of evaporating dish + solution $= b$ g
Mass of evaporating dish + solid $= c$ g
Mass of water $= (b - c)$ g
Mass of solid $= (c - a)$ g

Solubility of common salt in 100 g of water at t K. $= \dfrac{(c - a)}{(b - c)} \times 100$ g

In practice, this method offers many difficulties. Evaporation by steam is essential, as evaporation by burner causes splashing and great inaccuracy. When water has been expelled to the fullest possible extent by steam, the solid will have to be dried further, say at 390 K, but even then it is not easy to be sure of *complete* expulsion of water. This is especially so if the solute forms a hydrate with water. There is also the possibility of hydrolysis of the solute during evaporation; this is very common with chlorides.

In view of these difficulties, a chemical analysis may be preferable. In the case of common salt, the following is suitable. A small beaker (with watch-glass cover) is weighed. About 5 cm^3 of the saturated sodium chloride solution is added and the beaker is weighed again. The liquid is then diluted to, say, 100–120 cm^3 with distilled water, transferred, with the usual analytical precautions and washings, to a 250 cm^3 measuring flask, made up to the mark with distilled water and shaken. It is then titrated, in batches of 25.0 cm^3 by 0.1 M silver nitrate solution, with the usual potassium chromate indicator. The following illustrates the calculation:

Mass of beaker and cover-glass = 75.010 g
Mass of beaker, cover-glass, and solution = 80.363 g
Mass of solution = 5.353 g

Solution diluted to 250.0 cm^3

25.0 cm^3 of this dilute solution required 24.13 cm^3 (average) of 0.1 M $AgNO_3$ solution

$AgNO_3$ $\equiv$ NaCl
10 dm^3 of 0.1 M 58.5 g

$$\text{Mass of sodium chloride} = 58.5 \times \frac{24.13}{10\,000} \times \frac{250.0}{25.0}\ \text{g}$$
$$= 1.411\ \text{g}$$

$$\text{Mass of water} = (5.353 - 1.411)\ \text{g}$$
$$= 3.942\ \text{g}$$

$$\text{Solubility of common salt} = \frac{1.411}{3.942} \times 100\ \text{g}$$

$= 35.8$ g (in 100 g of water at t K.)

It must be admitted that the actual procedure in such cases depends (to a certain degree) on fore-knowledge of the result. For example, the choice of 'about 5 cm^3' of the saturated solution for dilution depends on knowledge that about one-quarter of the mass of the solution is common salt and the density of the solution is about 1.2 g cm^{-3}. This choice then provides about 1.5 g of sodium chloride in 250 cm^3 of diluted solution, which is about right for titration for 0.1 M $AgNO_3$ solution. In the absence of such knowledge, a trial run would probably be needed to acquire it.

Determination of solubilities at temperatures other than room temperature

The principles of the determination are the same as those of the method already given, but the saturated solution must be prepared at the required temperature and is then somewhat more difficult to manipulate and analyse.

A suitable volume of distilled water is heated to the temperature of the determination, placed in a thin-walled glass bottle with the finely divided solid and immersed in a thermostat at the appropriate temperature, t K.

The mixture is stirred, preferably mechanically, with excess of the solid continually present. Enough time must be allowed for saturation to be reached (not less than 2 hours in general). The mixture is then allowed to settle and some of the liquid is extracted by a dry pipette which has been heated to a temperature somewhat above t K and fitted with a filter to prevent the entrance of crystals of solid. The liquid is weighed (as before) in a pre-weighed vessel and analysed as before.

Solubility curves

If determinations of the solubility of a substance are made at suitable temperatures, a graph can be drawn from the results with temperatures along one axis and the solubility values along the other. Except in special cases, such *solubility curves* are continuous. Generally, solubility of solids increases with rise of temperature; in a few cases (e.g. calcium hydroxide), solubility decreases with rise of temperature; occasionally (e.g. calcium sulphate), solubility increases over a certain temperature range and then decreases. This is summarized in Figure 12.11. In some cases the solubility curves are discontinuous. This is usually caused by the fact that the compound in question can exist in two or more states of hydration.

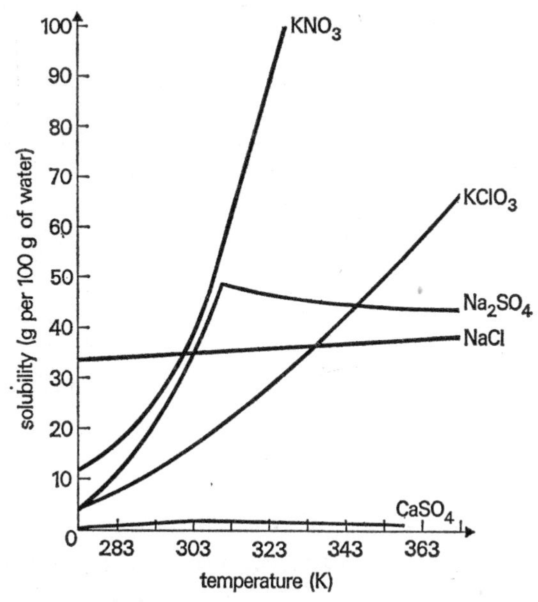

Figure 12.11. Solubility curves

It should be noted that most solids dissolve in water with *absorption* of heat; consequently, if a solution is in equilibrium with its solid solute at a certain temperature and the temperature is then raised, Le Chatelier's Principle (page 204) requires the equilibrium to shift so that the temperature tends to fall again. That is, heat must be absorbed by solution of more solute and the solubility therefore rises.

Solubilities of sparingly soluble compounds

The methods given above are not suitable for determining the solubilities of the so-called 'insoluble' salts, such as barium sulphate or silver chloride. For these, the following methods are available.

(i) *Conductance measurements*

A saturated solution of the compound (say silver chloride) is prepared in very pure water at the temperature in question (usually 298 K). The specific resistance (page 277) of the solution is determined and, from its reciprocal, the specific conductance is available. From this, the specific conductance of water is deducted to obtain the specific conductance of silver chloride. In such very dilute conditions, the silver chloride is taken as fully ionized so that the following relation holds:

$$\begin{array}{l}\text{molar conductance for} \\ \text{AgCl at infinite dilution}\end{array} = \frac{1000 \times \text{specific conductance of AgCl}}{\text{no. of mole of AgCl per dm}^3}$$

The molar conductance is the sum of the ionic mobilities of Ag^+ and Cl^- and is 138 ohm^{-1} cm^{-2}. The specific conductances of saturated silver chloride solution, and water, are 3.41×10^{-6} and 1.60×10^{-6} chm^{-1} cm^{-2} respectively at 298 K. By subtraction, the silver chloride figure is 1.81×10^{-6}. From this,

$$\text{respectively} \quad 138 = \frac{1000 \times (1.81 \times 10^{-6})}{\text{no. of mole of AgCl per dm}^3}$$

$$\text{solubility of AgCl in mol dm}^{-3} = 1000 \times \frac{(1.81 \times 10^{-6})}{138}$$

$$= 1.31 \times 10^{-5} \text{ (298 K)}$$

In grams per dm^3 this is $(1.31 \times 10^{-5}) \times 143.5 = 0.001\ 88$ g dm^{-3}. (AgCl = 143.5.)

(ii) *Colorimetric method*

This can be applied when a metal forms a suitably coloured salt. The case of lead sulphate is typical. A saturated solution of this salt is prepared at the relevant temperature. Hydrogen sulphide is passed through it, giving a brown liquid containing colloidal lead sulphide.

Lead ethanoate (acetate) solution of known concentration is then progressively diluted until, on passage of hydrogen sulphide, it matches the colour from the lead sulphate solution. The concentration of Pb^{2+} in the sulphate solution is then equal to the known concentration in the ethanoate solution. The solubility of the sulphate is then obtained by multiplying by the ratio $PbSO_4$:Pb.

(iii) *Radioactive method*

This method can be applied (say to lead sulphate) in the following way. Lead nitrate is dissolved in water and a known (very small) proportion of the same salt of a radioactive isotope of lead is added. All the lead is then precipitated as sulphate and the precipitate is thoroughly washed. The sulphate is then agitated with water till a saturated solution is obtained at the relevant temperature. A known mass of the filtered solution is evaporated to dryness and the amount of radioactive material is estimated in the residue by a Geiger counter. The proportion of this being already known, the total mass of solid can be calculated and hence its solubility.

Some further features of solubility curves

The behaviour of mixtures of potassium iodide and water

If pure water is cooled at a pressure of 101 300 $N m^{-2}$ (760 mmHg) and super-cooling is avoided, ice will separate as soon as the temperature of the water reaches 273 K. If a very dilute potassium iodide solution is similarly cooled, the freezing point will be a little depressed and ice will separate at a temperature a little below 273 K. If the concentration of potassium iodide is gradually increased, the freezing-point depression is accentuated and ice separates at progressively lower temperatures. This behaviour is expressed in the curve AB of Figure 12.12.

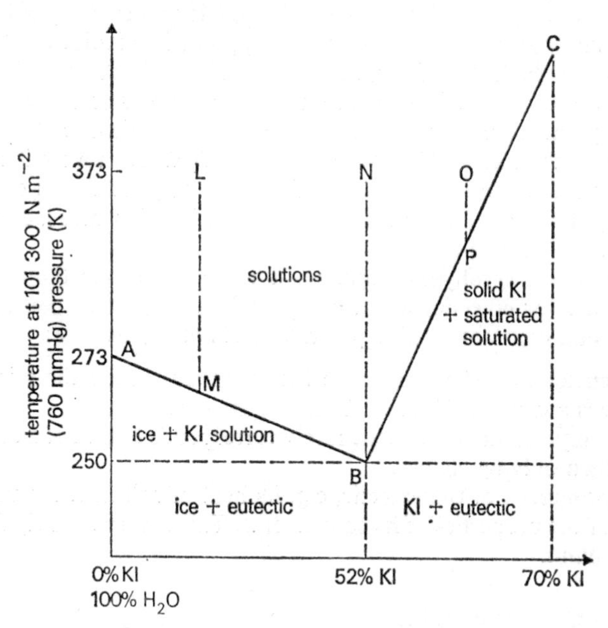

Figure 12.12. ***Cryohydrate of water-KI***

If a hot, concentrated solution of potassium iodide is cooled, it will finally reach a point at which crystallization of potassium iodide will begin. This point of crystallization will occur at lower temperatures for more dilute solutions and at higher temperatures for more concentrated solutions. The expression of this in graphic form is the curve BC, which is, in fact, the solubility curve of potassium iodide in water.

The situation can be illustrated by considering what happens when particular mixtures of water and potassium iodide are cooled.

If a solution of composition L (about 12 per cent KI) is cooled, ice begins to separate at the point M (about 268 K). As cooling is continued, more ice separates, and the composition of the solution follows the curve MB until the temperature falls to 250 K. By this time the solution has the composition B. It then remains constant in temperature while a mixture of composition B separates as solid crystals.

If a solution of composition N (or B), i.e. 52 per cent KI, is cooled, no separation of solid occurs at all until the temperature reaches 250 K. Then the temperature remains constant while a mixture crystallizes out which has the same composition as that of the solution, i.e. 52 per cent KI and 48 per cent H_2O. This mixture is called the *eutectic* between potassium iodide and water. If, as in this case, one of the constituents of the eutectic is water, the eutectic may also be called a *cryohydrate*. The eutectic is clearly that mixture of its constituents which has the lowest freezing point.

If a solution of composition O is cooled, potassium iodide crystals begin to separate at P. As they form, the composition of the solution follows the curve PB; then, at 250 K, the eutectic again separates as before.

At all points *above* the curves AB and BC, only *solutions* can exist. For this reason, the line ABC may be called the *liquidus*. In a similar way, at all points *below* the horizontal line of 250 K, only *solids* can exist. Consequently, this line may be called the *solidus*. In other areas, equilibrium mixtures can exist as marked.

As the name cryohydrate implies, it was at one time thought that a cryohydrate (or eutectic) was a compound because of its constant composition and constant melting point. This is not so, because:

1. The eutectic can be seen to be heterogeneous under the microscope, whereas a compound would be homogeneous.
2. The composition of a eutectic rarely corresponds to that of a compound and then only by chance.
3. The properties of the eutectic, e.g. its heat of solution, are simply the sum of the properties of its constituents. This is very unlikely for a true compound.

Other simple cases of eutectic formation

The typical feature of the water–KI system treated above is that the constituents form no compound. Many other similar cases occur, especially between metals. The case of zinc–cadmium is well known and is described in its main features by Figure 12.13.

If a mixture of composition A is cooled, it will remain liquid to the point B. Then solid zinc will begin to separate. The liquid will follow the curve BC in composition and will finally reach the composition of the eutectic, which will separate as solid at 543 K.

A liquid of the eutectic composition C will remain entirely liquid to 543 K if cooled, and will then deposit the solid eutectic mixture at that temperature.

A mixture of composition E will remain liquid when cooled until point D is reached. Then cadmium will separate as solid and the composition of the liquid will follow the curve DC. Then, at 543 K solid eutectic will separate.

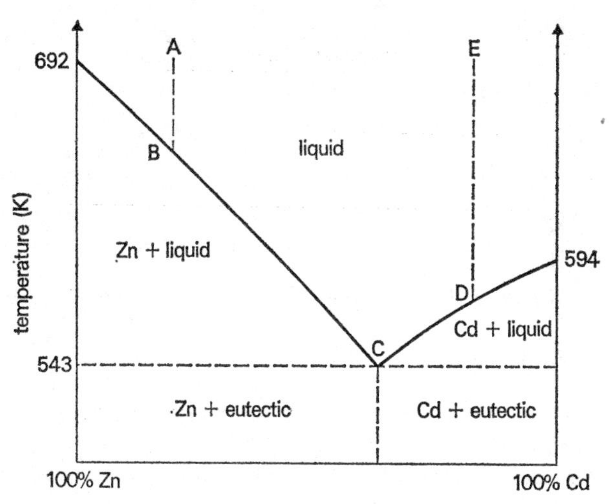

Figure 12.13. Eutectic of Zn-Cd

Compound formation

The formation of a compound produces a maximum in the freezing-point curve as in Figure 12.14. The freezing point and composition of the compound of X and Y (of the type X_mY_n) are represented at the point F. The addition of X *or* Y to this compound depresses the melting point of the compound, since both X and Y act as impurities with respect to the compound; consequently the freezing point is lowered on each side of F, producing the typical 'hump' effect (DFH) associated with compound formation.

Two eutectics occur in this case, of compositions represented by E_1 and E_2. E_1 is a eutectic between X and the compound; E_2 is a eutectic between the compound and Y. There is no eutectic between the materials X and Y. The following effects accompany the cooling of the various mixtures shown:

A. At temperature B, solid X begins to separate. The composition of liquid follows that of the curve B to E_1. When the point E_1 is reached, eutectic E_1 begins to separate and continues (at constant temperature) till all is solid.

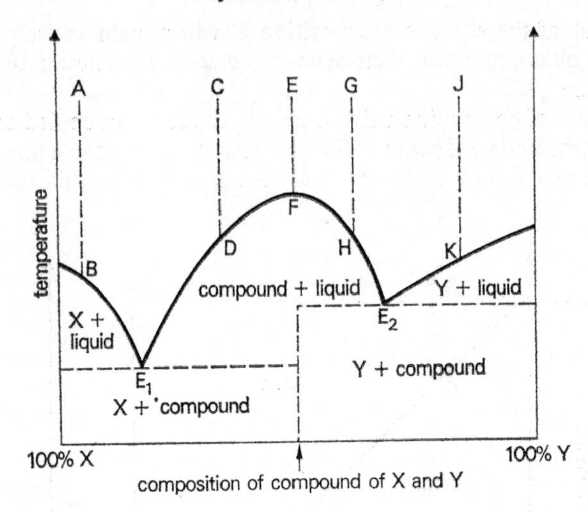

Figure 12.14. Eutectics with formation of compound

C. At temperature D, the compound begins to separate out solid. As it does so, the composition of the liquid follows the curve D to E_1. When the point E_1 is reached, eutectic E_1 begins to separate out solid and so on as in A.

E. This is the composition of the compound of X and Y. At F this compound begins to separate as pure crystals and continues to do so (at constant temperature) till all is solid.

G. At temperature H, the compound begins to separate as solid. The liquid follows the compositions represented by the curve H to E_2; when E_2 is reached, the eutectic E_2 begins to separate as solid and does so (at constant temperature) till all is solid.

J. At temperature K, solid Y begins to separate. As it does so, the liquid follows the compositions represented by the curve K to E_2. When the conditions of E_2 are reached, this eutectic begins to separate as solid and so on as in G.

Mixtures of the eutectic compositions E_1 and E_2 behave in practice like the compound, except that the appropriate eutectic separates out and the temperatures of solidification are lower.

The curve B E_1D F H E_2K is the liquidus above which (in the figure) only liquid exists; the horizontal dotted lines through E_1 and E_2 represent a solidus below which only solid exists. The other areas of the figure represent the conditions marked in them.

The metals magnesium and tin behave in accordance with the situation represented in Figure 12.14. The compound is Mg_2Sn (Mg, 28.8 per cent).

Much more complex cases of compound formation occur, with hydrates, in the solubility curves of compounds, usually salts. The case of iron(III) chloride is given in illustration in Figure 12.15. The curve shows iron(III) chloride forming four hydrates, $Fe_2Cl_6.xH_2O$, where x has the numerical values shown at the maxima of the curve.

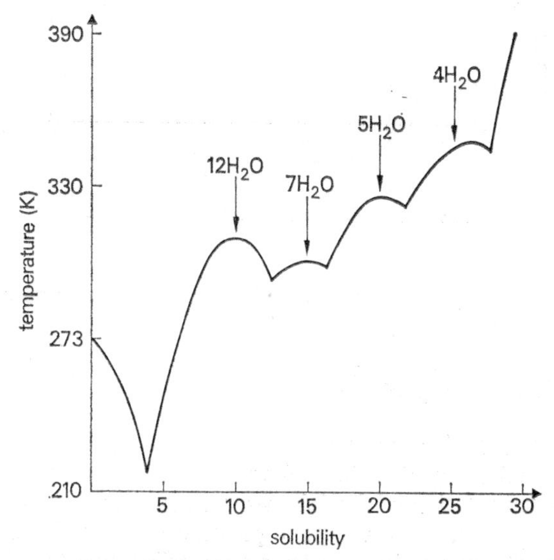

Figure 12.15. Hydrates of iron(III) chloride

3. *Vapour pressure of salt hydrates*

The behaviour of the three hydrates of copper(II) sulphate (which contain 5, 3, and 1 mole of water combined with one mole of copper sulphate) is typical of that of salt hydrates generally.

If the pentahydrate $CuSO_4.5H_2O$ is placed in an otherwise evacuated space at 323 K, it is found that it dehydrates gradually to the stage of $CuSO_4.3H_2O$. However, the vapour pressure remains steady at 6266 N m^{-2}(47 mmHg) so long as any of the pentahydrate remains.

As soon as all is converted to the trihydrate, a rapid fall of vapour pressure occurs to 4000 N m^{-2} (30 mmHg). It then remains steady at this value as dehydration to $CuSO_4.H_2O$ occurs, and there is no change of vapour pressure so long as any $CuSO_4.3H_2O$ is left.

As soon as $CuSO_4.H_2O$ is the *only* constituent of the solid, another rapid fall to 600 N m^{-2} (4.5 mmHg) occurs in the vapour pressure and this value is maintained so long as any monohydrate is present. These changes are summarized in Figure 12.16.

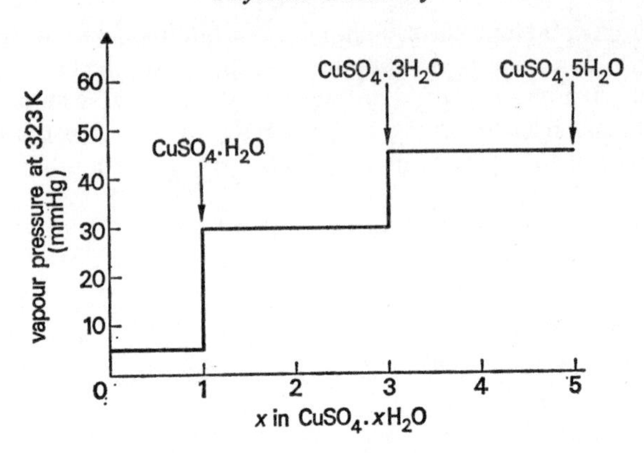

Figure 12.16. Hydrates of copper(II) sulphate

The phenomena of deliquescence and efflorescence are closely connected with vapour pressure of hydrates.

Deliquescence is the phenomenon by which a solid absorbs water-vapour from the air and passes into solution.

The explanation of this phenomenon is as follows. It is characteristic of a deliquescent solid that its saturated solution has a *low* vapour pressure, e.g. a saturated solution of the deliquescent compound, calcium chloride, at room temperature has a vapour pressure of only about 999 N m^{-2} (7.5 mmHg). Consequently, when an aqueous film from the air condenses on the solid salt, the vapour pressure of the solution formed is so low that the vapour pressure of water in the surrounding air is usually greater. Consequently more water condenses into the solution until finally the solid is entirely dissolved. Still more water may then be absorbed till the vapour pressure of the diluted solution is equal to that of water in the air. It will be obvious that if the aqueous vapour pressure in the air falls below 999 N m^{-2}, calcium chloride will not deliquesce. The vapour pressures of saturated solutions of most salts are greater than the usual aqueous vapour pressure in the atmosphere, which is usually about 1600 N m^{-2} (12 mmHg). Consequently, most salts are non-deliquescent.

Efflorescence is the loss of water of crystallization from a salt hydrate on exposure of it to air at room temperature.

The loss may be only partial, e.g. sodium carbonate decahydrate effloresces according to the equation:

$$Na_2CO_3.10H_2O(s) \rightarrow Na_2CO_3.H_2O(s) + 9H_2O(g)$$

The loss of the water is usually accompanied by a collapse of the crystal lattice, so that the crystals of the hydrate fall to powder.

Efflorescence occurs when the dissociation vapour pressure of the hydrate is greater than the partial pressure of aqueous vapour in the atmosphere. For example, the vapour pressure of sodium carbonate decahydrate is about 3200 N m^{-2} (24 mmHg) at room temperature. This exceeds the usual 1600 N m^{-2} pressure of aqueous vapour in the atmosphere, so that washing soda will usually effloresce. Sodium and iron(II) sulphates, $Na_2SO_4.10H_2O$ and $FeSO_4.7H_2O$, are also usually efflorescent. It is obvious that in very humid surroundings efflorescence may be suppressed because aqueous vapour pressure in the atmosphere balances that of the hydrate.

4. *Solubility of gases in liquids*

As far as is known at present, no gas has a high solubility in water unless it reacts chemically with the water. Gases such as nitrogen, hydrogen, and oxygen are typical of those which dissolve in water without chemical action; their solubilities are in the region of 2–4 volumes of gas in 100 volumes of water at room temperature. Gases of much higher solubility, of which chlorine, carbon dioxide, hydrogen chloride, and ammonia are representative, have solubilities in the region of 100–80 000 volumes in 100 volumes of water at room temperature. They all react with water, e.g.

$$CO_2(g) + H_2O(l) \rightleftharpoons H_2CO_3(aq) \rightleftharpoons 2H^+(aq) + CO_3^{2-}(aq)$$
$$Cl_2(g) + H_2O(l) \rightleftharpoons HCl(aq) + HClO(aq)$$

Determination of the solubility of a sparingly soluble gas

This is usually done by means of an absorption pipette (Figure 12.17). The experiment will be described for oxygen in water.

The gas pipette, P, has a three-way tap at E and an ordinary tap at R. Its capacity (v_1 cm^3) must be known. By suitably opening taps C and E to a source of oxygen connected at C, air is displaced from the copper (or lead) tube D. E and C are then closed. By raising A, B can be filled with mercury (air being expelled at C) and then, by connecting the oxygen supply at C, oxygen can be passed into B as A is lowered. The volume of oxygen is read off, a_1 cm^3, with levels of mercury equal in A and B. The pipette, P, full of water, is then connected; A is raised and C, E and R opened so that water runs out of P at R into a measuring cylinder, where the volume of water is measured, v_2 cm^3. The volume of water left in P is then $(v_1 - v_2)$ cm^3.

P is placed in a thermostat at the temperature required and is shaken repeatedly. As the gas dissolves, the tube A is adjusted to keep the levels equal in A and B (i.e. to maintain atmospheric pressure). When the level in B is constant, it is read off, a_2 cm^3.

The initial volume of oxygen (neglecting the volume of it in tube D, which remains constant) is a_1 cm^3. The final volume of oxygen is

$(a_2 + v_2)$ cm^3. That is, at the given temperature and pressure, $(a_1 - a_2 - v_2)$ cm^3 of oxygen are absorbed by $(v_1 - v_2)$ cm^3 of water.

The water used in the experiment must be air-free. Tube D is made of metal not rubber, because rubber is appreciably permeable to oxygen and is liable to distortion and change of capacity when shaken.

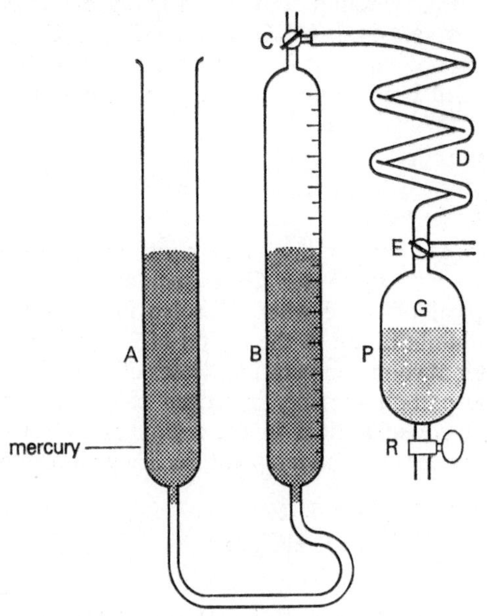

Figure 12.17. Solubility of a gas

Definition of solubility of a gas in a liquid

The solubility of a gas in a liquid is best expressed by means of an absorption coefficient; this can be defined as follows:

> The **absorption coefficient** of a gas, A, in a liquid, B, is the number of cm^3 of A, expressed at s.t.p., which saturate 1cm^3 of the liquid, B, at the temperature stated and at a pressure of one atmosphere.

Some absorption coefficients are given below for the solvent water:

Gas	*Absorption coefficient*	
	At 273 K	At 293 K
Hydrogen chloride	506	442
Carbon dioxide	1.71	0.88
Oxygen	0.049	0.028
Nitrogen	0.024	0.015

It will be noticed that all these gases become *less* soluble with rise of temperature. This is opposite to the behaviour of most solids. The dissolution of a gas in water is generally an exothermic process; consequently, if the temperature of a system of gas and liquid in equilibrium is *raised*, the equilibrium must shift so as to tend to *lower* it again (by Le Chatelier's Principle, page 204); that is, heat must be *absorbed* by expulsion of the gas from solution.

Henry's Law

Subject to certain qualifications mentioned below, Henry's Law states the effect of change of pressure on the solubility of a gas. The law takes the form:

> The mass of a given gas which saturates a constant volume of a given solvent is directly proportional to the pressure of the gas at equilibrium, provided that the temperature is constant and that the gas does not react chemically with the solvent.

Notice that Henry's Law does *not* apply to the aqueous behaviour of very soluble gases, which invariably react chemically with the water in which they 'dissolve'. Slight inaccuracies also arise in Henry's Law because gases do not obey Boyle's Law exactly.

If gaseous volume is kept constant at constant temperature, the mass of gas contained in that volume is directly proportional to the pressure. Consequently Henry's Law can also be expressed by the statement that, at constant temperature and provided that the gas does not react chemically with the solvent, the volume of a given gas which saturates a given volume of solvent is constant and independent of pressure.

The following figures illustrate the extent to which Henry's Law is obeyed by oxygen in water at 298 K. For perfect agreement with the law, the final column should show *constant figures*:

A Pressure ($N\ m^{-2}$)	B *Mass of oxygen* (g) *to saturate* 1 dm^3 *of water*	$\frac{B}{A} \times 10^5$
101 300	0.0408	5.37
81 330	0.0325	5.33
55 200	0.0220	5.31
40 000	0.0160	5.33
23 330	0.0095	5.43

Solution of gases from mixtures

The volume of a gas dissolved by water from a mixture is directly proportional to two factors and, therefore, to their product. These factors are:

1. The *absorption coefficient* of the gas.
2. The *partial pressure* of the gas in the mixture; this can be defined as follows:

> The **partial pressure** of a gas is that pressure which the gas would exert if it alone occupied the whole volume actually occupied by the mixture, temperature remaining constant.

From this it follows that the ratio of partial pressures in a mixture of gases is the same as the ratio of volumes.

The operation of these factors can be illustrated by their application to solution of air in water. Carbon dioxide is omitted because Henry's Law does not apply well to this gas.

	N_2	O_2	Ar
Volume composition of air per cent	78	21	1
Absorption coefficient of gas at 273 K.	0.024	0.049	0.053
Vols. dissolved in water are proportional to	$78 \times 0.024 = 1.87$	$21 \times 0.049 = 1.03$	$1 \times 0.053 = 0.053$
Total vol. of gas dissolved is proportional to	1.87	+ 1.03	+ 0.053
		= 2.95 (approx.)	

Consequently, the volume composition of air dissolved into water, or boiled out from it, is:

N_2	O_2	Ar
$\frac{1.87}{2.95} \times 100$ per cent	$\frac{1.03}{2.95} \times 100$ per cent	$\frac{0.053}{2.95} \times 100$ per cent
= 63.4 per cent	= 34.9 per cent	= 1.80 per cent

Determination of the solubility of a very soluble gas

The method of the gas-pipette is not suitable for this determination. A pyknometer is used (Figure 12.18). The pyknometer is weighed. A suitable volume of water is then introduced into it. With the pyknometer placed in a constant-temperature bath, the dry gas is passed until the solution is

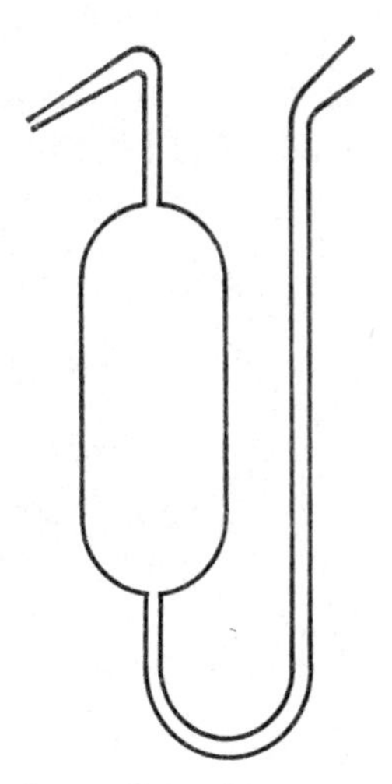

***Figure 12.18* Pyknometer**

saturated. The pyknometer is then sealed off by heating and drawing off its ends in a blowpipe flame. After cooling, the whole (including any parts drawn off during sealing) is weighed. This gives, by subtraction, the mass of solution. The solution is then analysed. It will generally be either alkaline or acidic, most of the very soluble gases being one or the other. If the liquid is acidic, the container is broken under excess of standard alkali and the excess is titrated. For alkaline liquids, excess standard acid is the absorbent.

5. *Mixtures of liquids*

When considering a mixture of two liquids (two components) in equilibrium with the mixture of the two vapours (two phases), there are three variables to consider: temperature, pressure, and the composition of either the liquid mixture or the vapour mixture.

If two completely miscible liquids behave *ideally*, the total vapour pressure of the mixture (at a given temperature) is equal to the sum of the partial pressures of the two constituents. This is because both constituents obey Raoult's Law of Vapour Pressure. They do not enter into combination with each other or cause dissociation or polymerization. This case is shown in Figure 12.19. The vapour pressure T, of the mixture is the sum of the vapour pressures PX and PY of the two constituents, and correspondingly for each point on AB. This ideal position is rarely, if ever, achieved in practice, but many cases occur in which it is approximately reached, so that the vapour pressures of mixtures of two such liquid components always lie between the vapour pressures of the two constituents (at the relevant temperature). No maximum or minimum of vapour pressure is shown. Correspondingly, the boiling-point curve at constant pressure shows no maximum or minimum (Figure 12.20).

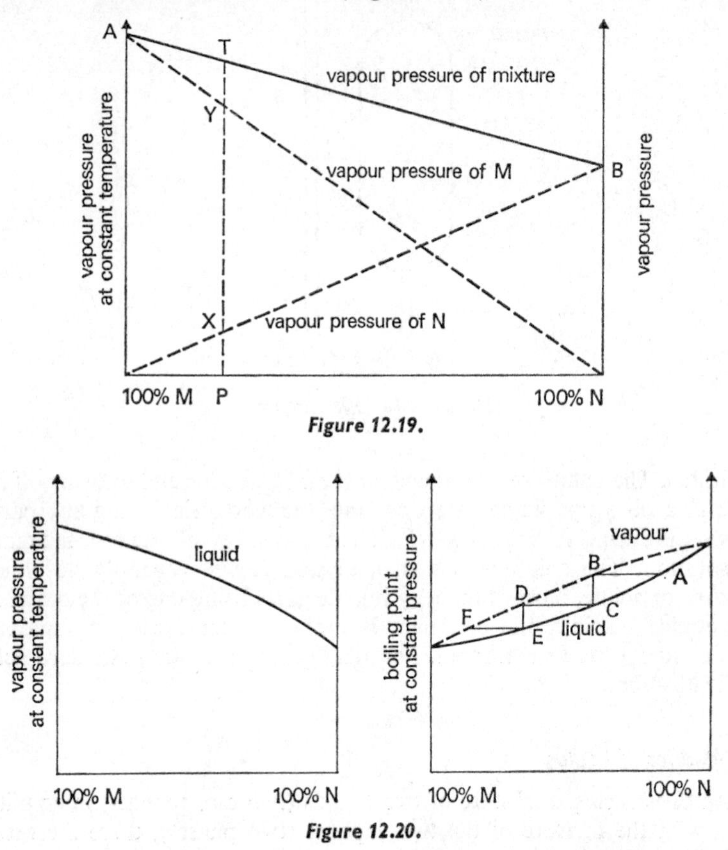

Figure 12.19.

Figure 12.20.

A lower vapour pressure than expected (negative deviation from Raoult's Law) is due to bond formation between the components of the mixture (see page 98). A higher vapour pressure than expected (positive deviation from Raoult's Law) indicates that bonds which existed between the molecules of one component have been broken when the components were mixed.

A mixture of this sort can be completely separated into its constituents by distillation. The dotted line in the boiling-point diagram represents the composition of vapour in equilibrium with the liquid mixtures. In Figure 12.20, liquid A is in equilibrium with vapour B, which is richer in the more volatile constituent, M, than A is. If mixture A is distilled, vapour B can be condensed to liquid C, which yields vapour D on distillation. This condenses to liquid E, giving vapour F and so on. It is evident that a succession of ideal distillations of this kind will finally yield pure M as vapour, which can be condensed to liquid. Ideal conditions will leave pure N in the distillation vessel and separation is theoretically complete.

By the use of a *fractionating column*, a fair approximation to the above ideal state of distillation can be reached. A typical fractionating column is shown in Figure 12.21 and consists of a central rod with discs at short

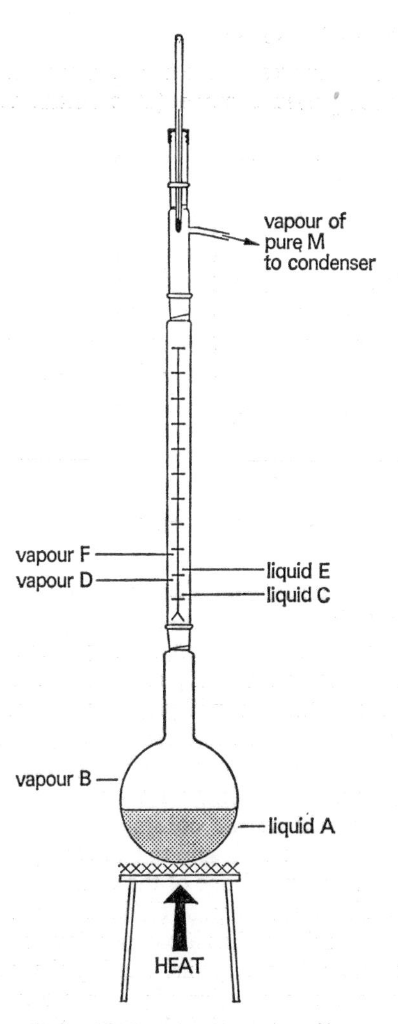

Figure 12.21. Fractionating column

intervals, the discs being almost as wide as the tube in which they are placed. Ideally, the ascending vapour should come into equilibrium with a liquid layer condensed on each disc. If so, the position corresponds to that of Figure 12.20 in the manner shown. The thermometer will register the boiling point of M until all of M has been delivered; then vapour of N will distil and the thermometer will register the boiling point of N.

Examples of mixtures which can be separated into their constituents in the above way are methanol–water; benzene–toluene; benzene–hexane.

Mixtures of minimum boiling point

If the vapour-pressure curves of the two constituents are both concave upwards, the vapour-pressure curve (at constant temperature) of the

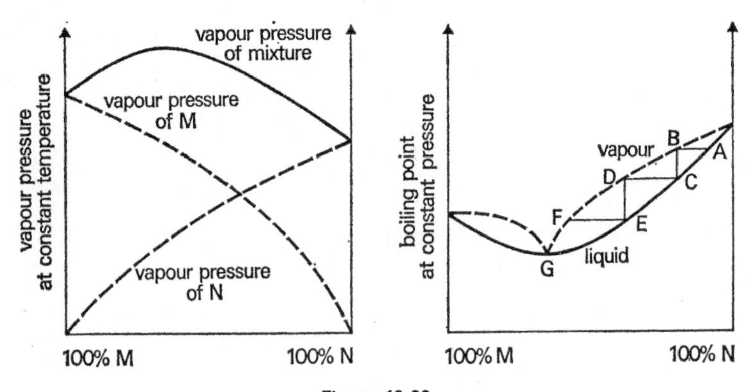

Figure 12.22.

mixture may show a maximum. Correspondingly (Figure 12.22), the boiling-point curve (at constant pressure) will show a minimum. If liquid A is distilled with a fractionating column, it will yield vapour, B, condensing to liquid, C, vaporizing to vapour, D, condensing to liquid, E, and so on. It is evident that the distillation will yield the mixture, G, which can be condensed, i.e. the mixture of minimum boiling point. Thus, complete separation of the constituents is not possible by distillation.

The mixture, G, is in equilibrium with vapour of its own composition and will distil unchanged at the constant, minimum boiling point at constant pressure. This mixture is said to be *azeotropic*. When the removal of such a mixture has exhausted one of the constituents, the other (excess) constituent will be left pure and can be distilled. A corresponding situation is found if the original mixture has a composition on the other side of G. Again, the azeotropic mixture distils, leaving the other constituent as residue; notice, again, that separation of pure samples of both constituents (from a given mixture) is not possible by distillation.

The most important case of this kind is that of ethanol-water. The azeotropic mixture (boiling point 351.1 K at 101 300 N m^{-2} (760 mmHg)) has 95.6 per cent of the alcohol and this is the limit of purification of dilute alcohol by distillation. Modern versions of the Coffey still can yield alcohol of this composition in a continuous process starting from dilute solutions of alcohol (about 8 per cent) produced by fermentation of starch.

Mixtures of maximum boiling point

If the vapour-pressure curves of the two constituents of a liquid mixture are both convex upwards, the vapour-pressure curve may show a *minimum*

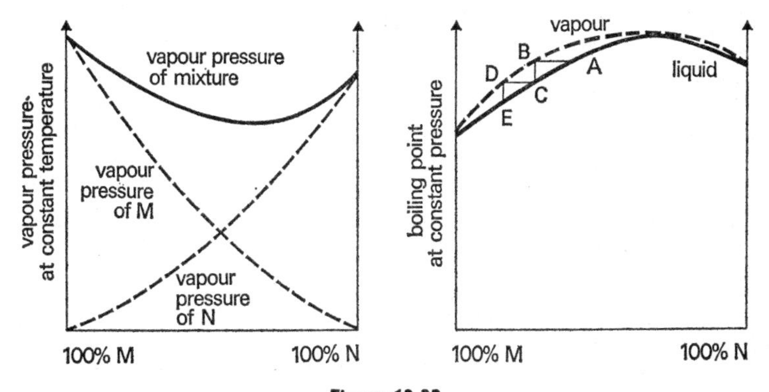

Figure 12.23.

(at constant temperature). Correspondingly (Figure 12.23), the boiling-point curve, at constant pressure, may show a *maximum.* In this case, a liquid with the composition of the azeotropic mixture distils unchanged at the maximum boiling point, being in equilibrium with vapour of its own composition. A mixture such as A distils vapour B, condensing to liquid C, vaporizing to D, and so on, delivering finally a liquid which tends to be pure M, or, at any rate, much richer in M than A was. The residual liquid becomes more concentrated in N until it reaches the composition of the azeotropic mixture, and then distils unchanged at the maximum boiling point. Similar changes occur on the other side of the azeotropic mixture, tending to deliver pure N as distillate and leave the azeotropic mixture as residue. Here again, complete separation of the constituents from a given mixture is impossible and any mixture, if distilled long enough, leaves the mixture of maximum boiling point.

All the mineral acids show behaviour of this kind with water. For hydrochloric acid, sulphuric acid, and nitric acid, the data are:

	Maximum boiling-point mixture	*Boiling point at* 101 300 N m^{-2} (760 mmHg)
Hydrochloric acid	20.2% Hcl	383 K
Sulphuric acid	98.7% H_2SO_4	611 K
Nitric acid	68.0% HNO_3	394 K

6. *Distillation of non-miscible liquids*

If two liquids are immiscible, their behaviour when distilled is quite simple. They operate quite independently and there are three phases present, and each liquid is in equilibrium with its own vapour. Each contributes its own separate vapour pressure at a particular temperature and the total vapour pressure of the mixture is the sum of the two contributions. The mixture boils when the total vapour pressure reaches the prevailing atmospheric pressure. The most important application of this situation is found in *steam distillation*, which is employed mainly in the purification of organic compounds.

The mixture for steam distillation is heated and steam is passed in from a can in which water is boiled. The issuing vapour, which (ideally) contains only water and the organic compound, is condensed by passage through a Liebig's condenser (see text-books of organic chemistry for details). The composition of the distilled mixture is given by the following calculation. Consider the case of a compound, X, insoluble in water and forming an aqueous mixture which boils at 370 K at 101 000 N m^{-2} (757 mmHg) pressure, the vapour pressure of water at 370 K being 96 000 N m^{-2} (720 mmHg). Let X have a relative molecular mass of 125. The vapour pressure of X at the boiling point is (101 000 − 96 000) N m^{-2} or 5000 N m^{-2}. From this, in the vapour which distils,

$$\frac{\text{vol. of X}}{\text{vol. of steam}} = \frac{\text{vapour pressure of X}}{\text{vapour pressure of steam}} = \frac{5000}{96\,000}$$

$$\frac{\text{mass of X in distillate}}{\text{mass of water in distillate}} = \frac{5000 \times (\text{mol. mass of X})}{96\,000 \times (\text{mol. mass of steam})} = \frac{5000 \times 125}{96\,000 \times 18} = \frac{1}{2.76}$$

That is, the percentage by mass of X in the distillate is

$$\frac{1}{3.76} \times 100 = 26.6 \text{ per cent}$$

The remainder is water.

As can be seen from the above calculation, to achieve a reasonable proportion of the compound in the final distillate, the compound should have the following characteristics:

1. It should have a considerable vapour pressure at temperatures near the usual boiling point of water.
2. It should have a fairly high relative molecular mass (usually about 90–140).

Steam distillation can be particularly useful when a compound is liable to decompose near its normal boiling point, as the process of steam distillation, in a similar manner to distillation under reduced pressure, results in the liquid boiling at a temperature below its normal boiling point.

Three-component Systems

The Partition Law (or Distribution Law)

The Partition Law is concerned with the distribution, between two solvents, of a material which is soluble in both of them. It can be stated as follows:

If a solute X distributes itself between two immiscible solvents A and B at constant temperature and X is in the same molecular condition in both solvents,

$$\frac{\text{concentration of X in A}}{\text{concentration of X in B}} = \text{a constant}$$

This constant is called the **partition coefficient** of X with respect to A and B.

Experimental illustration of the Partition Law

Equal volumes of water and tetrachloromethane are shaken in a separating funnel with a known mass of iodine. If the data are required not at room temperature but at (say) 298 K the funnel is placed in a thermostat at this temperature and left there for a considerable time, during which it is shaken vigorously at frequent intervals. This should ensure an equilibrium distribution of the iodine between the two solvents.

After time for settling, the lower (CCl_4) layer is run off and the aqueous layer (or a known fraction of it) is then titrated for iodine content by standard thiosulphate solution, say 0.01 M. The mass of iodine in the aqueous layer is given by the relation.

$$\begin{array}{ccc} I_2 & \equiv & 2Na_2S_2O_3 \\ 2 \times 127\ \text{g} & & 200\ \text{dm}^3\ 0.01\ \text{M soln.} \end{array}$$

and, from this and a knowledge of the volume of water used, the concentration of iodine in the aqueous layer is known. The total mass of iodine less the iodine in the aqueous layer gives the mass of iodine in the tetrachloromethane layer, and the concentration of iodine in this layer can be calculated.

The experiment can then be repeated with varying masses of iodine and varying proportions of water to tetrachloromethane. In all cases, the ratio of the iodine concentrations in tetrachloromethane and water should be (approximately) the same.

For example, suppose 1.00 g of iodine was shaken with 50 cm^3 of each solvent and 25.0 cm^3 of the aqueous layer required 4.50 cm^3 of 0.01 M sodium thiosulphate

$$\begin{array}{ccc} I_2 & \equiv & 2Na_2S_2O_3 \\ 2 \times 127\ \text{g} & & 200\ \text{dm}^3\ 0.01\ \text{M} \end{array}$$

$$\text{Mass of iodine in the total aqueous layer} = (2 \times 127) \times \frac{9.00}{200\,000}\ \text{g}$$

$$= 0.0114\ \text{g}$$

$$\text{Mass of iodine in tetrachlormethane} = (1.00 - 0.0114)\ \text{g}$$

$$= 0.9886\ \text{g}$$

$$\therefore \text{Partition coefficient} = \frac{0.9886/50}{0.0114/50} = \frac{86.7}{1}$$

Typical results obtained by varying the mass of iodine and the volume of solvents are: 87.9, 88.3, 87.5.

Ethereal extraction of organic compounds

The extraction (usually of organic compounds) from aqueous mixtures by ether is probably the most important application of the Partition Law. Ether is not, in fact, completely insoluble in water (which dissolves about 7 per cent of its mass of ether at room temperature), but the law can be applied approximately to cases of ethereal extraction, in calculations of which the following is typical:

Example A solution of 6 g of a substance X in 50 cm^3 of aqueous solution is in equilibrium, at room temperature, with an ethereal solution of X containing 108 g of X in 100 cm^3. Calculate what mass of X could be extracted by shaking 100 cm^3 of an aqueous liquid containing 10 g of X with (a) 100 cm^3 of ether, (b) 50 cm^3 of ether twice, at room temperature.

From the first data,

$$\text{partition coefficient of X between ether and water} = \frac{108/100}{6/50}$$

$$= \frac{108}{100} \times \frac{50}{6} = \frac{9}{1}$$

In extraction (a), let x g of X pass into ether. Then

$$\frac{x/100}{(10 - x)/100} = \frac{9}{1}$$

$$\frac{9(10 - x)}{100} = \frac{x}{100}$$

$$10x = 90$$

$$x = 9$$

i.e. 9 g of X pass into ether.

In extraction (b), for the first extraction, let a g of X pass into ether. Then

$$\frac{a/50}{(10 - a)/100} = \frac{9}{1}$$

$$\frac{9(10 - a)}{100} = \frac{a}{50}$$

$$11a = 90$$

$$a = \frac{90}{11}$$

$$= 8.2 \text{ g (approx.)}$$

For the second extraction, 1.8 g of X remain in the water. Let b g of X pass into ether. Then

$$\frac{b/50}{(1.8 - b)/100} = \frac{9}{1}$$

$$\frac{9(1.8 - b)}{100} = \frac{b}{50}$$

$$11b = 16.2$$

$$b = 1.5 \text{ g (approx-)}$$

The total extraction of X into ether is $(8.2 + 1.5)$ g $= 9.7$ g

It will be seen that the extraction is more efficient (9.7 g) if the same ether is used in two half-batches instead of all at once (9.0 g).

The above treatment of the Partition Law has assumed throughout that there is no difference between the molecular conditions of the solute in the two liquid layers. If such a difference arises, the situation becomes more complex. Two of the simpler possibilities are mentioned below.

1. *Solute ionized slightly in water and not at all in the other liquid*

In this case, the Partition Law equilibrium is maintained between the unionized molecules in both solvents. Since there are no ions in the non-aqueous layer, the question of ionic equilibrium cannot arise as between the two layers, but an ionic equilibrium is maintained, according to the Dilution Law, in the aqueous layer.

If C_1 is the total concentration of weak electrolyte, X, in the aqueous layer and the degree of ionization is α, the concentration of unionized molecules in the aqueous layer is $C_1(1 - \alpha)$. By the Partition Law,

$$\frac{C_2}{C_1(1 - \alpha)} = K \text{ (at constant temperature)}$$

where C_2 is the total concentration of X in the non-aqueous layer.

A case of this sort arises in the distribution of a weak acid (e.g. succinic acid or oxalic acid) between ether and water.

2. *Solute polymerized in the non-aqueous layer and in unimolecular condition in the aqueous layer*

Suppose a solute X is in the unimolecular condition in water and polymerized (entirely) as molecules X_n in the non-aqueous solvent. The results of experimental investigation show that in this case the Partition Law takes the form

$$\frac{\sqrt[n]{(\text{Conc. of X in non-aqueous layer})}}{\text{Conc. of X in water}} = \text{a constant (at constant temperature)}$$

The best-known cases of this are the equilibria between dimerized ethanoic acid, $(CH_3CO_2H)_2$, or benzoic acid (benzenecarboxylic acid), $(C_6H_5\,CO_2H)_2$, in benzene and the unimolecular forms of these acids in water. In these cases, $n = 2$. Figures for benzoic acid illustrate the position;

Concentration in water, C_1	Concentration in benzene, C_2	$\frac{\sqrt{C_2}}{C_1}$
0.0145	0.252	$\frac{\sqrt{0.252}}{0.0145} = 34.6$
0.0188	0.375	$\frac{\sqrt{0.375}}{0.0188} = 32.6$
0.0213	0.509	$\frac{\sqrt{0.509}}{0.0213} = 33.5$
0.0275	0.817	$\frac{\sqrt{0.817}}{0.0275} = 32.9$

The figures in the final column are approximately constant. Ionization of the acid in water is neglected.

3. *Solute in the form of a complex ion in the aqueous layer and in its norma state in the non-aqueous layer*

Partition may be used to determine the number of ammonia ligands co-ordinating with a copper(II) ion in aqueous solution. If the aqueous solution of the complex ion and excess ammonia is shaken with trichloromethane, equilibrium will be established and the total ammonia in each layer may be determined by titration with acid. The ammonia in the aqueous layer is made up of complex ammonia and free ammonia. If the partition coefficient of ammonia between the two solvents is known, the results of the analysis of the trichloromethane layer and the coefficient value may be used to calculate the concentration of free ammonia in the aqueous layer. Then by subtraction from the total ammonia in the aqueous layer it is possible to determine the complexed ammonia and hence the formula of the complex ion.

13. Energetics

All chemical reactions are accompanied by energy changes which are usually observed as heat changes. The chemicals involved (known as the system) may lose energy to the surroundings, such a reaction being known as an exothermic reaction. When the system absorbs energy from the surroundings it is known as an endothermic reaction.

The quantity of heat evolved or absorbed during a chemical reaction will depend on the following factors.

1. The amounts of substances involved will affect the magnitude of the heat change and it is usual to quote heat changes for specific molar quantities of reactants or products.

2. The physical states of the reactants and products must also be specified, as to convert a substance from one physical state to another will also involve an energy change. It is usual to indicate the state of each substance by the appropriate letter placed in brackets after the formula: (s) for solid, (l) for liquid and (g) for gas.

3. The temperature at which the experiment is carried out is either specified or it is indicated that it is the standard temperature which is taken as 298 K (room temperature, 25 °C).

4. The pressure at which the experiment is carried out is usually taken as the standard atmospheric pressure, 101 300 N m^{-2} (760 mmHg).

5. The magnitude of the heat change can also depend on whether the experiment is carried out at constant pressure (represented by ΔH) or at constant volume (represented by ΔU). It is usually more convenient to carry out experiments at a constant pressure.

Standard Enthalpy Change

The heat change at constant pressure under the standard conditions of 298 K and 101 300 N m^{-2} (760 mmHg) pressure is known as the standard enthalpy change and is represented by $\Delta H^{\ominus}$.

Sign convention

During an exothermic reaction energy is lost by the system and so the products must be at a lower energy state than the reactants (Figure 13.1(a)). The enthalpy change for such a reaction is said to be negative.

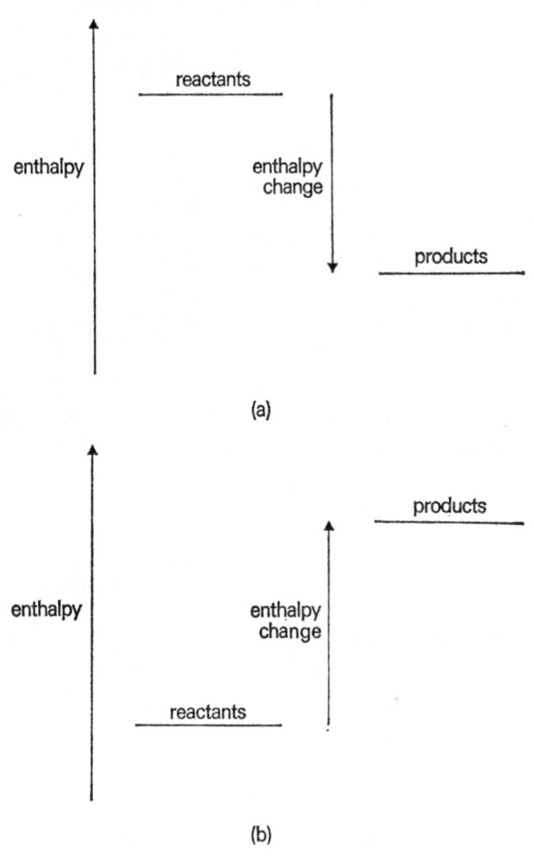

Figure 13.1. ***(a) Exothermic reaction and (b) endothermic reaction***

For an endothermic reaction the products appear to be at a higher energy state than the reactants and ΔH is positive as the system has gained energy from the surroundings (Figure 13.1(b)).

Relationship Between ΔH and ΔU

Liquids and solids show so little volume change in connection with thermochemical change that the issue of measurement at constant volume and constant pressure does not arise to any significant extent. With gases, however, the case is different. If observations are carried out at *constant volume*, no work can be done on, or by, the system. For an exothermic reaction carried out at *constant pressure*, a *diminution* of volume means that work is done *on* the system by the external pressure; that is, the heat change appears greater than it would at constant volume by the thermal equivalent of the work done. An *increase* of volume means that work is done *by* the expanding system against the external pressure, so the heat change appears less than it would at constant volume by the thermal equivalent of the work done. In quantitative terms, the work done is measured by the product $p\Delta V$, where p is the constant pressure and ΔV the change of volume. From the gas equation, $pV = nRT$, the work done is measured by RT for every change of 1 mole unit of gaseous volume (i.e. 22.4 dm^3), and is nRT for a change of n. R has the value of 8.3 $J\,K^{-1}\,mol^{-1}$ in thermal terms and T is the absolute temperature of the observations.

The relationship between ΔH and ΔU may be summarized by the equation

$$\Delta H = \Delta U + nRT$$

Consider the following example. 890 kJ are evolved when 1 mole of methane is burned in oxygen under standard conditions (101 325 $N\,m^{-2}$ and 298 K), i.e. $\Delta H = -890\ kJ\ mol^{-1}$. How much heat would be evolved if the same reaction were carried out at constant volume?

The equation for the reaction is

$$CH_4(g) + 2O_2(g) \rightarrow CO_2(g) + 2H_2O(l)$$

The volume of the water can be considered as negligible and there is a reduction from 3 moles of gas to 1 mole of gas, i.e. $n = -2$. The quantity of work done on the system when this contraction occurs is given by

$$\begin{aligned} nRT &= -2 \times 8.3 \times 298 \\ &= -4947\ J \\ &= -4.947\ kJ \end{aligned}$$

Using

$$\begin{aligned} \Delta H &= \Delta U + nRT \\ -890 &= \Delta U - 4.947 \\ \Delta U &= -885\ kJ\ mol^{-1} \end{aligned}$$

Experimental Determination of Enthalpy Changes

As previously indicated, when a value for an enthalpy change is quoted it must be absolutely clear under what conditions such a value is obtained. It is therefore necessary to have strict definitions for enthalpy terms which are in common use.

The standard enthalpy change for a reaction, $\Delta H^{\ominus}_{298}$, is equal to the heat evolved or absorbed when the molar quantities, represented by the equation, react at a pressure of 101 300 N m^{-2} (760 mmHg) and a temperature of 298 K, the substances being in their normal physical states for these conditions.

The standard enthalpy of formation, $\Delta H_f^{\ominus}$, is the heat change which occurs when 1 mole of a compound is formed from its elements, in their normal states, under the standard conditions of pressure and temperature (101 300 N m^{-2} and 298 K).

The standard enthalpy of combustion, $\Delta H_c^{\ominus}$, is the heat evolved when 1 mole of an element or compound is completely burned in oxygen under standard conditions.

The enthalpy changes for reactions involving liquids are most easily determined by insulating the reactants from the surroundings and recording the temperature change of the reactants. If the reaction does not involve organic solvents, a polystyrene beaker is a particularly convenient container for such a determination as it is a good insulator and, due to its small mass, it has a negligible heat capacity. The quantity of the heat change is calculated by multiplying the temperature change by the mass of the reactants and by their specific heat capacities. This value must then be multiplied up for the required molar quantities.

The alternative approach is to conduct the heat away from the reactants and to record the temperature of the surroundings. This method is particularly suitable for enthalpies of combustion.

Determination of heat of combustion

This determination is carried out in a *bomb calorimeter*, Figure 13.2. This instrument was originally devised by Berthelot. A typical example of it is made of steel. It is nickel plated on the outside and protected on the inside by a coating of non-oxidizable material (enamel or, better, gold or platinum). It has a tight-fitting, screw lid, working on to a lead washer. Material is burnt in a small platinum cup, C, in the calorimeter. Insulated platinum leads allow electric current to be passed through a thin spiral of iron wire, W, which glows and then fires the material.

A known mass of the material under investigation is placed in the platinum cup. Air is displaced by oxygen, which is allowed to reach a pressure of 20–25 atmospheres. The bomb calorimeter is then closed by a screw valve, and placed in water in a calorimeter fitted with a stirrer and an accurate thermometer, and well lagged to minimize heat losses. With the stirrer operating, temperature readings are taken at regular intervals of time. Current is then passed. The iron wire becomes red hot and fires the material, and combustion occurs. This brings a sudden rise of temperature,

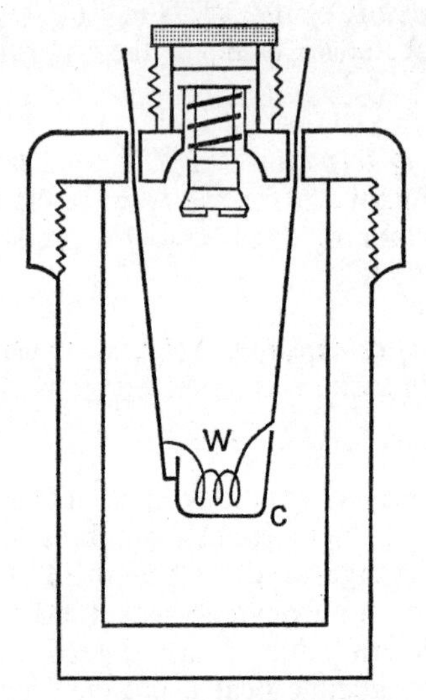

Figure 13.2. Bomb calorimeter

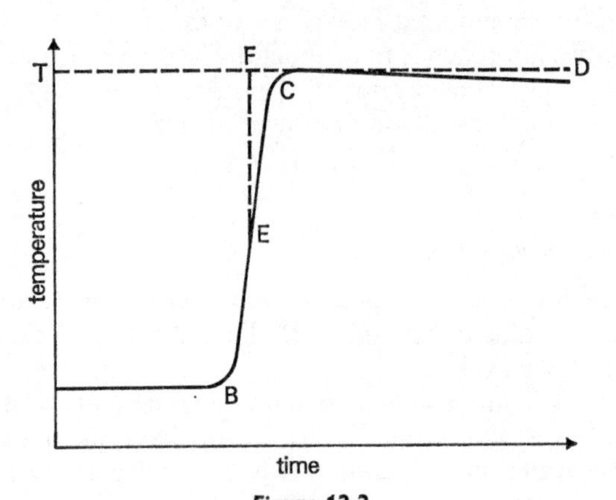

Figure 13.3.

which is registered on the temperature-time curve, Figure 13.3. Observations are continued until the system is cooling slowly and regularly, as in the part of the curve CD. If E is the mid-point of BC and EF is parallel to the temperature axis, the highest temperature reached can be taken as F (or T). The heat capacity of the whole system—bomb calorimeter, water, calorimeter, thermometer, stirrer, etc.—must be known.

Then, if the total heat capacity of the system is W, and the rise of temperature is t K, the total evolution of heat is Wt J. If the iron wire in burning produces x J, the heat evolved in the combustion of the material is $(Wt - x)$ J. Then, if the mass of the material used was a g, and its molecular mass M, the heat of combustion of the material is

$$\frac{M}{a}(Wt - x)$$

Hess's Law and Energy Cycles

This law is the essential basis of calculations in thermochemistry. It has both experimental and theoretical justification, as shown below. **Hess's Law** takes the form:

> The total energy change resulting from a chemical reaction is dependent only on the initial and final states of the reactants and is independent of the reaction route.

This law is an aspect of the general Law of Conservation of Energy or the First Law of Thermodynamics.

Suppose that a given chemical system changes from the state A to the state B by way of intermediate stages C and D, with liberation of x, y, and z J respectively in these stages; let the system also change from A to B by a single intermediate stage, E, with liberation of a and b J as follows;

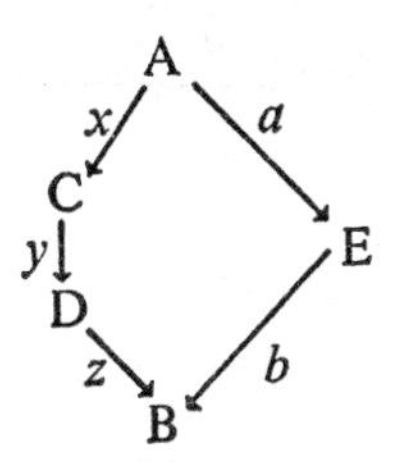

Hess's Law requires that $(a + b) = (x + y + z)$. If not, let $(a + b)$ be greater than $(x + y + z)$. It would then be possible to pass from A to B by E and gain $(a + b)$ J then reverse the process by D and C, giving back $(x + y + z)$ J. This would give a net gain on the circuit of $(a + b) - (x + y + z)$ J. This could be repeated indefinitely, yielding an unlimited supply of energy from no corresponding source. This is contrary to the Law of Conservation of Energy. Consequently, Hess's Law must be true.

Experimental illustration of Hess's Law is obtained by performing the same chemical change in two alternative ways and showing the heat change to be the same for both. The following is an example of this:

Conversion of 56 g of calcium oxide, CaO, into a dilute solution of the corresponding mass (111 g) of calcium chloride by methods A and B.

Method A

Convert the CaO to $Ca(OH)_2$.

$CaO(s) + H_2O(l) \rightarrow Ca(OH)_2(s)$
$\Delta H^{\ominus} = -63$ kJ

Convert the $Ca(OH)_2$ to a dilute solution.

$Ca(OH)_2(s) + H_2O(l) \rightarrow Ca(OH)_2(aq)$
$\Delta H^{\ominus} = -12.5$ kJ

Neutralize this with the corresponding amount of dilute HCl.

$Ca(OH)_2(aq) + 2HCl(aq) \rightarrow CaCl_2(aq) + 2H_2O(l)$
$\Delta H^{\ominus} = -117$ kJ

Total enthalpy change = $(-63 - 12.5 - 117) = -192.5$ kJ

Method B

Add the CaO directly to the appropriate mass of dilute HCl.

$CaO(s) + 2HCl(aq) \rightarrow CaCl_2(aq) + H_2O(l)$
$\Delta H^{\ominus} = -192.5$ kJ

Total enthalpy change = -192.5 kJ

The above example may be more conveniently arranged in the form of an energy cycle.

$$\begin{array}{ccc} CaO(s) + 2HCl(aq) & \xrightarrow{\Delta H_1^{\ominus}} & CaCl_2(aq) + H_2O(l) \\ \Delta H_2^{\ominus} \downarrow + H_2O(l) & & \Delta H_4^{\ominus} \uparrow + 2HCl(aq) \\ Ca(OH)_2(s) & \xrightarrow[H_2O(l)\ \Delta H_3^{\ominus}]{} & Ca(OH)_2(aq) \end{array}$$

$$\Delta H_1^{\ominus} = \Delta H_2^{\ominus} + \Delta H_3^{\ominus} + \Delta H_4^{\ominus}$$

Indirect Determination of Enthalpy Changes

The enthalpy changes for some reactions cannot be determined directly by experiment, but, by the application of Hess's Law, they can be calculated from other experimental results. For example, it is possible to determine experimentally the enthalpies of combustion $\Delta H_c^{\ominus}$ of methane, carbon, and hydrogen and these results may then be used to calculate the enthalpy of formation of methane which cannot be determined directly.

Given that $\Delta H_c^{\ominus}$ for methane is -890 kJ mol^{-1},

$\Delta H_c^{\ominus}$ for carbon is -393 kJ mol^{-1},

$\Delta H_c^{\ominus}$ for hydrogen is -286 kJ mol^{-1}, the equations associated with the above enthalpy changes are

$CH_4(g) + 2O_2(g) \rightarrow CO_2(g) + 2H_2O(l) \qquad \Delta H^\ominus = -890 \text{ kJ mol}^{-1}$

$C \text{ (graphite)} + O_2(g) \rightarrow CO_2(g) \qquad \Delta H^\ominus = -393 \text{ kJ mol}^{-1}$

$H_2(g) + \frac{1}{2}O_2(g) \rightarrow H_2O(l) \qquad \Delta H^\ominus = -286 \text{ kJ mol}^{-1}$

The equation for the formation of methane from its elements is

$$C \text{ (graphite)} + 2H_2(g) \rightarrow CH_4(g)$$

The following energy cycle can be constructed:

$$C \text{ (graphite)} + 2H_2(g) \xrightarrow{\Delta H_1^\ominus} CH_4(g)$$

$C \text{ (graphite)} + 2H_2(g) \xrightarrow[\Delta H_3^\ominus]{+2O_2(g)} CO_2(g) + 2H_2O(l)$

$CH_4(g) \xrightarrow[\Delta H_2^\ominus]{+2O_2(g)} CO_2(g) + 2H_2O(l)$

According to Hess's Law the total enthalpy change during the formation of the combustion products by one route is equal to the total change when they are formed by another route, i.e.

$$\Delta H_1^\ominus + \Delta H_2^\ominus = \Delta H_3^\ominus \qquad (1)$$

where $\Delta H_1^\ominus = \Delta H_f^\ominus$ the standard enthalpy of formation of methane,

$\Delta H_2^\ominus$ is the standard enthalpy of combustion of methane,

$\Delta H_3^\ominus$ is the enthalpy of combustion of carbon + 2 × the enthalpy of combustion of hydrogen.

Substituting into equation (1),

$$\Delta H_1^\ominus + (-890) = -393 + [2 \times (-286)]$$

$$\Delta H_f^\ominus \text{ for methane} = \Delta H_1^\ominus = -75 \text{ kJ mol}^{-1}$$

Standard Enthalpy of Atomization and Bond Energy Terms

The standard enthalpy of atomization of an element is the heat required to convert the element, in its normal state and under standard conditions, into 1 mole of atoms in the gas phase.

Note that enthalpies of atomization are quoted for 1 mole of gaseous atoms produced and not for 1 mole of the element in its normal state. The following example will emphasize this point.

Consider the conversion of hydrogen molecules to atoms.

$$H_2(g) \rightarrow 2H(g) \qquad \Delta H^\ominus = +436 \text{ kJ}$$

436 kJ mol^{-1} is known as the mean H—H bond energy, as it is the energy required to break 1 mole of H—H bonds. However, the standard enthalpy of atomization of hydrogen is 436/2 = +218 kJ mol^{-1} as this is the energy required to produce 1 mole of atoms in the gas phase.

Enthalpies of atomization can be used to calculate mean bond energies for bonds involving unlike atoms.

Mean C—H bond energy

Given that the standard enthalpy of formation of methane is −74.8 kJ mol^{-1} and the enthalpies of atomization of graphite and hydrogen are +715 and +218 kJ per mole of atoms in the gas phase, the following energy cycle can be constructed.

$$\text{C (graphite)} + 2H_2(g) \xrightarrow{\Delta H_1^\ominus} CH_4(g)$$

$$\text{C (graphite)} + 2H_2(g) \xrightarrow{\Delta H_2^\ominus} C(g) + 4H(g) \xrightarrow{\Delta H_3^\ominus} CH_4(g)$$

By Hess's Law,

$$\Delta H_1^\ominus = \Delta H_2^\ominus + \Delta H_3^\ominus$$

where $\Delta H_1^\ominus$ is the standard enthalpy of formation of methane,

$\Delta H_2^\ominus$ is the enthalpy of atomization of graphite + (4 × the enthalpy of atomization of hydrogen),

$\Delta H_3^\ominus$ is the enthalpy of formation of 1 mole of methane from gaseous atoms of carbon and hydrogen.

Substituting,

$$-74.8 = 715 + (4 \times 218) + \Delta H_3^\ominus$$
$$\Delta H_3^\ominus = -1661.8 \text{ kJ}$$

This is the energy evolved during the formation of 4 moles of C—H bonds from gaseous atoms of carbon and hydrogen. Therefore the mean C—H bond energy is

$$\frac{-1661.8}{4} = -415.4 \text{ kJ mol}^{-1}$$

Mean C—C bond energy

The standard enthalpy of formation of ethane is −84.6 kJ mol^{-1}. Using this value and the standard enthalpies of atomization of hydrogen and graphite from the previous example, it is possible to calculate the mean C—C bond energy from the following energy cycle.

$$2C\,(\text{graphite}) + 3H_2(g) \xrightarrow{\Delta H_1^\ominus} C_2H_6(g)$$

$$2C\,(\text{graphite}) + 3H_2(g) \xrightarrow{\Delta H_2^\ominus} 2C(g) + 6H(g) \xrightarrow{\Delta H_3^\ominus} C_2H_6(g)$$

By Hess's Law,

$$\Delta H_1^\ominus = \Delta H_2^\ominus + \Delta H_3^\ominus$$

where $\Delta H_1^\ominus$ is the standard enthalpy of formation of ethane,
$\Delta H_2^\ominus$ is (2 × standard enthalpy of atomization of graphite) + (6 × standard enthalpy of atomization of hydrogen),
$\Delta H_3^\ominus$ is the standard enthalpy of formation of ethane from gaseous atoms of carbon and hydrogen.

Substituting,

$$-84.6 = (2 \times 715) + (6 \times 218) + \Delta H_3^\ominus$$
$$\Delta H_3^\ominus = -2822.6 \text{ kJ mol}^{-1}$$

This is the energy evolved during the formation of 1 mole of C—C bonds and 6 moles of C—H bonds, i.e.

$$-2822.6 = 1(\text{C—C bond energy}) + (6 \times -415)$$

This gives a value for the mean C—C bond energy of -333 kJ mol^{-1}.

By calculations similar to those employed above it is possible to build up a table of mean bond energies. These bond energies strictly refer to the bond in a particular molecular environment (e.g. the C—H in methane as opposed to the C—H in trichloromethane) and care must be exercised when using them as it is common for bond strengths to be affected by the rest of an organic molecule. Despite these limitations, as will be seen in the following section, bond energy terms can be very useful.

Prediction of Enthalpy Changes from Bond Energy Terms

The enthalpy change for a reaction may be considered as the difference between the sum of the energies required to break bonds in the reactants and the sum of the energies evolved when new bonds are formed to give the products. Clearly a knowledge of bond energy terms makes it possible to predict enthalpy changes for reactions, e.g. assuming the following mean bond energies it is possible to predict a value for the standard enthalpy of hydrogenation of ethene (ethylene).

$$\text{C—C} = 345 \text{ kJ mol}^{-1}$$
$$\text{C=C} = 610 \text{ kJ mol}^{-1}$$
$$\text{C—H} = 415 \text{ kJ mol}^{-1}$$
$$\text{H—H} = 435 \text{ kJ mol}^{-1}$$

The equation for the hydrogenation of ethene is

$$H_2C{=}CH_2\,(g) + H_2\,(g) \rightarrow H_3C{-}CH_3\,(g)$$

This change could be considered to take place by the breaking of 1 mole of C=C bonds and 1 mole of H—H bonds which would require:

$$610 + 435 = 1045 \text{ kJ}$$

This is then followed by the forming of 1 mole of C—C bonds and 2 moles of C—H bonds. The heat evolved during the formation of these bonds would be

$$345 + (2 \times 415) = 1175 \text{ kJ}$$

The standard enthalpy of hydrogenation is likely to be

$$1045 - 1175 = -130 \text{ kJ mol}^{-1}$$

Delocalization Energy

Using the principles employed in the previous section it is possible to predict a value for the standard enthalpy of hydrogenation of benzene if its structure is assumed to consist of alternate double and single carbon–carbon bonds. Using the following mean bond energies:

$$\text{C=C} = 610 \text{ kJ mol}^{-1}$$
$$\text{C—C} = 345 \text{ kJ mol}^{-1}$$
$$\text{C—H} = 415 \text{ kJ mol}^{-1}$$
$$\text{H—H} = 435 \text{ kJ mol}^{-1}$$

The reaction may be represented by the equation

$$C_6H_6\,(l) + 3H_2\,(g) \rightarrow C_6H_{12}\,(l) \qquad (1)$$

This reaction could be considered to involve the breaking of 3 moles of C=C bonds and 3 moles of H—H bonds which would require

$$(3 \times 610) + (3 \times 435) = 3135 \text{ kJ}$$

This is then followed by the forming of 3 moles of C—C bonds and 6 moles of C—H bonds. The heat evolved during the formation of these bonds would be

$$(3 \times 345) + (6 \times 415) = 3525 \text{ kJ}$$

This gives a value of

$$3135 - 3525 = -390 \text{ kJ mol}^{-1}$$

for the standard enthalpy of hydrogenation of benzene.

The experimentally determined value is approximately -210 kJ mol^{-1}. The difference between these two values indicates that the actual structure of benzene is more stable than the alternate double and single bond structure by a factor equivalent to (390–210) = 180 kJ mol^{-1}. This extra stability is attributed to the delocalization of the bonding electrons over all six carbon atoms (see page 83).

Born-Haber Cycle and Lattice Energies

The Born-Haber energy cycle considers the energy changes which occur during the formation of an ionic crystal. The Born-Haber cycle for the formation of sodium chloride is given below.

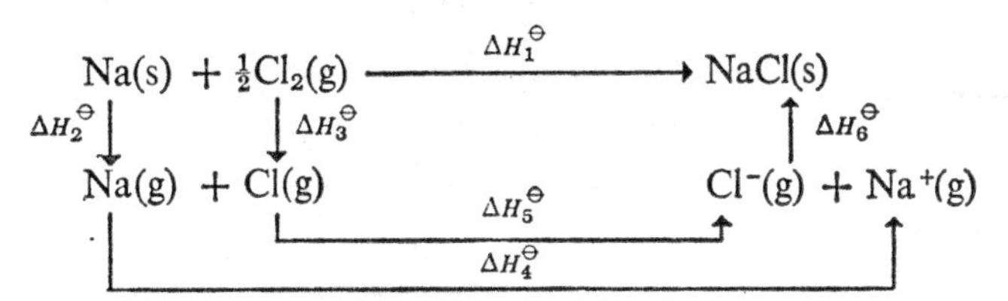

Starting with the elements in their standard states, the steps in the cycle and the enthalpy changes associated with them are:

1. The atomization of the elements to form gaseous atoms. These changes require energy and so the values for $\Delta H_2^\ominus$ and $\Delta H_3^\ominus$ will be positive. Remember standard enthalpies of atomization are quoted for 1 mole of gaseous atoms and it is not necessary to halve the chlorine value.

2. The gaseous metal atom is converted into a positive ion. This step needs energy and $\Delta H_4^\ominus$, which is the ionization energy for sodium, has a positive value.

3. The gaseous non-metal atom forms a negative ion. This usually results in the evolution of heat (O^- to O^{2-} being an exception) and $\Delta H_5^\ominus$, which is known as the electron affinity of chlorine, is negative.

4. The gaseous ions come together to form the crystal lattice. This step will be accompanied by evolution of heat and $\Delta H_6^\ominus$, which is the lattice energy, will be negative.

The lattice energy is defined as the standard enthalpy of formation of 1 mole of the crystal lattice from its ions in the gas phase.

The other half of the cycle is $\Delta H_1^\ominus$, which is the standard enthalpy of formation of the sodium chloride crystal, i.e. the enthalpy change which occurs when 1 mole of the compound is formed from its elements in their standard states.

Applying Hess's Law to the cycle:

$$\Delta H_1^\ominus = \Delta H_2^\ominus + \Delta H_3^\ominus + \Delta H_4^\ominus + \Delta H_5^\ominus + \Delta H_6^\ominus$$

All of these enthalpy changes except $\Delta H_6^\ominus$ can be determined either directly or indirectly from experimental results and hence the cycle can be used to calculate a value for the lattice energy (ΔH_6).

In the case of sodium chloride,

$\Delta H_1^\ominus = \Delta H_f^\ominus$ of NaCl (s) $= -411$ kJ mol^{-1}
$\Delta H_2^\ominus =$ standard enthalpy of atomization of sodium $= +108$ kJ mol^{-1}
$\Delta H_3^\ominus =$ standard enthalpy of atomization of chlorine $= +121$ kJ mol^{-1}
$\Delta H_4^\ominus =$ first ionization energy of sodium $= +493$ kJ mol^{-1}
$\Delta H_5^\ominus =$ electron affinity of chlorine $= -364$ kJ mol^{-1}
$\Delta H_6^\ominus =$ lattice energy for sodium chloride.

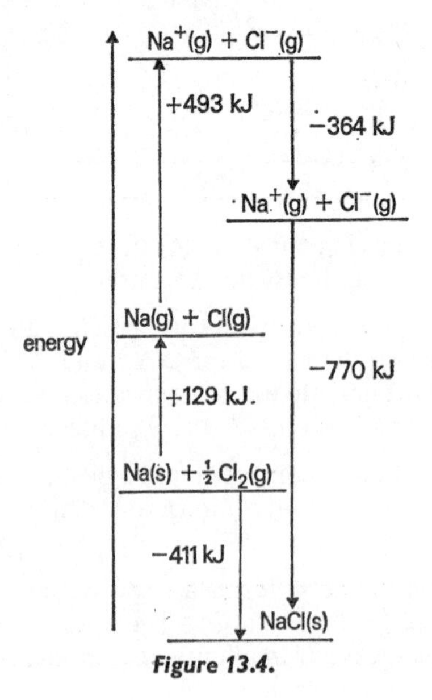

Figure 13.4.

Substituting,

$$-411 = 108 + 121 + 494 - 364 + \Delta H_6^{\ominus}$$
$$\Delta H_6^{\ominus} = -770 \text{ kJ mol}^{-1}$$

The lattice energy of sodium chloride is —770 kJ mol^{-1}. It is worth-while considering this energy cycle in the form of an energy level diagram, Figure 13.4.

If the ionic crystal is to be energetically stable with respect to its constituent elements in their standard states, then $\Delta H_f^{\ominus}$ must be negative. It is clear from Figure 13.4 that the main energy terms which determine the sign of $\Delta H_f^{\ominus}$ are the ionization energy of the metal atom and the lattice energy. It is mainly the relative magnitude of these two values which will determine whether or not the compound is formed exothermally from its elements in their standard states.

Theoretical Lattice Energies

Lattice energies determined by the Born-Haber cycle are based directly or indirectly on experimental results. They may be considered as the energy released when the ions in the gas phase come together to form 1 mole of the crystal, irrespective of the type of bonding in the resulting crystal. In other words, lattice energies can be calculated from the Born-Haber cycle even if the resulting lattice is not 100 per cent ionic and does involve some electron sharing.

If the lattice is considered to be 100 per cent ionic then it is possible to calculate a theoretical value for the lattice energy. This calculation must take into account the magnitude of the charges, their coming together from infinity, and therefore the variation of forces of attraction with distance of separation (inverse square law). In addition, using the geometry of the crystal, it is necessary to take into account both attractions between oppositely charged ions and repulsions between ions with like charges.

In the cases of salts such as sodium chloride there is good agreement between the theoretical and experimental values, indicating the validity of the assumption that they are 100 per cent ionic. Where there is an obvious discrepancy, as in the cases of silver chloride and zinc sulphide, it suggests that the bonding in the lattice is not entirely ionic and there is appreciable sharing of electrons (covalent bonding).

Standard Enthalpy of Solution and Standard Enthalpy of Hydration

The dissolving of an ionic compound in water can be either an exothermic or an endothermic process, which suggests that there are at least two energy terms involved. In fact, from the energy point of view, the process can be thought of as occurring in two stages. The first is the separation of the ions which will require energy equal in magnitude but opposite in sign to the lattice energy. The second stage is the hydration of the ions, which will be an exothermic process.

If the lattice energy is greater than the enthalpy of hydration, the enthalpy change associated with the dissolving of the compound to form a dilute solution will be endothermic. If the enthalpy of hydration is the greater, then the dissolving process will be exothermic.

For sodium chloride the following energy cycle may be drawn:

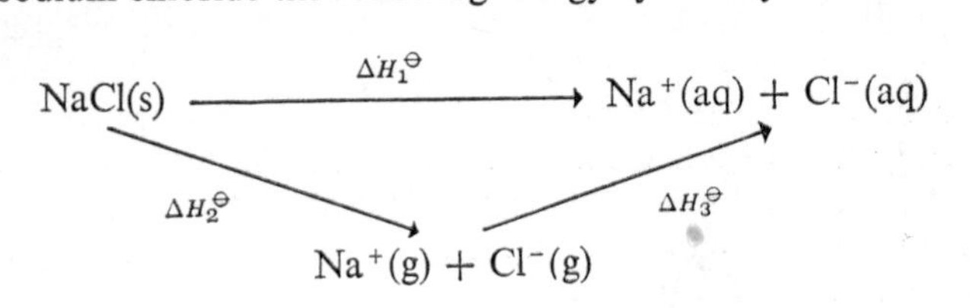

By Hess's Law,

$$\Delta H_1^{\ominus} = \Delta H_2^{\ominus} + \Delta H_3^{\ominus}$$

where $\Delta H_1^{\ominus}$ = the standard enthalpy of solution,

$\Delta H_2^{\ominus}$ = – (lattice energy),

$\Delta H_3^{\ominus}$ = the sum of the enthalpies of hydration of the two ions.

The standard enthalpy of solution of sodium chloride is +4 kJ mol^{-1} and the lattice energy is −770 kJ mol^{-1}. This means the enthalpy of hydration is −766 kJ mol^{-1}.

The balance between lattice energies and enthalpies of hydration accounts for the observation that certain hydrated compounds dissolve endothermally whereas the corresponding anhydrous compounds dissolve exothermally. The ions in the crystals are already partially hydrated so that when they are dissolved the lattice energy easily outweighs the enthalpy change associated with any further hydration. For example, anhydrous copper(II) sulphate dissolves to form a dilute solution with the evolution of 66.3 kJ mol^{-1}, whereas 1 mole of the hydrated copper(II) sulphate absorbs 11.4 kJ mol^{-1} when it is dissolved to form a dilute solution. The following energy cycle may be constructed;

$CuSO_4(s)$ ⟶ $Cu^{2+}(aq) + SO_4^{2-}(aq)$

$CuSO_4 . 5H_2O(s)$

From this it is possible to calculate that the enthalpy change associated with the formation of the hydrated crystals from the anhydrous salt is

$$66.3 + 11.4 = 77.7 \text{ kJ mol}^{-1}$$

i.e. $CuSO_4(s) + 5H_2O(l) \rightarrow CuSO_4{,}5H_2O(s) \quad \Delta H^{\ominus} = -77.7 \text{ kJ mol}^{-1}$

Endothermic Reactions, Entropy, and Free Energy

When a reaction, from the overall energy change point of view, is feasible, it is sometimes described as being energetically favourable. An exothermic reaction would appear to be energetically favourable as the system is losing heat energy to the surroundings and hence is moving to a lower energy state. However, this approach raises the question – Why do endothermic reactions occur? The existence of endothermic reactions suggests that enthalpy changes are not the sole factor to be considered. The other factor is given the name entropy and the change in entropy, ΔS, represents the change in the degree of disorder. It appears to be the natural tendency of systems to move to a more disordered or chaotic state. For example, if marbles of two different colours are shaken in a tray, one would expect a random, chaotic distribution of the two colours (high entropy) to result rather than the more ordered state of marbles of one colour in one half of the tray and marbles of the other colour in the other half (low entropy). Similarly, if two gases which do not react are placed in a container one would expect the gases to mix and achieve the highest state of entropy. In the context of a chemical reaction, if, for example, the number of moles of gas increases as the reaction proceeds, there will be an increase in entropy. In general the greater the number of particles which can move independently, the greater the entropy.

The overall term which incorporates both enthalpy change and entropy change is the change in **Gibb's free energy**. The standard free energy change, $\Delta G^{\ominus}$, is related to the standard enthalpy change and the standard entropy change by the equation

$$\Delta G^{\ominus} = \Delta H^{\ominus} - T\Delta S^{\ominus}$$

Thus the overall factor which determines whether or not a reaction is energetically feasible is the change in free energy. A negative free energy change for the complete reaction indicates an energetically favourable reaction. If the entropy change is small, for example when two moles of liquid produce two moles of two other, but fairly similar, liquids, then the enthalpy change may provide a good indication of the feasibility of the reaction, but it must be noted that the importance of the entropy change (as it is multiplied by the temperature on the Kelvin scale) increases with temperature.

Also it is important to note that overall free energy changes indicate energetically favourable reactions but give no indication of the rate of the reaction (see page 256).

14. Chemical Equilibria I: Molecular Equilibria

Reversible Reactions and the State of Dynamic Equilibrium

A reversible process, whether physical or chemical, in a closed system (i.e. one which will not allow any of the components to escape) and at a constant temperature, will eventually reach a state of equilibrium. The equilibrium will be dynamic and may be considered as a state at which the rate of the process in one direction (see page 254) is exactly balanced by the rate in the reverse direction. When a chemical reaction reaches this state of dynamic equilibrium, the equilibrium concentrations of the reactants and products will remain constant as long as the conditions are not changed.

Consider the homogeneous system

$$wA + xB \rightleftharpoons yC + zD$$

then, when the system is in a state of equilibrium, the equilibrium molar concentrations of the reactants and products are found to be related in the following way,

$$\frac{[\mathrm{C}]^y\,[\mathrm{D}]^z}{[\mathrm{A}]^w\,[\mathrm{B}]^x} = K_c$$

K_c, for a fixed temperature, is found to have a constant value and is known as the **equilibrium constant**. The square brackets indicate equilibrium concentrations measured in mol dm^{-3}. The relationship itself is known as the **Equilibrium Law**.

Application of the Equilibrium Law to Homogeneous Molecular Equilibria

1. *The reaction between ethanol and ethanoic acid*

The reaction is represented by the equation

$$CH_3CO_2H(l) + C_2H_5OH(l) \rightleftharpoons CH_3CO_2C_2H_5(l) + H_2O(l)$$

Suppose that *a* mole of ethanoic acid and *b* mole of the alcohol are mixed and allowed to come to equilibrium at constant temperature, so that, at equilibrium, *x* mole of ethyl ethanoate and water are formed. Then

$(a - x)$ mole of acid and $(b - x)$ mole of alcohol must remain. If the volume of the liquid is V dm^3, the molar concentrations of the reagents are

$$\text{ethanoic acid: } \frac{(a - x)}{V}$$

$$\text{alchohol: } \frac{(b - x)}{V}$$

$$\text{ester and water: each } \frac{x}{V}$$

It is advisable to employ a systematic method of setting out this information.

$$CH_3CO_2H(l) + C_2H_5OH(l) \rightleftharpoons CH_3CO_2C_2H_5(l) + H_2O(l)$$

Initial amounts (mol)	a	b	0	0
Equilibrium amounts (mol)	$a - x$	$b - x$	x	x
Equilibrium concentrations (mol dm^{-3})	$\frac{a - x}{V}$	$\frac{b - x}{V}$	$\frac{x}{V}$	$\frac{x}{V}$

The equilibrium law for this reaction is

$$K_c = \frac{[CH_3CO_2H]\,[C_2H_5OH]}{[CH_3CO_2C_2H_5]\,[H_2O]}$$

Substituting,

$$K_c = \frac{x^2/V^2}{(a - x)(b - x)/V^2}$$

$$= \frac{x^2}{(a - x)(b - x)}$$

A value for K_c for this reaction at a particular temperature may be determined by the following procedure.

Ethanoic acid and ethanol, both in a pure and dry state, are mixed to make a known volume, in the proportions of 1 mole of each (that is, 60 g of ethanoic acid to 46 g of ethanol). The containing flask is corked and kept in a thermostat at 298 K. At intervals, 2 cm^3 of the liquid are extracted and titrated at once by barium hydroxide solution (or carbonate-free sodium hydroxide solution) with phenolphthalein as indicator. This determines the ethanoic acid in the mixture. It will be found that the proportion of ethanoic acid present falls as esterification proceeds, finally reaching a constant minimum when equilibrium is attained. At this point, experiment shows that one-third of the original mole of ethanoic acid remains, i.e. two-thirds have been esterified. Applying these figures to the above equation, $a = b = 1$ and $x = \frac{2}{3}$; that is,

$$K_c = \frac{x^2}{(a - x)(b - x)} = \frac{(\frac{2}{3})^2}{\frac{1}{3} \cdot \frac{1}{3}} = 4$$

From these figures, the equilibrium constant of the reaction at 298 K is 4.

The equilibrium law states that for this particular temperature K_c for the ethanoic acid–ethanol reaction will always be 4. The implication of this is that if other initial concentrations of reactants are used the position of the dynamic equilibrium finally reached will be such that the equilibrium concentrations, when substituted into the equilibrium law, will still give a value of 4 for K_c. Suppose, now, that a mixture of 1 mole of ethanoic acid and 0.5 mole of ethanol is used. If the equilibrium law is correct, and if equilibrium is attained when n mole of ester (and water) have been formed,

$$\frac{n^2}{(1 - n)(0.5 - n)} = K_c = 4$$

This is a quadratic in n, which can be solved for n as follows:

$$\begin{aligned} n^2 &= 4(1 - n)(0.5 - n) \\ &= 4(0.5 - 1.5n + n^2) \\ &= 2 - 6n + 4n^2 \end{aligned}$$

i.e.

$$3n^2 - 6n + 2 = 0$$

From this,

$$\begin{aligned} n &= \frac{6 \pm \sqrt{(36 - 4 \times 3 \times 2)}}{6} \\ &= \frac{6 \pm \sqrt{12}}{6} \\ &= \frac{6 \pm 2\sqrt{3}}{6} \\ &= \frac{6 - 2 \times 1.732}{6} \\ &= 0.42 \end{aligned}$$

(The alternative value of n, taking the positive value of $2\sqrt{3}$, gives an impossible answer.) From this, equilibrium should be reached with

0.42 mole of ester and water, (1 − 0.42) or 0.58 mole of ethanoic acid and (0.50 − 0.42), or 0.08 mole of ethanol present. This can be tested by performing the actual experiment with initial proportions of 1 mole of ethanoic acid and 0.5 mole of ethanol. The order of agreement reached in experiments by Berthelot and St Gilles is shown in the table below.

The good agreement between the observed and calculated values for the ester justifies acceptance of the Equilibrium Law for this reaction.

Original proportions (mole)		*Ester produced at equilibrium (mole)*	
Ethanoic acid	*Ethanol*	*Observed by expt.*	*Calculated from $K_c = 4$*
1.00	0.080	0.078	0.078
1.00	0.280	0.226	0.232
1.00	0.330	0.293	0.311
1.00	0.500	0.414	0.423
1.00	0.670	0.519	0.528
1.00	1.500	0.819	0.785
1.00	2.000	0.858	0.845

2. *The reaction between hydrogen and iodine*

The classical example of the application of the Equilibrium Law to a reaction involving entirely gaseous reagents is Bodenstein's investigation of the synthesis of hydrogen iodide from its elements:

$$H_2(g) + I_2(g) \rightleftharpoons 2HI(g)$$

There is no change of total volume in this reaction, so that the results are independent of pressure (a fact which was first verified by Bodenstein). In the experimental work, he took a known mass of iodine (say 0.319 mole) and placed it in a glass bulb which was then sealed containing a known mass (say 0.206 mole) of hydrogen, actually under considerable pressure, which is convenient because of the very low density of the gas. The bulb was then heated at a *constant* temperature in the region of 723 K till equilibrium was attained. The bulb was then cooled *rapidly* to room temperature. The time of cooling was so short that no appreciable change could take place in the equilibrium as the temperature was falling; at room temperature, the velocity of reaction is so slow that the mixture can be kept indefinitely without appreciable change. This device of *rapid* cooling to a temperature at which reaction velocity is negligible is called *freezing the equilibrium* and is often resorted to on both the laboratory and industrial scales. In the present case, the 'frozen' equilibrium mixture was analysed for iodine and hydrogen iodide by absorption in excess of standard potassium hydroxide solution and the application of the usual methods of analysis. The iodine was found to be 0.134 mole and the hydrogen iodide 0.370 mole; from these figures, the hydrogen present at equilibrium was

$$0.206 - (0.319 - 0.134) \text{ mole} = 0.021 \text{ mole}$$

Applying the Equilibrium Law,

	$H_2(g)$ +	$I_2(g)$	$\rightleftharpoons$ $2HI(g)$	
Initially:	a	b	—	mole
At equilibrium:	$(a - x)$	$(b - x)$	$2x$	mole

If the volume at equilibrium is V dm^3,

$$K_c = \frac{(2x)^2/V^2}{[(a-x)/V]/[(b-x)/V]} = \frac{(2x)^2}{(a-x)(b-x)}$$

Notice that K_c is independent of V (and, therefore, of pressure).

Applying the above figures of experiment,

$$K_c = \frac{(0.370)^2}{0.021 \times 0.134} = 48.7$$

By using different proportions of hydrogen and iodine, other values of K_c can be calculated and should, if the Equilibrium Law is valid, be constant. For Bodenstein's figures, K_c shows some variation; but, in view of the experimental difficulties involved, the results are considered reasonably satisfactory ($K_c = 54 \pm 9$ approximately). The following calculation further illustrates the operation of the Equilibrium Law in this case.

Example In an experiment, 0.206 mole of hydrogen and 0.144 mole of iodine were heated (at 723 K) to equilibrium in the reaction $H_2 + I_2 \rightleftharpoons 2HI$. 0.258 mole of hydrogen iodide was formed. Calculate the equilibrium constant of the reaction to the nearest integer. Hence calculate the molar composition (at 723 K) of the equilibrium mixture obtained from heating 0.515 mole of hydrogen with 0.360 mole of iodine.

$$H_2(g) + I_2(g) \rightleftharpoons 2HI(g)$$

As 0.258 mole of hydrogen iodide was formed, $0.258/2 = 0.129$ mole of both hydrogen and iodine must have been used up. The iodine in the equilibrium mixture must, therefore, be $(0.144 - 0.129) = 0.015$ mole, and the hydrogen $(0.206 - 0.129) = 0.077$ mole.

$$K_c = \frac{[HI]^2}{[H_2][I_2]} = \frac{(0.258)^2}{0.077 \times 0.015} = 58$$

In the second part of the calculation, let there be $2x$ mole of hydrogen iodide produced at equilibrium. Then the number of mole of iodine in the equilibrium mixture is $(0.360 - x)$ and of hydrogen $(0.515 - x)$.

From this,

$$58 = \frac{(2x)^2}{(0.515 - x)(0.360 - x)}$$

$$4x^2 = 58(0.515 - x)(0.360 - x)$$

$$13.5x^2 - 12.7x + 2.68 = 0$$

$$x = \frac{12.7 \pm \sqrt{[(12.7)^2 - 4 \times 13.5 \times 2.68]}}{2 \times 13.5}$$

$$x = 0.32$$

From this, the equilibrium mixture is

Hydrogen iodide	0.64 mole
Iodine	$(0.360 - 0.32) = 0.04$ mole
Hydrogen	$(0.515 - 0.32) = 0.195$ mole

3. *Dissociation of phosphorus pentachloride (phosphorus(V) chloride)*

Phosphorus pentachloride dissociates according to the equation

$$PCl_5(g) \rightleftharpoons PCl_3(g) + Cl_2(g)$$

Suppose that a mole of the pentachloride are originally present and dissociate to form b mole each of PCl_3 and Cl_2, and to leave $(a - b)$ mole of PCl_5, all gaseous and in a volume V dm^3 at equilibrium at some definite temperature, t K. Applying the Equilibrium Law,

$$K_c = \frac{b/V \times b/V}{(a - b)/V} = \frac{b^2}{(a - b)V}$$

In this case, K_c involves V and so is not independent of pressure. If V increases, K_c must remain constant at constant temperature; consequently b^2 must increase and $(a - b)$ decrease to maintain the value of K_c. That is, *increase of volume increases the dissociation of the pentachloride*. This is the same as the assertion that decrease of pressure (at constant temperature) increases the dissociation, and *vice versa*.

Suppose that, to the above system in equilibrium, one of the products, say chlorine, is added with no change of volume or temperature. If the concentration of chlorine is raised to $(b + c)$ mole in V dm^3, this quantity must enter the numerator of the value of K_c above. To maintain the value of K_c, the b factor for PCl_3 must decrease and the $(a - b)$ factor for PCl_5 increase, i.e. the introduction of chlorine tends to suppress the dissociation of the pentachloride. The same is true if phosphorus trichloride is added in similar conditions.

When all the materials concerned are *gaseous*, their concentrations can be expressed in terms of pressure. For example, in the dissociation of phosphorus pentachloride:

$$PCl_5(g) \rightleftharpoons PCl_3(g) + Cl_2(g)$$

partial pressures at equilibrium: p_1 $\quad p_2$ $\quad p_3$

the equilibrium constant can have the alternative representations

$$K_c = \frac{[PCl_3][Cl_2]}{[PCl_5]}$$

$$K_p = \frac{p_2 \times p_3}{p_1}$$

The sum $(p_1 + p_2 + p_3)$ is the total pressure of the mixed vapours.

In terms of the total pressure, p, of the system, the partial pressure of a reactant is given by $p \times$ the mole fraction of the reactant present at equilibrium.

In the case of the dissociation of phosphorus pentachloride, the total number of moles present at equilibrium is

$$(a - b) + b + b = (a + b)$$

Therefore the partial pressures are given by

$$p_1 = p \times \frac{(a - b)}{(a + b)}$$

$$p_2 = p \times \frac{b}{(a + b)}$$

$$p_3 = p \times \frac{b}{(a + b)}$$

Substituting into the K_p form of the Equilibrium Law,

$$K_p = \frac{p^2b^2/(a + b)^2}{p(\mathrm{a} - b)/(a + b)}$$

$$= \frac{pb^2}{(a + b)(a - b)}$$

This expression can be used to make quantitative predictions of equilibrium concentrations for various total pressures.

4. *The equilibrium in the Haber synthesis of ammonia*

In the Haber process, ammonia is synthesized from its elements

	$N_2(g)$	+	$3H_2(g)$	$\rightleftharpoons$	$2NH_3(g)$
Initially:	1		3		— mole
At equilibrium:	$(a - x)$		$(b - 3x)$		$2x$ mole
	p_1		p_2		p_3 partial pressures

From this,

$$K_c = \frac{[NH_3]^2}{[N_2][H_2]^3}; \quad K_p = \frac{p_3^2}{p_1 \times p_2^3}$$

As in the above examples, K_c can be expressed in terms of a, b, x, and the total volume of the system, V. K_p can be expressed in terms of a, b, x, and p, the total pressure of the system.

Application of the Equilibrium Law to Heterogeneous Systems

In a general way, the Equilibrium Law can be applied only in homogeneous systems. With some appropriate assumptions, however, it can be applied to certain heterogeneous systems. Some of these are mentioned below.

1. *Action of heat on calcium carbonate*

When heated, calcium carbonate dissociates in the reversible reaction

$$CaCO_3(s) \rightleftharpoons CaO(s) + CO_2(g)$$

If the assumption is made that the solids $CaCO_3$ and CaO have definite vapour pressures and so participate in the equilibrium *in the vapour phase* (which is homogeneous), the Equilibrium Law can be applied, *to the vapour phase*, in the form

$$\frac{(\text{partial pressure of CaO}) \times (\text{partial pressure of } CO_2)}{(\text{partial pressure of } CaCO_3)} = \text{a constant}$$

The further assumption is then made that, *so long as any of the solid is present at all, no matter how little, its vapour pressure* (*at a given temperature*) *can be taken as constant*. Consequently, for a system in which $CaCO_3$ (solid), CaO (solid), and CO_2 are all present together in equilibrium, the partial pressures of both CaO and $CaCO_3$ can be taken as constant and the above expression becomes:

$$(\text{partial pressure of } CO_2) = \text{a constant} = K_p$$

This means that, by a deduction from the Equilibrium Law applied in this special way, we should expect that if calcium carbonate is heated in an originally evacuated container so that $CaCO_3$, CaO, and CO_2 are always present together, there is a definite pressure of carbon dioxide, (known as the *dissociation pressure* for calcium carbonate) for each temperature at which equilibrium is reached. This is, in fact, the case. Up to about 873 K, this equilibrium pressure of carbon dioxide is quite minute and the rate of decomposition of calcium carbonate is so slow (at atmospheric pressure) as to be negligible. At bright red heat, the equilibrium pressure of carbon dioxide is about one atmosphere and decomposition of calcium carbonate is rapid.

In a closed vessel of such a size that the equilibrium pressure of carbon dioxide can be reached with some calcium carbonate still left, complete decomposition of calcium carbonate cannot be achieved. In the open air, the partial pressure of carbon dioxide is negligibly small, the reverse action is negligible, and virtually complete decomposition of the carbonate is achieved.

2. *Reaction between iron and steam*

At red heat, iron reacts with steam, in a reversible reaction, to form its black oxide, iron(II) diiron(III) oxide, and hydrogen.

$$3Fe(s) + 4H_2O(g) \rightleftharpoons Fe_3O_4(s) + 4H_2(g)$$

Assuming, as in case 1, that the solids Fe and Fe_3O_4 have definite vapour pressures and participate in the equilibrium *in the vapour phase*, the Equilibrium Law can be applied *to the vapour phase* in the form

$$\frac{(\text{partial pressure of } Fe_3O_4) \times (\text{partial pressure of } H_2)^4}{(\text{partial pressure of Fe})^3 \times (\text{partial pressure of } H_2O)^4} = \text{a constant } (K_p)$$

Assuming again that so long as any of the solid is present its vapour pressure is constant, the partial pressures of Fe and Fe_3O_4 can be taken as constant in a system in equilibrium (at constant temperature) in which *all four* reagents are present together. The above expression can then be written

$$\frac{(\text{partial pressure of } H_2)^4}{(\text{partial pressure of } H_2O)^4} = \text{a constant } (K'_p)$$

or

$$\frac{(\text{partial pressure of } H_2)}{(\text{partial pressure of } H_2O)} = \text{a constant } (K''_p)$$

This means that, if iron and steam are heated together in an originale, evacuated vessel of such a size that equilibrium can be reached with Fly H_2O, Fe_3O_4, and H_2 *all present together*, there is a definite total equilibrium pressure for each temperature and it is made up of steam pressure and hydrogen pressure in a ratio characteristic of that temperature, and constant for it. Also, an exactly similar equilibrium position will be reached, for the same temperature, starting from the materials Fe_3O_4 and H_2, similar conditions applying.

If, however, steam is passed over iron (at red heat) in a continuous stream in an *open* tube, hydrogen is continually swept away by steam pressure as fast as it forms. Its effective concentration (or partial pressure) tends to zero, the reverse action is negligible and iron can, for practical purposes, be *fully* converted to its black oxide. (Notice, however, that the conversion of steam to hydrogen is far from complete.) If hydrogen is similarly passed over black iron oxide at red heat, the relations just stated for hydrogen and steam are reversed and the black oxide can be reduced, virtually quantitatively, to iron.

Qualitative Illustrations of Reversible Reactions and Chemical Equilibria

A **reversible reaction** may be defined as a reaction which will proceed in either direction if conditions are arranged appropriately.

The following are examples of reversible reactions which can be used to give qualitative illustration of chemical equilibrium.

1. *The bismuth chloride reaction*

$$BiCl_3(aq) + H_2O(l) \rightleftharpoons BiOCl(s) + 2HCl(aq)$$

If bismuth carbonate is treated carefully with dilute hydrochloric acid with shaking, it can be converted to a colourless liquid, bismuth chloride solution, containing a certain excess of hydrochloric acid.

If this solution is then poured into a moderate excess of water and stirred, a white precipitate will appear. This is bismuth oxychloride. In this case, the excess of water has pushed the equilibrium to the *right*, producing enough oxychloride to cause precipitation.

If the 'milky' liquid is then stirred while concentrated hydrochloric acid is added drop by drop, the white precipitate will eventually dissolve. In this case, the increased concentration of acid has pushed the equilibrium to the *left* (in the above equation), producing soluble bismuth chloride. By suitable alternate additions of water and acid the reaction can usually be reversed several times, though at greater dilutions the precipitate may be slow to appear.

2. *The thiocyanate reaction*

The reaction between iron(III) chloride and ammonium thiocyanate (both in solution) is represented by

$$Fe^{3+}(aq) + CNS^-(aq) \rightleftharpoons [Fe(CNS)]^{2+}(aq))$$

$[Fe(CNS)]^{2+}$ is very strongly coloured (blood-red) and the equilibrium position can be followed by observation of this colour.

Prepare a solution of ammonium thiocyanate (1 g in 100 cm^3) and of iron(III) chloride (1 g of the hydrate, $FeCl_3.6H_2O$, in water, 95 cm^3 and concentrated HCl, 5 cm^3). Add 2.5 cm^3 of each solution to 250 cm^3 of water in each of four beakers and stir; this should give a solution of a suitable intensity of red colour.

Keep one of the beakers for comparison of colour, adding 50 cm^3 of water. To the others, add:

To beaker (a), 50 cm^3 of the iron(III) chloride solution.
To beaker (b), 50 cm^3 of the ammonium thiocyanate solution.
To beaker (c), 50 cm^3 of water and 20 g of ammonium chloride and stir.

The ammonium chloride will quickly dissolve. It will be found that in (a) and (b) the red colour will intensify; i.e. increased concentration of either of the reagents on the left-hand side of the equation forces the equilibrium position to the *right*; in (c) the red colour will weaken, i.e. addition of the (colourless) right-hand side reagent forces the equilibrium position to the *left*.

Le Chatelier's Principle

The principle of Le Chatelier (sometimes ascribed to Le Chatelier and Braun jointly) is universal in its application in the field of physical events; here it will be treated mainly in its chemical applications.

Le Chatelier's Principle can be stated in the following way:

> If a system is in equilibrium and one of the factors affecting the equilibrium is changed, the equilibrium will shift so as to annul, or tend to annul, the effect of the change.

The principle is so general in its application that no better can be done than illustrate its working by characteristic examples, as below.

1. *Effects of temperature change*

Suppose the reversible reaction

$$A + B \rightleftharpoons C + D \text{ with heat absorbed}$$

operates (as shown) endothermally from left to right. Suppose that a system involving A, B, C, and D is in equilibrium at a certain temperature and the temperature is then *raised.* Le Chatelier's Principle requires that the equilibrium shall move in such a way as to tend to *reduce* the temperature again; that is, to absorb heat. This requires a shift of equilibrium to the *right*, producing higher concentrations of C and D. Thus, in all cases, an *endo*thermic reaction is favoured by *rise* of temperature.

Incidentally, the rise of temperature also increases the rate of reaction. Consequently it is usual to operate an endothermic reaction at the highest possible temperature, e.g. that of the electric arc, and not to employ a catalyst. Examples of endothermic reactions promoted by high temperature are:

$$\left.\begin{array}{lll} N_2(s) + O_2(s) & \rightleftharpoons 2NO(l) & \Delta H^\ominus + 180\text{ kJ} \\ C(s) + 2S(s) & \rightleftharpoons CS_2(l) & \Delta H^\ominus + 79\text{ kJ} \\ 3O_2(g) & \rightleftharpoons 2O_3(g) & \Delta H^\ominus + 285\text{ kJ} \end{array}\right\} \text{reading from left to right}$$

Reading the reaction in the reverse direction,

$$D + C \rightleftharpoons A + B$$

with heat evolved, it is exothermic when operating from left to right. If the temperature of an equilibrium system containing A, B, C and D is *lowered*, Le Chatelier's Principle requires the equilibrium to shift so as to tend to *raise* the temperature again; that is, to evolve heat. This requires a shift of equilibrium to the *right* as the equation is written above. Thus, an *exothermic* reaction is favoured by *lowering* of temperature.

Incidentally, the fall of temperature reduces the reaction rate, which is a disadvantage to industrial production. For this reason, a *catalyst* is usually employed in industrial exothermic reactions. Then a temperature can be used which is low enough, with the help of the catalyst, to give both a reasonably favourable equilibrium and a reasonably rapid rate of attainment of the equilibrium. Important examples of this are:

Reaction		Catalyst	Temperature
$N_2(g) + 3H_2(g) \rightleftharpoons 2NH_3(g)$	$\Delta H = -92$ kJ	Iron	723 K
$2SO_2(g) + O_2(g) \rightleftharpoons 2SO_3(g)$	$\Delta H = -778$ kJ	V_2O_5 or Pt	723 K

Since the great majority of industrial reactions are exothermic, catalysts are very frequently used.

2. *Effects of pressure change*

Since liquids and solids are only very slightly compressible, the effects of pressure change are significant only in the case of gaseous reagents.

Consider a reaction

$$nA + mB \rightleftharpoons C$$

in which A, B, and C are all gaseous and C is produced with a diminution in the number of molecules, that is, a decrease of volume. Let the system containing A, B, and C be in equilibrium (at constant temperature), and then let the pressure on the system be *increased*. Le Chatelier's Principle requires the equilibrium to shift so as to tend to *decrease* the pressure again. This requires the gaseous system to take up a smaller volume and so exert lower pressure; that is, the formation of C is favoured, and the equilibrium moves from left to right as the reaction is written above. In general, a reaction which proceeds with *decrease* of volume is favoured by *high* pressure. The converse of this is also true. If, however, a reaction proceeds with no volume change, the associated equilibria are independent of pressure.

An important industrial case, in which very high pressure (at least 200 atmospheres) is employed, is the synthesis of ammonia (Haber's process).

$$\underset{1\ \text{vol.}}{N_2(g)} + \underset{3\ \text{vol.}}{3H_2(g)} \rightleftharpoons \underset{2\ \text{vol.}}{2NH_3(g)} \quad \text{(at constant temp. and pressure)}$$

Sulphur dioxide is converted to trioxide with diminution of volume:

$$\underset{2\ \text{vol.}}{2SO_2(g)} + \underset{1\ \text{vol.}}{O_2(g)} \rightleftharpoons \underset{2\ \text{vol.}}{2SO_3(g)} \quad \text{(at constant temp. and pressure)}$$

This oxidation is favoured by high pressure, but, in practice, the yield at atmospheric pressure is such that higher pressure is not necessary.

Absorptions of gas to form a solid product are favoured by high pressure, e.g.

$$\begin{array}{lccc} & NaOH(s) + & CO(g) & \rightarrow HCO_2Na(s) \\ \text{Vols.} & \text{negligible} & 22.4\ \text{dm}^3 & \text{negligible} \quad \text{(at s.t.p.)} \end{array}$$

Polymerizations of gases are favoured by high pressure because the polymerization reduces the number of molecules and, further, the polymer may occupy relatively negligible volume as liquid or solid, e.g.

$$\underset{\text{ethyne}}{3C_2H_2(g)} \rightarrow \underset{\text{benzene}}{C_6H_6(l)}$$

$$\underset{\text{ethene}}{nC_2H_4(g)} \rightarrow \underset{\text{polyethene}}{(C_2H_4)_n(s)}$$

Reactions in which there is no volume change, and in which the equilibria are independent of pressure, include

$$N_2(g) + O_2(g) \rightleftharpoons 2NO(g)$$
$$H_2(g) + I_2(g) \rightleftharpoons 2HI(g)$$

3. *Changes of concentration*

Consider a reversible homogeneous reaction:

$$A + B \rightleftharpoons C + D$$

Suppose the system is in equilibrium with A, B, C, and D all present. Then let the concentration of A be increased by addition of A from outside the system. Le Chatelier's Principle requires the equilibrium to shift so that the concentration of A is reduced; i.e. the reaction will move from left to right, reducing the concentration of B and increasing that of C and D.

Corresponding effects are produced by an increase in the concentration of B; increase of concentration of C or D favours the reverse reaction by similar reasoning. Examples of these principles in operation were provided by the bismuth oxychloride and iron(III) thiocyanate reactions earlier in this chapter.

Le Chatelier's Principle and solubility

Most solids which dissolve in water do so with *absorption* of heat. If an aqueous solution is in equilibrium with the corresponding solid, and the temperature is then *raised*, Le Chatelier's Principle requires that the equilibrium must shift so as to tend to *lower* the temperature, i.e. to absorb heat. To do this, more solid must dissolve. This explains why most solids show increased solubility with rise of temperature.

Gases, however, usually dissolve in water with *evolution* of heat and, by the reverse of the above reasoning, should become *less* soluble with rise of temperature. This, in fact, they do. If a gas is in equilibrium with its aqueous solution (at constant temperature) and the pressure is then raised, Le Chatelier's Principle requires the equilibrium to shift so as to tend to reduce the pressure. To do this, the gas must tend to occupy a smaller volume, i.e. to dissolve further in the water. This, in fact, it does, the mass of gas dissolved in a given mass of water at constant temperature being directly proportional to the pressure (Henry's Law, see page 165).

Le Chatelier's Principle and the Equilibrium Law

Le Chatelier's Principle can be used to make qualitative predictions about the change in position of an equilibrium when any of the conditions temperature, pressure, or concentration are varied. The Equilibrium Law makes both qualitative and quantitative predictions about the effect of changing either the pressures or the concentrations but it cannot be used alone to predict the effect of changes in temperature. The sign of the enthalpy change, together with Le Chatelier's Principle, will give some qualitative indication of the effect of temperature but quantitative predictions of the effect of changes in temperature on an equilibrium constant and hence on the position of equilibrium can only be made by using the relationship between the free energy change and the equilibrium constant:

$$\Delta G = -RT \ln K$$

Again the free energy change, ΔG, (see page 193) is seen to be of more fundamental importance than the enthalpy change.

It is important to realize that all predictions from Le Chatelier's Principle, the Equilibrium Law, and free energy changes for the complete reaction refer to the position of equilibrium and not to the rate of attainment of equilibrium.

15. Chemical Equilibria II: Ionic Equilibria

The **Dilution Law of Ostwald** is an attempt to apply the Equilibrium Law to ionic solutions and so to obtain a mathematical expression that will state accurately the behaviour of a given electrolyte in solutions of varying concentration at constant temperature.

Consider ethanoic acid as a typical weak, binary electrolyte (i.e. an electrolyte ionizing only slightly and into two ions per molecule). The ionic situation is

	$CH_3CO_2H(aq)$	$\rightleftharpoons CH_3CO_2^-(aq)$	$+ \; H^+(aq)$
Originally, before ionization:	1 mole	—	—
At equilibrium in V dm³ of solution:	$(1 - \alpha)$ mole	α	α mole

where α is the degree of ionization of the acid.

By the Equilibrium Law at constant temperature.

$$K_a = \frac{[CH_3CO_2^-][H^+]}{[CH_3CO\ H]}$$

Substituting

$$K_a = \frac{\alpha^2/V^2}{(1 - \alpha)/V}$$

$$= \frac{\alpha^2}{(1 - \alpha)V}$$

where K_a is the *dissociation constant* of the electrolyte.

(By convention, the concentrations of the materials on the right-hand side of the ionic equation are put into the *numerator* of the final fraction. V is the volume of solution in dm³ which contains 1 mole of the original electrolyte.)

This final fraction is the mathematical expression of the Dilution Law and, so far, rests on pure theory. To test it, the degree of ionization of ethanoic acid must be determined over a wide range of dilutions; the results must be applied to the left-hand side of the Dilution Law fraction above to see if they do, in fact, give a constant value as the Dilution Law requires. Experimental results obtained for ethanoic acid are shown in the table below.

The degree of ionization is determined by electrical conductance. It will be observed that the final column shows reasonably constant figures after V has reached a value of about 2 dm^3 and the solution is 0.5 M, or less, in concentration. This general position is found to apply for all weak electrolytes, i.e. the Dilution Law is a satisfactory expression of the ionization behaviour of weak electrolytes in dilute solution. K is used to represent the dissociation constant of weak electrolytes in general, whereas the use of K_a is confined to the dissociation of weak acids.

At 298 K throughout

Vol. V (dm³), containing 1 mole of ethanoic acid	*Degree of ionization* α	$\frac{\alpha^2}{(1 - \alpha)V}$
1.977	0.00570	1.65×10^{-5}
5.374	0.00981	1.81×10^{-5}
10.753	0.0138	1.80×10^{-5}
24.875	0.0216	1.92×10^{-5}
63.26	0.0336	1.85×10^{-5}

For *very* weak electrolytes, i.e. those with very little ionization, an approximate form of the Dilution Law can be used. In this case, α is so small that $(1 - \alpha)$ does not differ appreciably from unity and can be taken as unity with no significant error. So, for very weak electrolytes, the expression

$$\frac{\alpha^2}{(1 - \alpha)V} = K$$

can be written

$$\frac{\alpha^2}{V} = K$$

and this is the Dilution Law in approximate form.

From this form, it follows that $\alpha^2 = KV$, or $\alpha = \sqrt{K}\sqrt{V}$. Since the square root of K is a constant, it follows that, for *very* weak electrolytes the degree of ionization is directly proportional to the square root of the volume of solution containing 1 mole of the electrolyte. If preferred, the equation for the dissociation of weak electrolytes may be determined using concentrations rather than dilutions. For example if the initial concentration of ethanoic acid is c (mol dm^{-3}) and the degree of dissociation is α,

$$CH_3CO_2H(aq) \rightarrow CH_3CO_2^-(aq) + H^+(aq)$$

Before ionization:	c	0	0	mol dm^{-3}
At equilibrium:	$c(1-\alpha)$	$c\alpha$	$c\alpha$	mol dm^{-3}

Substituting in the Equilibrium Law,

$$K_a = \frac{c^2\alpha^2}{c(1-\alpha)}$$

$$= \frac{c\alpha^2}{(1-\alpha)}$$

The following example shows a characteristic use of the Dilution Law.

Example Ethanoic acid has a degree of ionization of 1.4 per cent in decimolar solution at 298 K. Calculate its degree of ionization in one-hundredth molar solution at the same temperature.

$$CH_3CO_2H(aq) \rightleftharpoons CH_3CO_2^-(aq) + H^+(aq)$$

At equilibrium:	$(1-\alpha)$	α	α mole

If the volume containing 1 mole of acid is V dm^3,

$$\frac{\alpha^2}{(1-\alpha)V} = K_a$$

Applying the data for decimolar solution,

$$K_a = \frac{(0.014)^2}{(1-0.014)\times 10} = 0.000\,019\,9$$

$$= 1.99 \times 10^{-5}\ \text{mol dm}^{-3}$$

If α_1 is the degree of ionization of ethanoic acid in one-hundredth molar solution,

$$\frac{\alpha_1^2}{(1-\alpha_1)\cdot 100} = 0.000\,019\,9$$

Rearranging this expression, we have

$$\alpha_1^2 + 0.001\,99\alpha_1 - 0.001\,99 = 0$$

$$\alpha_1 = 0.0487$$

That is, the degree of ionization of ethanoic acid in 0.01 M solution at 298 K is 0.0487 or 4.87 per cent.

Using the approximate form of the Dilution Law,

$$\frac{\alpha}{\alpha_1} = \frac{\sqrt{10}}{\sqrt{100}}$$

$$\begin{aligned}\alpha_1 &= \alpha\sqrt{10} \\ &= 0.014 \times 3.16 \\ &= 0.044.\end{aligned}$$

We have seen that the Dilution Law expresses quite well the behaviour of weak electrolytes in dilute solution. It breaks down completely, however, when applied to strong electrolytes. The case of potassium chloride illustrates this fact.

All measurements at 291 K

Vol. (dm^3) containing 1 mole of KCl	*Degree of ionization apparent*	$\frac{\alpha^2}{(1-\alpha)V}$
1	0.757	2.35
5	0.831	0.815
50	0.923	0.222
200	0.958	0.108
1000	0.980	0.0485

It will then be seen that the required 'constant' of the third column varies progressively from 2.35 to less than 0.05, falling to about one-fiftieth of the initial value. Several attempts have been made, notably by Debye and Hückel, to produce alternative equations for the behaviour of strong electrolytes, but they lie outside the scope of this book.

Acidity, Alkalinity, and Neutralization

The ionic state of water

Water has a solvent action for a very wide variety of materials and is, in consequence, difficult to purify. Water is purified by running it through columns containing cation and anion exchange resins. During this process the cations in the water are replaced by hydrogen ions and the anions by hydroxide ions. The H^+ ions and the OH^- ions then combine to form water.

The progress of the purification can be followed by measurement of the electrical conductance of the water. Starting at a comparatively high level (caused by impurity), it will be found to fall gradually to a very low value, which cannot be reduced by further efforts at purification. This constant electrical conductance is believed to be genuinely that of water itself and to be the result of an ionization which is usually represented as

$$H_2O(l) \rightleftharpoons H^+(aq) + OH^-(aq)$$

though the hydrogen ion is actually hydrated (as the hydroxonium ion H_3O^+; see also page 75) and the ionization is more properly written as

$$2H_2O(l) \rightleftharpoons H_3O^+(aq) + OH^-(aq)$$

Except for special reasons, the simpler equation will generally be used. Applying the Equilibrium Law,

$$K = \frac{[H^+]\,[OH^-]}{[H_2O]}$$

Due to the ionization being very slight, the equilibrium concentration of the water does not differ appreciably from the original concentration and so it is regarded as being a constant and is incorporated in the equilibrium constant. The expression now becomes

$$K_w = [H^+]\,[OH^-]$$

The very small conductance of pure water (specific conductance 3.84×10^{-8} ohm^{-1} at 294 K) can be shown to correspond to the following concentrations of hydrogen ion H^+, and hydroxide ion, OH^-, in the water:

$$[H^+] = [OH^-] = 1 \times 10^{-7} \text{ at } 298 \text{ K}$$

(The square brackets indicate concentrations in mol dm^{-3}, the usual convention.) From these figures, the product of the hydrogen and hydroxide ion concentrations is given by

$$\begin{aligned} \text{at 298 K } [H^+][OH^-] &= (1 \times 10^{-7}) \times (1 \times 10^{-7}) \\ &= 1 \times 10^{-14} \text{ mol}^2\text{ dm}^{-6} \\ &= K_w \end{aligned}$$

This quantity, K_w of the value 1×10^{-14}, is called the *ionic product* of water and is a very important constant. It is always maintained in an aqueous liquid at 298 K. In pure water, it is maintained by *equal* concentrations of H^+ and OH^- ions, both being 1×10^{-7} mol dm^{-3} and this situation defines a *neutral* solution. That is, a neutral aqueous solution is one in which the concentrations of hydrogen ion, H^+, and hydroxide ion, OH^-, are equal at the value of 1×10^{-7} mol dm^{-3} at 298 K.

At any other temperature K_w will have a different value and the H^+ and OH^- concentrations of a neutral solution will be different. In fact, as the reaction between hydrogen ions and hydroxide ions is exothermic, a rise in temperature will cause an increase in the concentration of the hydrogen ions in a neutral solution.

However, at 298 K the value of K_w has to be maintained at 1×10^{-14}, so that any variation in the concentration of one of the ions must be compensated for by a change in the concentration of the other ion. For example, suppose pure water is acidified by hydrogen chloride till the concentration of HCl is 10^{-2} mol dm^{-3}. At this concentration it can be considered as fully ionized, so that $[H^+] = 10^{-2}$. This H^+ participates in the equilibrium of water:

$$[H^+][OH^-] = K_w = 1 \times 10^{-14} \text{ mol dm}^{-3}$$

Consequently $[OH^-]$ falls to 10^{-12} mol dm^{-3}. The product K_w is still maintained, but by *unequal* concentrations of H^+ and OH^- ion. The preponderance of H^+ is characteristic of an *acidic* solution. Similarly, if pure water is made one-thousandth molar in potassium hydroxide $[OH^-] = 10^{-3}$, K_w (1×10^{-14}) is maintained by a fall of H^+ concentration to 10^{-11} mol dm^{-3}. The preponderance of OH^- is characteristic of an *alkaline* solution.

It is important to notice that, however acidic an aqueous solution may be, it still retains enough OH^- to satisfy the requirement $K_w = [H^+][OH^-] = 1 \times 10^{-14}$. Similarly, however alkaline an aqueous solution may be, it still retains enough H^+ to satisfy this requirement.

The nature of neutralization

This is best illustrated by an example. Suppose 0.2 M hydrochloric acid and 0.2 M sodium hydroxide solution are mixed, perfectly and instantaneously, in equal volumes so that both become 0.1 M. At the moment of mixing, assuming complete ionization of the two strong electrolytes, $[H^+] = [OH^-] = 10^{-1}$. That is, for the moment, the product of $[H^+][OH^-] = 10^{-2}$. This is very much in excess of the permissible value of K_w, the ionic product of water, which is 1×10^{-14}. Consequently, OH^- and H^+ are at once withdrawn from the solution as unionized molecules of water until the value of K_w is again attained. This process of readjusting the disturbed ionic product of water is the process of neutralization.

$$H^+(aq) + OH^-(aq) \rightleftharpoons H_2O(l)$$

The reaction is reversible but lies so far to the right at equilibrium that for all practical measurements neutralization is quantitatively complete; this is the assumption made in acid-alkali titrations in the laboratory.

The hydrogen ion index, pH

The hydrogen ion index is a device which has been adopted for convenience in stating the nature of a solution with respect to acidity. It is defined in the following way:

> If the concentration of hydrogen ion, H^+, in a solution is 10^{-x} mol dm^{-3}, then the pH of the solution is x.

i.e.

$$pH = -\lg [H^+] \text{ (or } -\lg_{10} [H^+])$$

This leads to the following conclusions.

Neutrality Pure water shows the situation (at 298 K),

$$[H^+] = 10^{-7} \quad (\text{Also } [OH^-] = 10^{-7})$$

From this, pH in pure water is 7. That is, *a neutral aqueous solution at 298 K has a pH of* 7.

Acidity Consider, say, 0.01 M HCl. The acid may be taken as fully ionized, i.e. $[H^+] = 10^{-2}$. From this, the pH of the solution is 2. Similarly, all other *acidic solutions have a pH less than* 7.

Alkalinity Since alkalinity is associated with the ion, OH^-, it might be thought that it would be indicated by a hydroxide ion index, pOH. However, we have seen that an aqueous liquid always shows the relation (at 298 K):

$$K_w = [H^+][OH^-] = 1 \times 10^{-14} \text{ mol}^2 \text{ dm}^{-6}$$

From this, it follows that pH + pOH = 14; consequently, pOH is always related to pH in the same solution by the constant relation

$$pOH = 14 - pH$$

For this reason, it is quite convenient to state both acidity and alkalinity in terms of the one index, pH. Consider 0.001 M potassium hydroxide solution. If the alkali is fully ionized, $[OH^-] = 10^{-3}$. That is, pOH is 3 and pH is (14 − 3) or 11 for the solution. From this, we see that *an alkaline solution has a pH greater than* 7.

pH of aqueous liquid	*Nature of liquid*
less than 7	acidic
7	neutral
greater than 7	alkaline

It is very important to be quite clear about the following position. Since the value of pH is a logarithmic index with the negative sign omitted, an *increase* in hydrogen ion concentration (i.e. in acidity) is accompanied by a *decrease* in the pH of the solution. For example, an increase in hydrogen ion concentration from 10^{-5} to 10^{-2} mol dm^{-3} means a decrease in pH from 5 to 2. Consequently, low values of pH indicate high acidity. Conversely, high values of pH mean low acidity and values of pH greater than 7 indicate alkaline solution, increasing in alkalinity as the pH rises.

The value pH = 0, for example, indicates a solution with a concentration of 10° mole of H^+ dm^{-3}. Since $10^{\circ} = 1$, such a solution is a molar acid with the acid fully ionized. Correspondingly, pH = 14 indicates a solution in which pOH = 0, i.e. a molar solution of a fully ionized strong alkali.

The following examples illustrate the calculation of the pH of a solution.

Example 1 Calculate the pH of a decimolar solution of ethanoic acid, in which the acid is 1.4 per cent ionized.

If the decimolar acid is fully ionized, the H^+ concentration is 10^{-1} mol dm^{-3}. But as the degree of ionization is 0.014, the actual H^+ concentration is $10^{-1} \times 0.014 = 0.0014$ mol dm^{-3}, i.e.

$$
\begin{aligned}
[H^+] &= 0.0014 \\
&= 10^{3.146} \qquad (\lg 1.4 = 0.146) \\
&= 10^{-3+0.146} \\
&= 10^{-2.854}
\end{aligned}
$$

i.e. $$\text{pH} = 2.85 \text{ (to 3 sig. figs.)}$$

(Notice that the logarithmic index has to be made completely negative by manipulation of the negative characteristic and the positive mantissa of the logarithm.)

Example 2 Calculate the pH of a hundredth-molar solution of ammonium hydroxide, in which the degree of ionization of the electrolyte is 0.043.

If the alkali is fully ionized, the OH^- ion concentration is 10^{-2} mol dm^{-3}. But, as the degree of ionization is 0.043, the actual OH^- concentration is $10^{-2} \times 0.043 = 0.00043$ mol dm^{-3}, i.e.

$$
\begin{aligned}
[OH^-] &= 0.00043 \\
&= 10^{4.634} \qquad (\lg 4.3 = 0.634) \\
&= 10^{-4+0.634} \\
&= 10^{-3.366}
\end{aligned}
$$

From this,

$$
\begin{aligned}
\text{pOH} &= 3.37 \quad \text{(to 3 sig. figs.)} \\
\text{pH} &= 14 - 3.37 = 10.63
\end{aligned}
$$

Acids and Bases

Development of the idea of acidity

Some vague notion of the more obvious properties of acids has probably existed for a very long time because of their occurrence in sour and unripe fruits and the production of ethanoic acid by the souring of wine (the oxidation of ethanol). At any rate, by the middle of the seventeenth century the properties of acids were recognized as including a sour taste, corrosive action, ability to change the colours of certain vegetable products (i.e. indicators, such as litmus), ability to precipitate sulphur from potassium sulphide, and ability to react with alkalis to produce neutral substances. Alkalis were recognized as possessing detergent properties, the ability to dissolve fats and oils (saponification) and also sulphur, to reverse the action on indicators (above) and 'neutralize' acids. Later, the ability of certain (strong) acids to dissolve metals with liberation of hydrogen was recognized.

Towards the end of the eighteenth century, Lavoisier produced the *oxygen theory of acids*. This was an unjustified generalization from the fact that many acids can be produced by the combination of non-metallic oxides with water, e.g. H_2SO_4, H_2SO_3, HPO_3, H_3PO_4, H_3AsO_3, and so contain oxygen. Lavoisier extended this experience wrongly to produce the theory that oxygen is the essential acid-producing element. The name *oxygen* is, in fact, from *oxus* (sour) and *gennao* (to produce). Davy's proof that chlorine is an element, containing no oxygen, and that hydrochloric acid is produced by the combination of hydrogen with chlorine finally disproved the oxygen theory, though Berthollet had earlier recognized the absence of oxygen in the acids HCN and H_2S.

In the early nineteenth century, it was gradually recognized that the element essential to an acid is *hydrogen*, but that the hydrogen must be of a particular type, that is, hydrogen which is capable of being replaced by a metal (with formation of a salt), e.g.

$$Zn + H_2SO_4 \rightarrow ZnSO_4 + H_2 \text{ (direct replacement)}$$

$$CuO + H_2SO_4 \rightarrow CuSO_4 + H_2O \text{ (indirect replacement)}$$

At this time, an acid could have been roughly defined as a compound which turns blue litmus to red and contains hydrogen which can be replaced, directly or indirectly, by a metal.

With the introduction of ionic ideas (about 1880) *replaceable hydrogen* was recognized as hydrogen which is capable of ionizing, so that an acid could be defined as a compound which, in water, yields hydrogen ions, e.g.

$$HCl \rightleftharpoons H^+ + Cl^-$$
$$H_2SO_4 \rightleftharpoons 2H^+ + SO_4^{2-}$$

It is now recognized that the simple hydrogen ion (or proton) does not exist in aqueous solution but is solvated to give the *hydroxonium ion*, H_3O^+. Subject to the discussion which follows, we may say that the ability to produce this ion in water is recognized as the essential property of an acid, e.g.

$$HCl + H_2O \rightleftharpoons H_3O^+ + Cl^-$$
$$H_2SO_4 + 2H_2O \rightleftharpoons 2H_3O^+ + SO_4^{2-}$$
$$CH_3CO_2H + H_2O \rightleftharpoons H_3O^+ + CH_3CO_2^-$$

Essential nature of acid and base

From the ionic point of view an acid can be regarded as a substance which has a tendency to lose one or more protons (i.e. hydrogen ions) per molecule. Conversely, a base can be regarded as a substance which has a tendency to gain one or more protons per molecule or ion. This definition of acids and bases was first put forward by Brønsted and Lowry in 1922. The relation can be expressed in the equations

$$\text{acid} \rightleftharpoons H^+ + \text{base}$$
$$H^+ + \text{base} \rightleftharpoons \text{acid}$$

That is, for every acid, there exists a base, which is produced when the acid loses a proton (or H^+). The acid and base which stand in this relation to one another are said to be *conjugates*. (For examples, see below.) An acid which loses a proton readily from its molecule is said to be a *strong* acid; an acid which loses a proton with difficulty from its molecule is said to be a *weak* acid. It is obvious that a *strong* acid must have a *weak* conjugate base, and *vice versa*; for example:

Strong acid *Weak base*

$$HCl \rightleftharpoons H^+ + Cl^-$$
$$H_2SO_4 \rightleftharpoons 2H^+ + SO_4^{2-}$$
$$HNO_3 \rightleftharpoons H^+ + NO_3^-$$

Weak acid *Strong base*

$$CH_3CO_2H \rightleftharpoons H^+ + CH_3CO_2^-$$
$$H_2O \rightleftharpoons H^+ + OH^-$$
$$H_2C_2O_4 \rightleftharpoons 2H^+ + C_2O_4^{2-}$$

In practice, however, the free proton never exists in solution but is always solvated. When water is the solvent, the H^+ ion (or proton) is hydrated to H_3O^+, which is known as the *hydroxonium ion*. The water molecule which hydrates the proton in this way is acting as a proton acceptor, i.e. as a *base*, so the essential relation can be expressed in equations such as

$$HCl + H_2O \rightleftharpoons H_3O^+ + Cl^- \quad (1)$$
$$CH_3CO_2H + H_2O \rightleftharpoons H_3O^+ + CH_3CO_2^- \quad (2)$$
$$\text{acid}_1 + \text{base}_2 \rightleftharpoons \text{acid}_2 + \text{base}_1$$

In equation (1), HCl is a very strong acid, losing a proton with ease; the conjugate base, Cl^-, is very weak. In dilute solution, the equilibrium lies so far to the right, that for practical purposes ionization is complete.

In equation (2), ethanoic acid is weak, losing a proton with difficulty; the conjugate base, $CH_3CO_2^-$, is strong. In decimolar solution, the equilibrium lies so much to the left that only fourteen molecules of ethanoic acid per thousand are ionized. The ionization increases with dilution (according to the Dilution Law) so that with dilution the acid becomes stronger.

This raises the point that the terms *strong* and *concentrated* are now used, in connection with acids and bases, with quite distinct meanings.

> A **strong** acid is one which very readily loses a proton from its molecule and so (in dilute solution at least) tends to be highly ionized.
>
> A **concentrated** acid is one in which the proportion of acid to water is very high.
>
> The opposite of a strong acid is a **weak** acid, and this term is applied to an acid which loses a proton with difficulty and which, at any significant concentration, is only slightly ionized.
>
> The opposite of a concentrated acid is a **dilute** acid, and the term is applied to a liquid in which the proportion of acid to water is low.

An example of a *concentrated, weak* acid is glacial ethanoic acid. The percentage of water in it is almost nil and the acid loses a proton with difficulty. An example of a *dilute, strong* acid is bench hydrochloric acid. The proportion of hydrochloric acid is low (about 7 per cent) and the acid loses a proton very readily. These terms are used correspondingly with bases.

Water in relation to acidic and basic behaviour

Water can show *basic* behaviour by accepting a proton and does this when acids ionize in aqueous solution, e.g.

$$HCl + H_2O \rightleftharpoons H_3O^+ + Cl^-$$

The molecule H_2O accepts a proton to become the hydroxonium ion, H_3O^+.

Water can also show *acidic* behaviour by losing a proton (H^+) to produce the base, OH^-. This occurs, for example, when ammonia gas reacts with water producing an alkaline solution.

$$NH_3 + H_2O \rightleftharpoons NH_4^+ + OH^-$$

Water supplies a proton to the NH_3 molecule to produce the ion, NH_4^+.

Having proton-accepting (*protophilic*) and proton-donating (*protogenic*) properties, water is called an *amphiprotic* solvent.

It should be noted that the hydroxyl bases must now be considered as only one group of bases among many, all of which show, in varying degrees, the property of combining with protons. Some of these bases were shown a little earlier and included materials such as Cl^-, SO_4^{2-}, NO_3^-, $C_2O_4^{2-}$ and $CH_3CO_2^-$ as well as OH^-. From the modern point of view, the principal distinguishing feature about the hydroxyl base OH^- is its very great strength as a base. The equilibrium position of its proton-accepting reaction

$$OH^- + H^+ \rightleftharpoons H_2O \quad (\text{or } OH^- + H_3O^+ \rightleftharpoons 2H_2O)$$

lies so far to the right (as written above) that for practical purposes of measurement in titration it is regarded as irreversible. This reaction is the very familiar phenomenon of acid-alkali neutralization when read from left to right, but is merely one special case of its kind, i.e.

$$\text{base} + \text{proton} \rightleftharpoons \text{acid}$$

The following definitions can now be formally stated:

An **acid** is a substance which shows a tendency to lose a proton; if the tendency is marked, the acid is said to be strong; if the tendency is slight, the acid is said to be weak.

A **base** is a substance which shows a tendency to gain a proton; if the tendency is marked, the base is said to be strong; if the tendency is slight, the base is said to be weak.

An **alkali** is a substance which is soluble in water and produces the hydroxide base, OH^-, in the solution.

Neutralization is the special case of the reaction

$$\text{base} + \text{proton} \rightleftharpoons \text{acid}$$

in which the hydroxyl base, OH^-, is involved, i.e.

$$OH^-(aq) + H^+(aq) \rightleftharpoons H_2O(l)$$

It must be noted that in an alkaline solution the hydroxyl base, OH^-, is always accompanied by an equivalent concentration of a cation, such as Na^+, K^+, Ca^{2+}, or NH_4^+. When the base undergoes neutralization, combining with protons supplied by the participating acid, the conjugate base of the acid (an acidic ion such as Cl^-, SO_4^{2-}, or NO_3^-) is left in solution, together with the cation associated with the base. This aggregation of ions in the liquid constitutes the *salt*, which is normally regarded as a product of neutralization. For example,

$$Na^+ + OH^- + H^+ + Cl^- \rightarrow Na^+ + Cl^- + H_2O$$

The effect of this reaction is to substitute a cation for some (or all) of the hydrogen of an acid which is capable of being lost as protons. In this sense a salt is a by-product of neutralization, the essential of which is proton-gain by the base, OH^-. A salt can be defined in the following way:

A **salt** is a compound produced when a metallic cation is substituted for some (or all) of the hydrogen of an acid which is capable of being lost as protons. When the substitution is complete, the salt is said to be *normal*; when incomplete, the salt is said to be an *acid salt*; for example:

Acid	*Acid salts*	*Normal salts*
H_2SO_4	$Na^+HSO_4^-$	$(Na^+)_2SO_4^{2-}$
H_2CO_3	$K^+HCO_3^-$	$(K^+)_2CO_3^{2-}$
H_2SO_3	$Na^+HSO_3^-$	$(Na^+)_2SO_3^{2-}$

One important aspect of the Brønsted-Lowry view of acids and bases is that it can be applied to non-aqueous solvents. For instance, the dissolving of hydrogen chloride in liquid ammonia may be compared to its dissolving in water.

$$HCl + NH_3 \rightleftharpoons NH_4^+ + Cl^-$$
$$HCl + H_2O \rightleftharpoons H_3O^+ + Cl^-$$

In fact a solution of ammonium chloride in liquid ammonia does have acidic properties. In addition, a solution of sodium amide in liquid ammonia exhibits basic properties in a similar manner to a solution of sodium hydroxide in water.

$$NaNH_2 \rightarrow Na^+ + NH_2^-$$
$$NaOH \rightarrow Na^+ + OH^-$$

So sodium amide neutralizes ammonium chloride in liquid ammonia solution.

Comparison of the strengths of acids

It has already been mentioned that a strong acid is one which readily loses a proton and is highly ionized in dilute aqueous solution. A weak acid is one which loses a proton with difficulty and is never more than slightly ionized in any solution of a usual working concentration, say from 2 M to 0.1 M. The great majority of acids are weak, e.g. ethanoic, oxalic, tartaric, citric, and other organic acids. The common strong acids are the three mineral acids, hydrochloric, sulphuric, and nitric. Hydrobromic acid, HBr, and perchloric acid are also strong. The division between strong and weak acids is not rigid. Certain acids exist, e.g. sulphurous acid, H_2SO_3, which are moderately ionized at the usual working dilutions and cannot be classified as definitely strong or weak.

It must be remembered that no *absolute* comparison of strengths of acids can be made. The relative strengths of two given acids in aqueous solution may vary with dilution and with temperature. For example, at decimolar concentration and room temperature, hydrochloric acid is almost completely ionized, while ethanoic acid is about 1.4 per cent ionized. At 0.001 M, hydrochloric acid scarcely alters its strength at all because ionization was already almost complete, but ethanoic acid alters its ionization to about 14 per cent, so becoming considerably stronger. This means that comparisons of strength can be made only at stated dilutions. Since ionization also changes with temperature, this condition should also be specified.

As ionization of a weak acid increases with dilution, it is obvious that the acid becomes stronger as it becomes more dilute; in theory, at infinite dilution all acids become equally strong because all become fully ionized.

Methods for comparing strengths of acids

The following are some of the methods available for comparing strengths of acids:

1. *On a logarithmic scale*

The dissociation constant of a weak electrolyte is sometimes expressed on a logarithmic scale, which for an acid is denoted by $pK_a = -\lg K_a$, and for a base, $pK_b = -\lg K_b$. The advantage of this scale is that the rather cumbersome numbers are converted to relatively simple ones. For example, for ethanoic acid,

$$\begin{aligned} K_a &= 1.99 \times 10^{-5} \text{ mol dm}^{-3} \\ \lg K_a &= \bar{5}.2989 \\ &= -5 + 0.2989 \\ &= -4.7011 \\ pK_a &= -\lg K_a = 4.7 \end{aligned}$$

The pK_a values of some common weak acids are given below:

chloroethanoic acid	$pK_a = 2.9$
methanoic (formic) acid	$pK_a = 3.8$
ethanoic acid	$pK_a = 4.7$
propanoic acid	$pK_a = 4.9$
carbonic acid	
1st dissociation	$pK_a = 6.4$
2nd dissociation	$pK_a = 10.3$

A large pK_a value (which means a small K_a value) indicates a weak acid. Similar terms for bases, K_b and pK_b, may be used to give an indication of their comparative strengths.

The explanation of why a particular acid is stronger than another is related to the composition and structure of the acids concerned. It is profitable to consider two acids which are only slightly different in structure and attempt to relate this difference to a comparison of their pK_a values. Ethanoic and chloroethanoic acid only differ by the substitution of a chlorine atom for a hydrogen atom.

```
   H    O            H    O
   |   //            |   //
H—C—C           Cl—C—C
   |   \             |   \
   H    O—H          H    O—H
```

The difference in structure results in a significant difference in pK_a, which can be explained in terms of the more electronegative chlorine atom transmitting its attraction for electrons throughout the molecule and so increasing the bond polarity (see page 97) of the O—H bond and hence the ease of ionization of the hydrogen.

2. *By observations of standard enthalpy of neutralization*

Let two acids, of which the strengths are to be compared, be A and B. The enthalpy of neutralization of 1 mole of hydrogen ions from the acid A at a given dilution by 1 mole of hydroxide ions from sodium hydroxide in suitable dilution is measured (or obtained from records). Let this be x (J). Similar data is obtained from acid B. Let this be y (J).

Then the heat given out by mixing A and B in the same solution, so that they each provide 1 mole of hydrogen ions, with sufficient sodium hydroxide solution to provide 1 mole of hydroxide ions is measured. Let this be z(J). In this experiment the acids may be considered as competing for the base. If acid A neutralizes n mole of the hydroxide ions and acid B neutralizes $(1 - n)$ mole, the following relation must hold:

$$nx + (1 - n)y = z$$

From this n can be calculated. Then:

$$\frac{\text{Strength of A}}{\text{Strength of B}} = \frac{n}{1 - n}$$

This method is unsuitable for acids of roughly similar strengths because for such acids, the heat changes are usually similar.

Constant values of enthalpy of neutralization for strong acids and bases

When accurate thermal observations began to be made, it was noticed that the enthalpy of neutralization of 1 mole of H^+ ions from *strong* acid by 1 mole of OH^- ions from any *strong* base, both being in dilute solution, was almost constant at a value close to 57.3 kJ. Writing typical equations in the molecular manner:

$$NaOH(aq) + HCl(aq) \rightarrow NaCl(aq) + H_2O(l) \qquad \Delta H = -57.3 \text{ kJ}$$
$$KOH(aq) + HNO_3(aq) \rightarrow KNO_3(aq) + H_2O(l) \qquad \Delta H = -57.3 \text{ kJ}$$
$$NaOH(aq) + HBr(aq) \rightarrow NaBr(aq) + H_2O(l) \qquad \Delta H = -57.7 \text{ kJ}$$

This result was very surprising because these reactions, in molecular presentation, appear quite different. The ionic theory, however, requires this result for the following reasons. The acid and base, being 'strong', are both fully ionized (or very nearly so) in dilute solution. The salt produced is also a strong electrolyte (whatever its individual identity) and so is fully ionized. Consequently the only change occurring in all these neutralizations is the formation of 1 mole of water from its ions. This is a constant change, requiring a constant evolution of heat; as:

$$Na^+(aq) + OH^-(aq) + H^+(aq) + Cl^-(aq) \rightarrow Na^+(aq) + Cl^-(aq) + H_2O(l) \quad \Delta H = -57.3\ kJ$$

$$K^+(aq) + OH^-(aq) + H^+(aq) + NO_3^-(aq) \rightarrow K^+(aq) + NO_3^-(aq) + H_2O(l) \quad \Delta H = -57.3\ kJ$$

In both cases the real change is $H^+ + OH^- \rightarrow H_2O \quad \Delta H = -57.3$ kJ.

If, however, the acid concerned is weak, i.e. only slightly ionized, the enthalpy of neutralization differs from $\Delta H = -57.3$ kJ. The reason is that the process of neutralization takes place in two stages:

1. The completion of the ionization of the weak acid, which is accompanied by a heat change, e.g. for ethanoic acid:

$$CH_3CO_2H(aq) \rightleftharpoons CH_3CO_2^-(aq) + H^+(aq) \quad \Delta H = X\ kJ$$

2. The neutralization proper, as with sodium hydroxide:

$$Na^+(aq) + OH^-(aq) + H^+(aq) + CH_3CO_2^-(aq) \rightarrow CH_3\ CO_2^-(aq) + Na^+(aq) + H_2O(l) \quad \Delta H = -57.3\ kJ$$

The net heat change is, therefore, $(-57.3 + X)$ J. For sodium hydroxide + ethanoic acid, the experimental figure is -56.1 kJ. That is

$$-57.3 + X = -56.1$$

so that

$$X = +\ 1.2\ kJ$$

Similar considerations apply for a weak base.

Salt Hydrolysis

Salt hydrolysis is essentially the reversal of neutralization. Consider the reaction between an acid, HA, and a base, BOH, neither being strong. The equilibrium between them can be written

$$\underbrace{H^+ + A^-}_{\substack{\rightleftharpoons \\ \text{HA} \\ \text{acid}}} + \underbrace{B^+ + OH^-}_{\substack{\rightleftharpoons \\ \text{BOH} \\ \text{base}}} \rightleftharpoons \underbrace{B^+ \ + \ A^-}_{\text{salt}} + \underbrace{H_2O}_{\substack{\rightleftharpoons \\ H^+ + OH^- \\ \text{water}}}$$

Read from left to right, this process is *neutralization*; read from right to left, it is *salt hydrolysis*. The following discussion will show that salt hydrolysis is appreciable except when the salt is derived from a strong acid and a strong base. There are four possible cases as below. One example of each will be given in detail and others in outline only.

1. *Salt of a weak acid and a strong base*

(*a*) *Sodium ethanoate* This salt is derived from the weak acid, ethanoic, and the strong base, sodium hydroxide. In sodium ethanoate solution, the equilibria set up are

$$\begin{array}{ccccc} & & CH_3CO_2^-(aq) & & Na^+(aq) \\ & & + & & \\ H_2O(l) & \rightleftharpoons & H^+(aq) & + & OH^-(aq) \\ & & \downarrow\uparrow & & \\ & & CH_3CO_2H(aq) & & \end{array}$$

The salt (a strong electrolyte) can be considered fully ionized. Since sodium hydroxide is a strong base, it can be considered as remaining completely ionized, with no formation of 'molecules' of NaOH. Ethanoic acid, however, is weak. As shown above, it must form some unionized molecules, using H^+ derived from water. This disturbs the equilibrium of water ($[H^+][OH^-] = K_w = 1 \times 10^{-14}$ at 298 K). To restore K_w, more water ionizes. This puts OH^- into excess, causes the pH of the solution to rise above 7 and the solution to react alkaline.

This is a general situation.

A solution of a salt derived from a weak acid and a strong base always contains unionized molecules of the acid and reacts alkaline.

Further examples of this are given below and can be fully argued on the same lines as for sodium ethanoate.

(b) *Sodium carbonate*

$$\begin{array}{ccccc} & & 2Na^+(aq) & & CO_3^{2-}(aq) \\ & & & & + \\ 2H_2O(l) & \rightleftharpoons & 2OH^-(aq) & + & 2H^+(aq) \\ & & & & \downarrow\uparrow \\ & & & & H_2CO_3(aq) \end{array}$$

(c) *Potassium cyanide*

$$\begin{array}{ccccc} & & K^+(aq) & & CN^-(aq) \\ & & & & + \\ H_2O(l) & \rightleftharpoons & OH^-(aq) & + & H^+(aq) \\ & & & & \downarrow\uparrow \\ & & & & HCN(aq) \\ & & & & \downarrow\uparrow \; (A) \\ & & & & HCN(g) \end{array}$$

The general situation here is similar to that of sodium ethanoate solution; there is, however, the additional factor that hydrogen cyanide, HCN, is extremely volatile. Consequently, if the solution is boiled, the vapour of hydrogen cyanide is expelled into the air. The equilibrium (A) tends to be displaced continually downwards. This causes further hydrolysis to replace the hydrogen cyanide in the solution, which is again expelled. This continues until eventually only K^+ and OH^- ions remain (with the trace of H^+ to maintain K_w); that is, the final solution is one of potassium hydroxide. Corresponding reactions occur with sodium cyanide.

2. *Salt of a strong acid and a weak base*

(a) *Ammonium chloride* The salt is derived from the strong acid, hydrochloric, and the weak base, ammonium hydroxide. In ammonium chloride solution, the situation is

$$\begin{array}{ccccc} & & NH_4^+(aq) & & Cl^-(aq) \\ & & + & & \\ H_2O(l) & \rightleftharpoons & OH^-(aq) & + & H^+(aq) \\ & & \downarrow\uparrow & & \\ & & NH_4OH(aq) & & \end{array}$$

Like all salts, ammonium chloride is a strong electrolyte and can be taken as fully ionized. Hydrochloric acid is a strong acid, so there is no formation of HCl molecules. Ammonium hydroxide, however, is a weak base so that molecules of NH_4OH are formed, as shown, by use of the OH^- ion from water. Withdrawal of OH^- ion disturbs the ionic equilibrium of water. ($[H^+][OH^-] = K_w = 1 \times 10^{-14}$ at 298 K) and, to restore K_w, water ionizes further. This puts H^+ into excess, the pH of the solution falls below 7 and it reacts acidic.

This is a general situation.

A solution of a salt from a strong acid and a weak base always contains unionized molecules of the base and reacts acidic.

Further examples are stated in outline below and can be fully argued as in the case of ammonium chloride.

(*b*) *Ammonium sulphate*

$$\begin{array}{ccccc} & & 2NH_4^+(aq) & & SO_4^{2-}(aq) \\ & & + & & \\ 2H_2O(l) & \rightleftharpoons & 2OH^-(aq) & + & 2H^+(aq) \\ & & \downarrow\uparrow & & \\ & & 2NH_4OH(aq) & & \end{array}$$

Salts of organic bases, such as aniline, behave like ammonium salts. Taking aniline hydrochloride as the example, the situation is

$$\begin{array}{ccccc} & & C_6H_5NH_3^+(aq) & & Cl^-(aq) \\ & & + & & \\ H_2O(l) & \rightleftharpoons & OH^-(aq) & + & H^+(aq) \\ & & \downarrow\uparrow & & \\ & & C_6H_5NH_3OH(aq) & & \end{array}$$

(*c*) *Iron (III) chloride* (and *aluminium chloride*)

$$\begin{array}{ccccc} & & Fe^{3+}(aq) & & 3Cl^-(aq) \\ & & + & & \\ 3H_2O(l) & \rightleftharpoons & 3OH^-(aq) & + & 3H^+(aq) \\ & & \downarrow\uparrow & & \\ & & Fe(OH)_3(aq) & & \end{array}$$

The case of aluminium chloride is similar, substituting Al for Fe.
In these cases, hydrolysis may be so marked that the solution will dissolve magnesium with liberation of hydrogen.

$$Mg(s) + 2H^+(aq) \rightarrow Mg^{2+}(aq) + H_2(g)$$

(*d*) *Copper*(II) *sulphate*

$$\begin{array}{ccccc} & & Cu^{2+}(aq) & & SO_4^{2-}(aq) \\ & & + & & \\ 2H_2O(l) & \rightleftharpoons & 2OH^-(aq) & + & 2H^+(aq) \\ & & \downarrow\uparrow & & \\ & & Cu(OH)_2(aq) & & \end{array}$$

The above explanation of the hydrolysis of iron(III) chloride, aluminium chloride, and copper(II) sulphate is consistent with that used for ammonium salts but it is somewhat simplified as it ignores the hydration of the ions. The cations in these solutions are heavily hydrated and exist as $[Fe(H_2O)_6]^{3+}$, $[Al(H_2O)_6]^{3+}$, and $[Cu(H_2O)_6]^{2+}$. It is the reaction of these hydrated ions with water which causes their solutions to be acidic. In the case of iron(III) chloride the following sequence of reactions occurs, culminating in the precipitation of colloidal hydrated iron(III) hydroxide.

$$\begin{array}{llll} Fe(H_2O)_6^{3+} & + H_2O \rightleftharpoons & Fe(H_2O)_5OH^{2+} & + H_3O^+ \\ Fe(H_2O)_5OH^{2+} & + H_2O \rightleftharpoons & Fe(H_2O)_4(OH)_2^+ & + H_3O^+ \\ Fe(H_2O)_4(OH)_2^+ & + H_2O \rightleftharpoons & Fe(H_2O)_3(OH)_3 & + H_3O^+ \end{array}$$

Similar equilibria are considered to be set up in aqueous solutions of aluminium chloride and copper(II) sulphate.

3. *Salt of a weak acid and a weak base*

(a) *Ammonium ethanoate* This salt is derived from the weak base, ammonium hydroxide, and the weak acid, ethanoic acid. The salt is a strong electrolyte and can be taken as fully ionized.

$$\begin{array}{ccccc} & & NH_4^+(aq) & & CH_3CO_2^-(aq) \\ & & + & & + \\ H_2O(l) & \rightleftharpoons & OH^-(aq) & + & H^+(aq) \\ & & \downarrow\uparrow & & \downarrow\uparrow \\ & & NH_4OH(aq) & & CH_3CO_2H(aq) \end{array}$$

Since the acid and base are both weak, unionized molecules of both must be formed as shown, by utilizing the ions H^+ and OH^- derived from water. The removal of these ions disturbs the ionic product of water ($[H^+][OH^-] = K_w = 1 \times 10^{-14}$ at 298 K. To restore K_w, water ionizes further. In this particular case, the acid and base are about equally weak (K is about 2×10^{-5} for both) so that the concentrations of H^+ and OH^- remain about equal in the solution in spite of hydrolysis. So ammonium ethanoate is strongly hydrolysed in solution, but the solution remains almost neutral.

In general, if the acid from which the salt is derived is stronger than the base, the solution tends to react acidic, and *vice versa*.

(b) *Aluminium sulphide* This salt, Al_2S_3, can be made by heating the two *dry* elements together. The corresponding acid, H_2S, and base, $Al(OH)_3$, are, however, so weak that the salt is completely hydrolysed in water, with liberation of hydrogen sulphide and precipitation of aluminium hydroxide.

$$\begin{array}{ccccc} & & 2Al^{3+}(aq) & & 3S^{2-}(aq) \\ & & + & & + \\ 6H_2O & \rightleftharpoons & 6OH^-(aq) & + & 6H^+(aq) \\ & & \downarrow\uparrow & & \downarrow\uparrow \\ & & 2Al(OH)_3(s) & & 3H_2S(g) \end{array}$$

Because of this hydrolysis, the mixing of solutions of a soluble sulphide and an aluminium salt precipitates aluminium *hydroxide*, not aluminium sulphide as might be expected.

Because of similar hydrolysis effects, aluminium carbonate is unknown. The mixing of sodium carbonate solution and a solution of an aluminium salt precipitates aluminium hydroxide and liberates carbon dioxide.

$$\begin{array}{ccccc} & & 2\,Al^{3+}(aq) & & 3CO_3^{2-}(aq) \\ & & + & & + \\ 6H_2O & \rightleftharpoons & 6OH^-(aq) & + & 6H^+(aq) \\ & & \downarrow\uparrow & & \downarrow\uparrow \\ & & 2\,Al(OH)_3(s) & & 3H_2CO_3(aq) \rightleftharpoons 3H_2O(l) + 3CO_2(g) \end{array}$$

The same is true for a iron(III) salt and sodium carbonate. The reactions are similar, substituting Fe for Al.

4. *Salt of a strong acid and a strong base*

A typical case of this kind is sodium chloride. The situation in a solution of this salt is

$$\begin{array}{ccccc} & & Na^+(aq) & & Cl^-(aq) \\ H_2O(l) & \rightleftharpoons & OH^-(aq) & + & H^+(aq) \end{array}$$

The acid and base are both strong so that no formation of molecules, NaOH or HCl, occurs. The ionic equilibrium of water remains undisturbed, there is no hydrolysis, and the solution remains neutral. The salts of sodium and potassium hydroxide with any of the three mineral acids show this situation.

Qualitative detection of hydrolysis in a salt solution

A simple method of detecting hydrolysis in a solution of a salt is to employ a *universal indicator*. This is a mixture of indicators in solution so chosen that it will register a definite change of colour for each change of one unit (sometimes half a unit) of pH. The indicator is added to the salt solution in a proportion corresponding to the maker's instructions (usually 4 drops to 20 cm^3.) The pH can then be estimated by comparison with a set of standard tubes of indicator in solutions of known pH. Any departure from pH 7 indicates hydrolysis, acid hydrolysis for pH less than 7, alkaline hydrolysis for pH greater than 7. This method cannot be used in coloured salt solutions.

Quantitative treatment of salt hydrolysis

The general reaction for the hydrolysis of a salt of a weak acid, HA, and a strong base is

$$A^-(aq) + H_2O(l) \rightleftharpoons OH^-(aq) + HA(aq)$$

The *hydrolysis constant*, K_h, for this case is written as

$$K_h = \frac{[OH^-][HA]}{[A^-]}$$

The concentration of H_2O is virtually constant. By introducing the quantity $[H^+]$ in both numerator and denominator, we can write

$$K_h = [OH^-][H^+] \times \frac{[HA]}{[H^+][A^-]}$$

$$= K_w \times \frac{1}{K_a}$$

$$= \frac{K_w}{K_a}$$

where K_w is the ionic product of water and K_a is the dissociation constant of the acid.

In a similar way, the hydrolysis constant for the hydrolysis of the salt of a weak base and a strong acid is

$$K_h = \frac{K_w}{K_b}$$

It can also be shown that if the salt is derived from a weak base and a weak acid, its hydrolysis constant is given by

$$K_h = \frac{K_w}{K_a\, K_b}$$

Nature and Choice of Indicators

Nature of an indicator

An indicator is a very weak acid or a very weak alkali. Since the great majority of indicators are in the former class, the main discussion of indicators will be stated in these terms. As a very weak acid, the indicator is of the general type, HA, and ionizes as $HA(aq) \rightleftharpoons H^+(aq) + A^-(aq)$ with the balance of the equilibrium very much to the *left* of the equation. It is a vital feature of an acidic indicator that there shall be a marked difference of colour between the undissociated molecule, HA, and the anion, A^-. This colour change is often associated with a change (accompanying ionization) from a benzene-type organic structure to a quinone-type structure, i.e.

from [benzene ring] to [quinonoid ring: =C< with HC=CH above and HC=CH below, >C=]

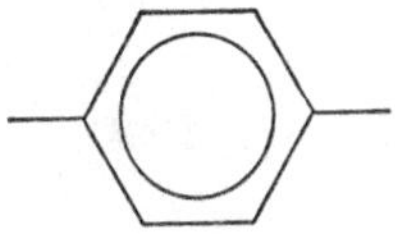

HC CH
HC CH

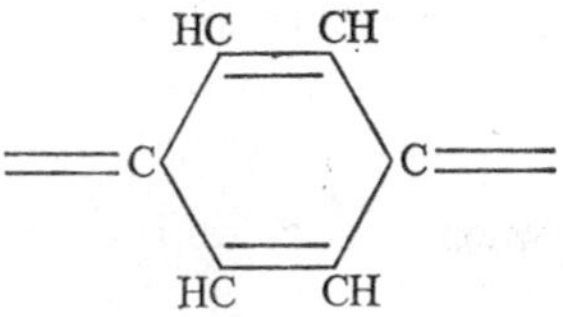

As the indicator is a very weak electrolyte, its ionization is governed by the Dilution Law, so that

$$\frac{[H^+][A^-]}{[HA]} = K_a$$

If the indicator is in an acidic solution, a relatively high concentration of H^+ is also present in the solution and participates in the above equilibrium. The high concentration of H^+ tends to increase K_a. To maintain the value of K_a, $[A^-]$ must decrease and [HA] must increase. That is, the colour of the undissociated molecule is seen, e.g. colourless for the indicator, phenolphthalein.

If the indicator is placed in an alkaline solution (say sodium hydroxide solution), it will form its salt, which is highly ionized.

$$HA + Na^+(aq) + OH^-(aq) \rightarrow Na^+(aq) + A^-(aq) + H_2O$$

Consequently, the colour of the solution changes to that of the anion, A^-, e.g. purple for phenolphthalein.

A few indicators are basic, i.e. of the type, BOH(aq), ionizing very slightly as $BOH(aq) \rightleftharpoons B^+(aq) + OH^-(aq)$. In this case, there must be a marked colour difference between the undissociated molecule, BOH, and the cation, B^+. By use of the Dilution Law,

$$\frac{[B^+][OH^-]}{[BOH]} = K_b$$

and an argument as above, it can be shown that, in alkaline solution (excess of OH^-), the colour of BOH predominates. In acidic solution, the colour of B^+ is seen as a result of salt formation, e.g.

$$BOH(aq) + H^+(aq) + Cl^-(aq) \rightarrow B^+(aq) + Cl^-(aq) + H_2O(l)$$

Change point of indicators

Rearrangement of the equilibrium expression for an indicator which is a weak acid gives

$$[H^+] = \frac{K_a\,[HA]}{[A^-]}$$

The intermediate colour of a two colour indicator will occur when

$$[HA] = [A^-]$$

i.e. when

$$\frac{[HA]}{[A^-]} = 1$$

and

$$[H^+] = K_a$$

Each indicator has its own particular K_a value. Consequently, different indicators change colour at different $[H^+]$ concentrations (and pH values).

In general, it is found that *a change of about two units of pH is necessary to produce the full colour change in a given indicator*. The change-points of a few of the indicators which are in most frequent use are given in the table on page 231.

The pH values are stated to the nearest integer. It will be seen that these indicators cover the whole range of pH values from 10 to 3, i.e. the whole range which is significant for ordinary acid–alkali titration. Litmus is little used because, being a natural product, it is variable in quality. Bromocresol purple (change-point, pH 5–7) covers roughly the same range as litmus near the true neutral point of pH 7.

It is important to note that the 'neutral' point as shown by most indicators, i.e. the mid-point of their pH range for complete colour change, is not the true neutral point of pH 7. For example, the 'neutral' of phenolphthalein is about pH 9 and, in the absolute sense, is appreciably alkaline;

Indicator	*Colours*	*pH*
phenolphthalein	purple colourless	10 8
litmus	blue red	8 6
methyl red	yellow pink	6 4
methyl orange	yellow pink	5 3

the 'neutral' of methyl orange is about pH 4 and is appreciably acidic. Thus a saturated solution of carbon dioxide or carbonic acid, pH about 6.3, is quite strongly 'acidic' to phenolphthalein (change-point pH 10–8) but is 'alkaline' to methyl orange (change-point pH 5–3). These varying change-points are a valuable feature of indicators, making for flexibility in their use.

Choice of indicators for titration

The graphs in Figure 15.1 show approximately the change of pH in solutions when acid–alkali titration is taking place. The graphs cover all the four possible cases—strong and weak acids and strong and weak alkalis. The change-points of the three most important indicators are included for reference.

An efficient indicator is one which, in the titration for which it is employed, will change colour to give a *sharp* end-point. That is, it will move over its full range of colour for the addition of a single drop of the titrating acid (or alkali) from the burette. With the axes disposed as in Figure 15.1 this means that the change-point of the indicator must appear on a vertical part of the curve of pH change, i.e. a part where a maximum change of pH occurs on the vertical axis for a minimum change of titrating liquid on the horizontal axis. The end-point shown by the indicator will then be sharp. This consideration dictates the choice of indicator for each of the four possible cases stated below.

1. *Strong acid–strong alkali* Inspection of Figure 15.1 shows that the combined strong acid–strong alkali curves are almost vertical for the entire pH range 3–11. This range covers the change-points of all three indicators shown. All of these are suitable for use and will show almost

identical titration figures at the end-point. This covers cases like HCl–NaOH, HNO_3–KOH, H_2SO_4–NaOH.

2. *Strong acid–weak alkali* Inspection of Figure 15.1 shows that the combined strong acid–weak alkali curve is almost vertical over the range pH 3–7, only. The indicators, methyl orange and methyl red, have their change-points on this range and either is suitable, preferably methyl red. The most important case here is that of HCl–NH_4OH.

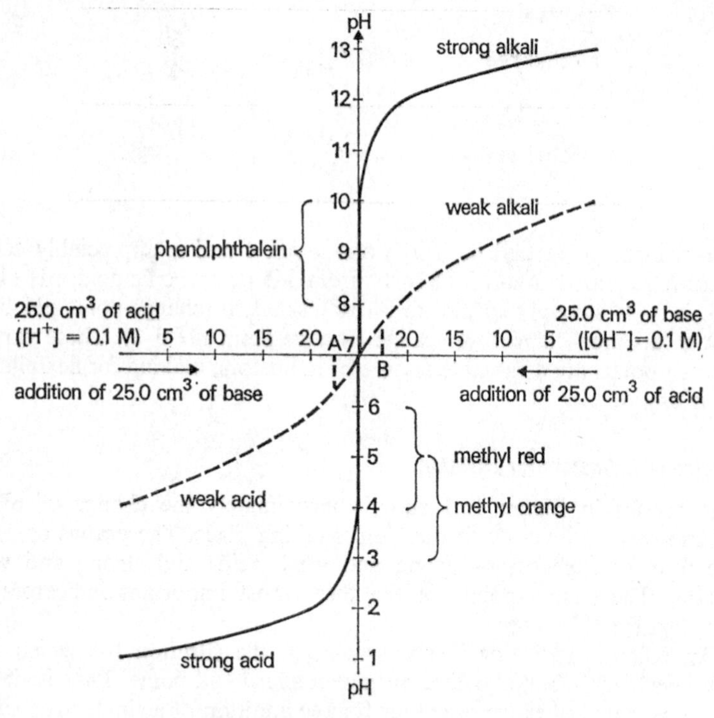

Figure 15.1. Change of pH in neutralization

3. *Weak acid–strong alkali* Figure 15.1 shows that the combined weak acid–strong alkali curves are almost vertical over the pH range 7–11, only. Phenolphthalein has its change-point on this range and is a suitable indicator. This covers the titration of weak organic acids, such as ethanoic or oxalic, with sodium or potassium hydroxide.

4. *Weak acid–weak alkali* Figure 15.1 shows that no part of the combined weak acid–weak alkali curve is vertical. Consequently, no indicator will give a sharp end-point and titration of weak acid by weak alkali is not possible with an indicator depending on colour change. The graph

shows that the pH is changing most rapidly near the neutral point of pH 7. But even here the necessary change of pH by two units, which covers the colour range of an indicator, will require an addition of titrating liquid over the range A–B. This is several cm^3 and at no point is the colour change sharp. Titration of pairs like ethanoic acid–ammonia, or oxalic acid–ammonia is, therefore, impossible with any acid–alkali indicator.

Titration of salts of weak acids by strong acids

The most important case of this kind is the titration of a soluble carbonate by mineral acid, usually sodium carbonate solution by standard hydrochloric acid. This is not really a neutralization of the type

$$H^+(aq) + OH^-(aq) \rightleftharpoons H_2O(l)$$

except in so far as the carbonate is hydrolysed. It is mainly the displacement of the weak acid, carbonic acid, from its salt by the strong mineral acid, with subsequent decomposition of the weak acid to carbon dioxide and water.

$$\begin{array}{rl} CO_3^{2-}(aq) + 2H^+(aq) \rightarrow & H_2CO_3(aq) \\ & \quad\Downarrow\!\Uparrow \\ & H_2O(l) + CO_2(g) \end{array}$$

Consequently, as long as any carbonate remains to react with the hydrochloric acid, the highest acidity attainable is that of a saturated solution of carbon dioxide, i.e. about pH 6.3. If methyl orange is used as indicator (change-point, pH 5–3), the solution appears to be alkaline and the indicator remains yellow. As soon as all the carbonate has reacted and a slight excess of hydrochloric acid appears, pH decreases rapidly, covering the change-point of methyl orange, which alters its colour to orange (or pink) to give the end-point.

A similar case is the titration of a soluble borate by dilute hydrochloric acid. Here, the weak acid is boric acid and the reaction is

$$B_4O_7^{2-}(aq) + 2H^+(aq) + 5H_2O(l) \rightarrow 4H_3BO_3(aq)$$

Buffer solutions

A buffer solution is one which is made up to have a particular hydrogen ion index, pH, and to retain that value of pH in spite of possible accidental contamination by acid or alkali.

A typical buffer solution for acidic values of pH, say 4–7, contains a weak acid and its sodium or potassium salt. A mixture of ethanoic acid and sodium ethanoate is commonly used. (Suitable quantities are quoted later.) This buffer solution operates in the following way. The sodium ethanoate which, like all salts, is a strong electrolyte, can be taken as fully ionized, while the weak ethanoic acid is slightly ionized. In the buffer solution the ionization of the acid is partially suppressed by the common ion effect of ethanoate ion, so it appears weaker than in a corresponding, purely aqueous, solution. Water is also ionized in the usual way ($K_w = 1 \times 10^{-14}$ at 298 K), but its ions play no considerable part in this case. The ionic situation is

From sodium ethanoate: $Na^+(aq) \quad CH_3CO_2^-(aq)$

$$CH_3CO_2H(aq) \rightleftharpoons H^+(aq) + CH_3CO_2^-(aq)$$

The salt, sodium ethanoate, provides a large reserve of ethanoate ion. The weak acid, ethanoic, provides a large potential reserve of hydrogen ion, which it can realize by ionizing as required. If the solution is contaminated by acid, i.e. additional H^+, the reserve ethanoate ion immediately reduces the effective concentration of the H^+ to negligible proportions by combining with it to form ethanoic acid molecules.

$$H^+(aq) + CH_3CO_2^-(aq) \rightleftharpoons CH_3CO_2H(aq)$$

In the conditions of the buffer solution, the equilibrium position in this reaction lies overwhelmingly to the right.

If the solution is contaminated by alkali, i.e. additional OH^-, it is immediately combined with H^+ from the ethanoic acid,

$$H^+(aq) + OH^-(aq) \rightleftharpoons H_2O(l)$$

and reduced to negligible concentration. Reserve ethanoic acid can then ionize to restore the situation to a point different to a negligible extent from the original one. For example, the addition of 1 cm^3 of 0.01 M HCl to 1 dm^3 of a sodium ethanoate–ethanoic acid buffer solution of pH 3.70 produces no change in the pH which can be expressed in three significant figures. The same addition of acid to water alters the pH of the liquid from 7.0 to 5.0.

Correspondingly, a buffer solution on the alkaline side (pH 7–11) can be made with a weak base and one of its salts. A mixture of ammonium hydroxide and ammonium chloride illustrates the principle, but is not much used in practice because of the volatility of ammonia. The salt provides a large source of the ion, NH_4^+, and the base a large potential source of OH^- ion.

From NH_4Cl $\quad NH_4^+(aq) \quad Cl^-(aq)$

$$NH_4OH(aq) \rightleftharpoons NH_4^+(aq) + OH^-(aq)$$

Contamination by alkali, i.e. OH^- ion, is taken up by formation of molecules of the weak base.

$$OH^-(aq) + NH_4^+(aq) \rightleftharpoons NH_4OH(aq)$$

Contamination by acid, i.e. H^+ ion, is taken up by the OH^- ion from the base, which then ionizes to restore the original situation almost exactly.

$$H^+(aq) + OH^-(aq) \rightleftharpoons H_2O(l)$$

The following approximate calculations illustrate the position quantitatively. Suppose a buffer solution contains 0.02 mole of ethanoic acid ($K_a = 1.8 \times 10^{-5}$) and 0.2 mole of sodium ethanoate per dm^3. For the acid,

$$K_a = \frac{[CH_3CO_2^-][H^+]}{[CH_3CO_2H]} = 1.8 \times 10^{-5}\ mol\ dm^{-3}\ \text{(by the Dilution Law)}$$

i.e.

$$[H^+] = 1.8 \times 10^{-5} \times \frac{[CH_3CO_2H]}{[CH_3CO_2^-]}$$

taking the sodium ethanoate as fully ionized, $[CH_3CO_2^-] = 0.2$, and ethanoate ion from the acid as relatively negligible. The ionization of the acid is slight in the presence of its salt by common ion effect, so that, approximately, $[CH_3CO_2H] = 0.02$. Substituting these figures,

$$[H^+] = 1.8 \times 10^{-5} \times \frac{0.02}{0.2}$$

$$= 1.8 \times 10^{-6}$$

$$= 10^{0.26} \times 10^{-6} \qquad \text{(since lg 1.8 = 0.26 approx.)}$$

$$= 10^{-5.74}$$

That is $\quad pH = 5.74$

This combination (0.2 mole of sodium ethanoate and 0.02 mole of ethanoic acid per dm^3) gives, therefore, a buffer solution of pH 5.74. Other mixtures used as buffer solutions are citric acid and its sodium salt, sodium carbonate and hydrogencarbonate, boric acid and borax, the two sodium phosphates, Na_2HPO_4 and NaH_2PO_4.

Sparingly Soluble Electrolytes

Solubility product

This very important concept concerning saturated solutions of sparingly soluble electrolytes will be discussed first in relation to the special case of silver chloride, and then generalized later.

Suppose a quantity of pure water is taken at some definite temperature (usually 298 K). If silver chloride is added to the water, with stirring, and plenty of time and unlimited silver chloride are available, a point will ultimately be reached at which the solution becomes saturated with silver chloride so that no more can dissolve. An equilibrium is then reached of the type

$$AgCl(s) \rightleftharpoons Ag^+(aq) + Cl^-(aq)$$

Applying the Equilibrium Law,

$$K = \frac{[Ag^+(aq)]\,[Cl^-(aq)]}{[AgCl(s)]}$$

The AgCl being a solid means that [AgCl(s)] can be regarded as constant, so that

$$K_{sp} = [Ag^+(aq)]\,[Cl^-(aq)]$$

In this expression, K_{sp} is a constant and is called the *solubility product* of silver chloride. Note that the ionic concentrations in this solubility product equation relate to a saturated solution at a definite temperature.

The important factor about the solubility product of a compound is that it defines the point at which the compound is about to precipitate, because it relates to conditions in a *saturated* solution. In the case of silver chloride at 298 K experimental results show that 1 dm^3 of saturated solution contains 1×10^{-5} mole of the salt. Since the mole produces 1 mole of Ag^+ and Cl^-, it follows that

$$[AgCl] = [Ag^+] = [Cl^-] = 1 \times 10^{-5} \text{ mol dm}^{-3}$$

From this, the solubility product of silver chloride is

$$[Ag^+][Cl^-] = 1 \times 10^{-10} \text{ mol}^2 \text{ dm}^{-6}$$

This value of the solubility product is significant in the following way. Suppose equal volumes of 0.05 M silver nitrate solution and 0.05 M sodium chloride solution are mixed thoroughly and instantaneously, so that each of the salts momentarily becomes 0.01 M by mutual dilution. Then, for a moment only,

$$[Ag^+] = [Cl^-] = 1 \times 10^{-2} \text{ mol}^2 \text{ dm}^{-6}$$

so that the product $[Ag^+][Cl^-]$ is 1×10^{-4}. This is greatly in excess of the solubility product (1×10^{-10}) of silver chloride. At once, Ag^+ and Cl^- are withdrawn from solution as a precipitate, and precipitation continues till the amounts of these ions left in solution are just sufficient to reach the solubility product of silver chloride. This is so small that precipitation is almost complete.

Notice, however, that it is the *product* of the ionic concentrations which is significant. Suppose we have a saturated solution of silver chloride, which the situation (at 298 K in) is $[Ag]^+ = [Cl^-] = 1 \times 10^{-5}$, so that

$$[Ag^+][Cl^-] = K_{sp} = 1 \times 10^{-10} \text{ mol}^2 \text{ dm}^{-6}$$

Then let hydrogen chloride be dissolved in the liquid up to a concentration of, say, 0.01 M, so that $[Cl^-] = 1 \times 10^{-2}$. This chloride ion participates in the solubility product relation of silver chloride though it is not derived from this salt. The chloride ion concentration from silver chloride being relatively negligible, the ionic concentrations in solution must satisfy the requirement

$$[Ag^+][Cl^-] = [Ag^+][10^{-2}] = K_{sp} = 1 \times 10^{-10}$$

That is, at equilibrium, $[Ag^+] = 1 \times 10^{-8}$; so, by dissolving the stated amount of hydrogen chloride in the solution, the concentration of Ag^+ is reduced from 1×10^{-5} to 1×10^{-8} mol dm^{-3} by precipitation of the corresponding quantity of silver chloride. This is an aspect of the *common ion effect*. The mass of silver chloride precipitated per dm^3 is $(10^{-5} - 10^{-8})$ [AgCl]. Since 10^{-8} is relatively negligible, this mass is $10^{-5} \times 143.5$ g, i.e. 0.001 435 g of silver chloride per dm^3.

General expression for solubility product

Suppose a compound A_mB_n ionizes in the following way:

$$A_mB_n \rightleftharpoons mA^{n+} + nB^{m-}$$

Each ion operates separately in the solubility product relation, so that the solubility product of A_mB_n is given by

$$[A^{n+}]^m\,[B^{m-}]^n = K_{sp}$$

and the square brackets indicate equilibrium concentrations measured in mol dm^{-3}.

Examples are:

$$Ag_2CrO_4(s) \rightleftharpoons 2Ag^+(aq) + CrO_4^{2-}(aq) \quad K_{sp} = [Ag^+]^2[CrO_4^{2-}]$$

$$PbCl_2(s) \rightleftharpoons Pb^{2+}(aq) + 2Cl^-(aq) \quad K_{sp} = [Pb^{2+}][Cl^-]^2$$

Some typical examples of solubility product values are given in the following table:

Compound	*Solubility product at 298 K*
AgCl	1×10^{-10}
Ag_2CrO_4	1×10^{-12}
ZnS	1×10^{-24}
CoS	2×10^{-27}
PbS	4×10^{-28}
$Mg(OH)_2$	1.2×10^{-11} (291 K)

Applications of the concept of solubility product

The first two applications are concerned with a scheme of qualitative analysis in which metal ions are separated first into groups. These groups are known as analytical groups and their numbers bear no relation to the numbers of the groups in the Periodic Table.

1. *Precipitation of metallic sulphides in qualitative analysis*

Two groups of insoluble sulphides occur—those of Group II, which are precipitated from a solution acidified with hydrochloric acid, and those of Group IV, which are precipitated by hydrogen sulphide in the presence of ammonia, i.e. by ammonium sulphide. The important members of these groups are:

Group II	Group IV
HgS, PbS, CuS, CdS, Bi_2S_3	NiS, CoS, MnS, ZnS

Differentiation into these groups depends on the following facts.

Hydrogen sulphide is a very weak electrolyte. Considering its ionization,

$$H_2S(aq) \rightleftharpoons 2H^+(aq) + S^{2-}(aq)$$

the situation is governed by the Dilution Law which requires that

$$\frac{[H^+]^2\,[S^{2-}]}{[H_2S]} = K_a$$

where K_a is the dissociation constant of the electrolyte and has a value in the neighbourhood of 1×10^{-22}. It is obvious from this that the concentration of S^{2-} ion in a solution of hydrogen sulphide is always very low. In Group II conditions, hydrochloric acid is present, supplying a large concentration of H^+. This participates in the above equilibrium, tending to raise the value of K_a. To maintain K_a at its correct level, the concentration of S^{2-} must fall, and of H_2S must rise, i.e. the ionization of H_2S is considerably suppressed, so that the concentration of S^{2-} is reduced much below its already low value in solution. This is an example of the 'common ion effect'.

The sulphides of Groups II and IV all have low solubility products, but the solubility products of the Group II sulphides are lower than those of the Group IV sulphides. Typical cases are:

Group II	Group IV
Copper(II) sulphide	Zinc sulphide
$[Cu^{2+}][S^{2-}] = 3 \times 10^{-42}\ mol^2\,dm^{-6}$	$[Zn^{2+}][S^{2-}] = 1 \times 10^{-24}\ mol^2\,dm^{-6}$

In Group II conditions (HCl present), the concentration of S^{2-} is so low that the higher solubility products of Group IV sulphides cannot be reached and these sulphides do not precipitate. The much lower solubility products of Group II sulphides can, however, be reached and (momentarily) exceeded; consequently, the Group II sulphides precipitate.

In Group IV conditions (NH_4OH present) the H^+ concentration is reduced due to its reaction with the OH^- ions. This results in increased ionization of the hydrogen sulphide. The concentration of S^{2-} is relatively high. Consequently the solubility products of Group IV sulphides are reached and (momentarily) passed and the Group IV sulphides precipitate.

Cadmium sulphide is a borderline case. It is allowed for in the analysis tables as a Group II sulphide. If, however, the concentration of hydrochloric acid is too high, the concentration of S^{2-} may be reduced to a point at which the solubility product, $[Cd^{2+}]\ [S^{2-}] = K_{sp}$ cannot be reached. In this case, cadmium sulphide will not precipitate in Group II. This is why a sample of the solution should be *well diluted* and subjected to passage of H_2S before it goes to Group III. The dilution ensures precipitation of cadmium sulphide, which might otherwise be missed.

The differences between the sulphides of Groups IIA and IIB are considered in connection with the chemistry of arsenic, antimony, bismuth, and tin, in *Inorganic Chemistry*, the companion volume to this book.

2. *Precipitation of metallic hydroxides in qualitative analysis*

After the precipitation of the Group II sulphides discussed above, the remaining metals with 'insoluble' hydroxides include Fe, Al, Cr, Ni, Co, Zn, and Mn. If, however, the precipitating agent used is ammonium hydroxide *in the presence of ammonium chloride*, only the hydroxides of Fe, Al, and Cr precipitate. They constitute Group III. This is explained as follows.

Ammonium hydroxide is a weak base, i.e. only feebly ionized at the usual dilution (2M – 4 M). By the Equilibrium Law, applied to the ionization $NH_4OH(aq) \rightleftharpoons NH_4^+(aq) + OH^-(aq)$,

$$\frac{[NH_4^+][OH^-]}{[NH_4OH]} = K_b = 2 \times 10^{-5}\ \text{mol dm}^{-3}\ \text{(at room temperature)}$$

If the salt NH_4Cl is present, it is fully ionized and provides a high concentration of the ion, NH_4^+. This participates in the above equilibrium. To maintain the value of K_b, the concentration of OH^- must fall and the concentration of NH_4OH must rise. That is, the weak base becomes weaker and the concentration of OH^-, small at any time, is depressed to a very low value, an example of the 'common ion effect'.

The 'insoluble' hydroxides named above all have low solubility products. The Group III hydroxides, however, are much less soluble than the rest. Consequently, the low concentration of OH^- available in the presence of ammonium chloride suffices to reach the solubility products of these hydroxides and they precipitate. The higher solubility products of the hydroxides of Ni, Co, Zn, and Mn are not attained and these hydroxides do not precipitate. Manganese hydroxide is a borderline case. If the concentration of ammonium chloride is rather low, this hydroxide may precipitate partially in Group III. Manganese will then follow the same course as iron in Group III. Some manganese will, however, always go forward and appear as sulphide in Group IV.

Solubility of salts of weak acids in dilute strong acids

There are many cases in which the salt of a weak acid is only very slightly soluble in water, but dissolves readily in a dilute mineral acid. Calcium oxalate is a typical case and will be used in illustration.

If calcium oxalate is stirred in water, it will dissolve to the point at which, in the solution, $[Ca^{2+}]\,[C_2O_4^{2-}] = K_{sp}$, where K_{sp} is the solubility product of the salt at the temperature in question. K_{sp} is very small so that, of a moderate quantity of calcium oxalate, only a very little will dissolve and the rest will remain in suspension. When hydrochloric acid is added, the ionic situation is

$$\begin{array}{lcccc} \text{From calcium oxalate:} & & Ca^{2+}(aq) & & C_2O_4^{2-}(aq) \\ & & & & + \\ 2HCl(aq) & \rightleftharpoons & 2Cl^-(aq) & + & 2H^+(aq) \\ & & & & \downarrow\!\uparrow \\ & & & & H_2C_2O_4(aq) \end{array}$$

Oxalic acid is weak; consequently unionized molecules of the acid are produced by utilizing oxalate ion of the salt and hydrogen ion from hydrochloric acid. The withdrawal of $C_2O_4^{2-}$ ion in this way reduces the product $[Ca^{2+}][C_2O_4^{2-}]$ momentarily below the value, K_{sp}. To restore this value, calcium oxalate passes from the suspended material into solution. More formation of $H_2C_2O_4$ molecules occurs, drawing more calcium oxalate into solution. If sufficient hydrochloric acid is available, this continues till all the calcium oxalate is dissolved. The free $C_2O_4{}^{2-}$ ion is then insufficient to reach the solubility product of calcium oxalate.

Calcium phosphate provides a similar case, being only sparingly soluble in water and readily soluble in dilute hydrochloric acid. The relevant equations are

$$[Ca^{2+}]^3\,[PO_4^{3-}]^2 = K_{sp} \text{ (very small)}$$

$$\begin{array}{lcccc} \text{From calcium phosphate:} & & 3Ca^{2+}(aq) & & 2PO_4^{3-}(aq) \\ & & & & + \\ 6HCl(aq) & \rightleftharpoons & 6Cl^-(aq) & + & 6H^+(aq) \\ & & & & \downarrow\!\uparrow \\ & & & & 2H_3PO_4(aq) \\ & & & & \text{(a weak acid)} \end{array}$$

Another application of solubility product ideas occurs in the following connection. Many cases are found in which materials (usually salts) are almost insoluble in water but pass readily into solution in water containing certain other dissolved chemicals. One of these will be discussed in detail; others will be stated in outline which can be filled in by similar development.

Why silver chloride is almost insoluble in water but readily soluble in ammonia solution

If silver chloride (about 5 g) is put into water (about 100 cm^3), the salt will dissolve until the product of the ionic concentrations in solution reaches the solubility product of silver chloride, i.e. until

$$[Ag^+][Cl^-] = K_{sp} = 1 \times 10^{-10} \text{ mol}^2 \text{ dm}^{-6} \text{ at } 298 \text{ K}$$

This value is so small that almost all the silver chloride will remain as a solid. If, however, ammonia is added, the solid material will quickly 'dissolve'. The reason for this is as follows.

When ammonia is added, a complex cation is formed.

$$Ag^+(aq) + 2NH_3(aq) \rightleftharpoons [Ag(NH_3)_2]^+(aq)$$

The silver ion for this complex ion formation is taken from the solution and no longer participates in the solubility product equilibrium of silver chloride. Consequently, K_{sp} is no longer attained in the solution. To restore the value of K_{sp}, silver chloride passes from the precipitate into solution as Ag^+ and Cl^- ions. The ion, Ag^+, at once forms a more complex ion with the ammonia, which in turn causes more silver chloride to dissolve in the attempt to restore the value of K_{sp}. This continues (if sufficient ammonia is present) until all the silver chloride precipitate has dissolved and the concentration of Ag^+ in the solution is so small that the solubility product of silver chloride cannot be reached. The silver chloride is then said to have 'dissolved' in ammonia though, in fact, a definite chemical reaction has occurred. The final 'solution' contains the silver almost entirely as the ion, $[Ag(NH_3)_2]^+$.

Most of the cases in which a material is almost insoluble in water but 'soluble' when the water contains another chemical are explained by complex ion formation as above. The complex is sometimes on the cation, sometimes on the anion. The following is a selection of such cases stated in outline. Each can be fully argued as in the case of ammonia and silver chloride above.

1. *Silver cyanide: almost insoluble in water; readily soluble in potassium cyanide solution*

The solubility of silver cyanide in water reaches its limit when $[Ag^+][CN^-] = K_{sp}$ (very small). When potassium cyanide is present, a complex anion is formed:

$$\underbrace{Ag^+(aq) + CN^-(aq)}_{\text{from AgCN}} + \underbrace{CN^-(aq)}_{\text{from KCN}} \rightleftharpoons [Ag(CN)_2]^-(aq)$$

2. *Copper(II) hydroxide: almost insoluble in water; readily soluble in excess of aqueous ammonia*

The solubility of copper hydroxide reaches its limit in water when $[Cu^{2+}][OH^-]^2 = K_{sp}$ (very small). When ammonia is present, a complex cation is formed:

$$Cu^{2+}(aq) + 4NH_3(aq) \rightleftharpoons [Cu(NH_3)_4]^{2+}(aq)$$

3. *Mercury(II) iodide: almost insoluble in water; readily soluble in excess of potassium iodide solution*

The solubility of Mercury(II) iodide in water reaches its limit when $[Hg^{2+}][I^-]^2 = K_{sp}$ (very small). When potassium iodide is present, a complex anion is formed:

$$\underbrace{Hg^{2+}(aq) + 2I^-(aq)}_{\text{from } HgI_2} + \underbrace{2I^-(aq)}_{\text{from KI}} \rightleftharpoons [HgI_4]^{2-}(aq)$$

4. *Iodine: sparingly soluble in water; readily soluble in a concentrated solution of potassium iodide*

Iodine dissolves in water till the equilibrium

$$I_2(s) \rightleftharpoons I_2(aq)$$

is satisfied. This equilibrium lies very much to the left and very little iodine dissolves (relative to the amount of water). Note that no solubility product issue arises here; the concept of solubility product is not applicable to an element.

When potassium iodide is present, a complex anion is formed.

$$I_2(aq) + I^-(aq) \rightleftharpoons I_3^-(aq)$$

5. *Lead chloride: sparingly soluble in cold water; considerably more soluble in concentrated acid*

The solubility of lead chloride in water reaches its limit when $[Pb^{2+}][Cl^-]^2 = K_{sp}$ (small in cold water). In concentrated hydrochloric acid, a complex anion is formed:

$$\underbrace{Pb^{2+}(aq) + 2Cl^-(aq)}_{\text{from } PbCl_2} + 2Cl^-(aq) \rightleftharpoons [PbCl_4]^{2-}(aq)$$

Adequate presentation of each of these cases requires a fully developed argument on the lines shown for the case of silver chloride and ammonia earlier.

16. Chemical Equilibria III: Redox

Originally the term *oxidation* was applied solely to cases in which oxygen was gained, and *reduction* to those in which oxygen was lost, e.g.

Oxidation: $PbS(s) + 2O_2(g) \rightarrow PbSO_4(s)$
Reduction: $2KClO_3(s) \rightarrow 2KCl(s) + 3O_2(g)$

The two processes could occur together, e.g.

$$CuO(s) + H_2(g) \rightarrow Cu(s) + H_2O(l)$$

The copper(II) oxide was reduced; the hydrogen was oxidized.

With the coming of electrolysis in the early nineteenth century, oxygen was frequently found as an anode product, opposed to hydrogen as a cathode product, as in the electrolysis of dilute sulphuric acid or sodium hydroxide solution. For this reason, oxygen and hydrogen appeared to be rather special chemical opposites, so that loss of hydrogen came to be considered as comparable to gain of oxygen. Loss of hydrogen was then classed as oxidation and gain of hydrogen as reduction. For example, in the reaction

$$H_2S(g) + Cl_2(g) \rightarrow 2HCl(g) + S(s)$$

the hydrogen sulphide was said to be oxidized by loss of hydrogen and the chlorine to be reduced by gain of hydrogen. No oxygen is involved.

With the further growth of chemical knowledge, many cases were recognized in which a metallic element exercised two valencies (possibly more) and formed two (or more) series of compounds. Iron, forming the iron(II) series (with iron *di*valent) and the iron(III) series (with iron *tri*valent), is a well-known case. Here the conversion of iron(II) oxide, FeO, to iron(III) oxide, Fe_2O_3, is a clear case of oxidation by gain of oxygen. Since all iron(II) compounds correspond to iron(II) oxide and show iron with valency 2, and all iron(III) compounds correspond to iron(III) oxide and show iron with a valency 3, the conversion of *any* iron(II) compound to *any* iron(III) compound came to be regarded as oxidation (and *vice versa* for reduction), e.g.

$$2FeCl_2(s) + Cl_2(g) \rightarrow 2FeCl_3(s)$$

(iron(II) chloride oxidized to iron(III) chloride)

$$Fe_2(SO_4)_3(aq) + H_2S(g) \rightarrow 2FeSO_4(aq) + H_2SO_4(aq) + S(s)$$

(iron(III) sulphate reduced to iron(II) sulphate)

In such cases, oxidation involves an increase (and reduction a decrease) in the valency of the metal. A similar case is the tin(II), tin(IV) interrelation.

$$\text{tin(II) compound} \underset{\text{reduction}}{\overset{\text{oxidation}}{\rightleftharpoons}} \text{tin(IV) compound}$$

In 1880, ionic ideas entered chemistry. It was then quickly seen that in cases like the iron(II)–iron(III) relation, oxidation involved an increase in the proportion of the electronegative constituent of a compound, i.e. the anion. This can be seen from the examples

$$\left.\begin{array}{c}\text{iron(II) chloride}\\ Fe^{2+}2Cl^-\\ \text{tin(II) chloride}\\ Sn^{2+}2Cl^-\end{array}\right\} \underset{\text{reduction}}{\overset{\text{oxidation}}{\rightleftharpoons}} \left\{\begin{array}{c}\text{iron(III) chloride}\\ Fe^{3+}3Cl^-\\ \text{tin(IV) chloride}\\ Sn^{4+}4Cl^-\end{array}\right.$$

In the late nineteenth century, therefore, oxidation may be said to have included the following ideas:

1. Increase of oxygen content.
2. Decrease of hydrogen content.
3. Increase in the proportion of the anion, with increase in the valency of the metal present.

Since ideas of atomic structure entered chemistry about forty years ago, oxidation and reduction have come to be regarded mainly as electronic phenomena in the following sense:

Oxidation is the process of electron-loss.
Reduction is the process of electron-gain.

These two processes are complementary and must occur together.

An **oxidizing agent** is an electron-acceptor.
A **reducing agent** is an electron-donor.

As the following discussion shows, this much simplified concept includes all the older ideas. In the light of it, consider the oxidation of iron(II) chloride to iron(III) chloride by chlorine.

As a molecular equation: $2FeCl_2(aq) + Cl_2(aq) \rightarrow 2FeCl_3(aq)$
Ionically: $2Fe^{2+}(aq) + Cl_2(aq) \rightarrow 2Fe^{3+}(aq) + 2Cl^-(aq)$

This ionic equation summarizes two processes.

1. Each iron(II) ion *loses* one electron. It is oxidized and so acts as a reducing agent. The valency of iron increases from 2 to 3.

$$2Fe^{2+}(aq) - 2e^- \rightarrow 2Fe^{3+}(aq)$$

2. The electrons are *accepted* by chlorine atoms. The chlorine is reduced and so acts as an oxidizing agent. The proportion of the anion, Cl^-, is increased.

$$Cl_2(aq) + 2e^- \rightarrow 2Cl^-(aq)$$

These ideas are quite general. The following are common oxidizing agents. Notice how, in each case, they operate by accepting electrons (supplied by a reducing agent).

Oxidizing agents

Chlorine: $Cl_2(aq) + 2e^- \rightarrow 2Cl^-(aq)$
Bromine: $Br_2(aq) + 2e^- \rightarrow 2Br^-(aq)$
Iodine: $I_2(aq) + 2e^- \rightarrow 2I^-(aq)$

Potassium permanganate (in acidic solution):

$$MnO_4^-(aq) + 8H^+(aq) + 5e^- \rightarrow Mn^{2+}(aq) + 4H_2O(l)$$

Potassium dichromate (in acidic solution):

$$Cr_2O_7^{2-}(aq) + 14H^+(aq) + 6e^- \rightarrow 2Cr^{3+}(aq) + 7H_2O(l)$$

Iron(III) salts: $Fe^{3+}(aq) + e^- \rightarrow Fe^{2+}(aq)$

Mercury(II) salts (two stages of reduction—to mercury(I) salts and to mercury)

$$2Hg^{2+}(aq) + 2e^- \rightarrow Hg_2{}^{2+}(aq)$$
$$Hg_2^{2+}(aq) + 2e^- \rightarrow 2Hg(l)$$

Hydrogen ions: $2H^+(aq) + 2e^- \rightarrow H_2(g)$
Hydrogen peroxide: $H_2O_2(l) + 2H^+(aq) + 2e^- \rightarrow 2H_2O(l)$
Manganese(IV) oxide (in the presence of acid):

$$MnO_2(s) + 4H^+(aq) + 2e^- \rightarrow Mn^{2+}(aq) + 2H_2O(l)$$

The following are common reducing agents. Notice how, in each case, they operate by donating electrons (which are accepted by an oxidizing agent).

Reducing agents

Iron(II) salts: $Fe^{2+}(aq) \rightarrow Fe^{3+}(aq) + e^-$
Tin(II) salts: $Sn^{2+}(aq) \rightarrow Sn^{4+}(aq) + 2e^-$
Metals: $X(s) \rightarrow X^{n+}(aq) + ne^-$

Hydrogen sulphide: $H_2S(aq) \rightleftharpoons 2H^+(aq) + S^{2-}(aq)$;
$S^{2-}(aq) \rightarrow S(s) + 2e^-$
Sulphurous acid: $SO_3^{2-}(aq) + H_2O(l) \rightarrow SO_4^{2-}(aq) + 2H^+(aq) + 2e^-$

Iodides: $2I^-(aq) \rightarrow I_2(aq) + 2e^-$
Oxalates: $C_2O_4^{2-}(aq) \rightarrow 2CO_2(aq) + 2e^-$
Sodium thiosulphate: $2S_2O_3^{2-}(aq) \rightarrow S_4O_6^{2-}(aq) + 2e^-$

These oxidizing and reducing agents do not necessarily all interact with each other. However, when observations indicate that such a reaction has occurred, or redox potentials (see page 248) are used to predict that a reaction is likely to occur, the complete ionic equation for the reaction is obtained by multiplying the two half equations by appropriate numbers and then adding them together so that the electrons cancel out. For example, when hydrogen sulphide is bubbled into a solution of an iron(III) salt, a yellowish precipitate is observed. This indicates that the sulphide ions have been oxidized to sulphur and it is likely that the iron(III) ions have been reduced to iron(II) ions. The two ion–electron half equations are

$$Fe^{3+}(aq) + e^- \rightarrow Fe^{2+}(aq)$$
$$S^{2-}(aq) \rightarrow S(s) + 2e^-$$

The complete ionic equation is obtained by multiplying the first of these equations by two and then adding it to the second equation. The equation for the reaction is

$$2Fe^{3+}(aq) + S^{2-}(aq) \rightarrow 2Fe^{2+}(aq) + S(s)$$

Oxidation Numbers

As an alternative to the ion–electron half equation method of interpreting redox reactions, a system based on oxidation numbers may be used. It has the advantage that it can also be applied to substances which do not exist in the form of ions.

The oxidation number of an element indicates the oxidation state of that element in a particular compound. An element in an uncombined state has an oxidation number of zero. When in the form of a simple ion in a compound its oxidation number is equal to the charge on the ion.

In the reaction

$$2Na(s) + Cl_2(g) \rightarrow 2NaCl(s)$$

the oxidation number of sodium changes from 0 to +1 and that of chlorine changes from 0 to −1. The increase in the oxidation number of sodium when it is oxidized is exactly balanced by the decrease in oxidation number of chlorine when it is reduced.

When assigning oxidation numbers to elements in covalent compounds or more complex ions, it is necessary to fix arbitrarily the oxidation numbers of certain elements.

Rules for assigning and using oxidation numbers

1. The oxidation number of an uncombined element is zero.
2. The algebraic sum of the oxidation numbers of the constituents of a compound is zero.
3. The algebraic sum of the oxidation numbers of the constituents of an ion is equal to the charge on the ion.
4. In order to apply the system to covalent compounds and more complex ions the following invariable oxidation numbers are adopted:
   F is -1,
   O is -2 except when combined with fluorine and in peroxides,
   H is $+1$ except in metal hydrides.
5. The total change in oxidation number during a redox reaction is zero. The increase in oxidation number of the species being oxidized is exactly balanced by the decrease in oxidation number of the species being reduced.

The following examples will serve to illustrate the system.

Compound or ion	*Oxidation numbers of constituent elements*	
$CaCl_2$	Ca $+2$	Cl -1
MgO	Mg $+2$	O -2
P_4O_{10}	P $+5$	O -2
H_2O	H $+1$	O -2
ClO_3^-	Cl $+5$	O -2
H_2O_2	H $+1$	O -1
F_2O_2	F -1	O $+1$
NaH	Na $+1$	H -1

Oxidation numbers may be used to balance redox equations. An acidified solution containing dichromate ions will oxidize iron(II) ions to iron(III), the dichromate itself being reduced to chromium(III). The oxidation number of the chromium is being reduced and that of iron(II) is being increased. The total increase must be equal to the total reduction.

The oxidation number of chromium in (i) $Cr_2O_7^{2-}$ is $+6$
(ii) Cr^{3+} is $+3$

The oxidation number of iron in (i) Fe^{2+} is $+2$
(ii) Fe^{3+} is $+3$

Total reduction $= 6$

Increase $= 1$

Therefore, for every mole of $Cr_2O_7^{2-}$ which is reduced, 6 mole of Fe^{2+} must be oxidized and the balanced equation is

$$Cr_2O_7^{2-}(aq) + 6Fe^{2+}(aq) + 14H^+(aq) \rightarrow 2Cr^{3+}(aq) + 6Fe^{3+}(aq) + 7H_2O(l)$$

The oxidation numbers of H and O do not change during the reaction.

Oxidation numbers may also be used to predict the oxidation state of a product. For example, it can be shown by experiment that 1 mole of bromate ions will oxidize acidified iodide ions to produce 3 mole of iodine molecules. The increase in oxidation number of the iodine must be balanced by the decrease in oxidation number of the bromine. When $6I^-$ changes to $3I_2$, the change in oxidation number is from -1 to 0 for each I, which gives a total increase of 6. In BrO_3^- the bromine has an oxidation number of $+5$ and if it is to decrease by 6 it must change to an oxidation state of -1, i.e. to Br^-. The complete equation is

$$BrO_3^-(aq) + 6I^-(aq) + 6H^+(aq) \rightarrow 3I_2(aq) + Br^-(aq) + 3H_2O(l)$$

Electrode Potentials

A metal dipping into a solution containing its ions establishes an equilibrium with its ions. For zinc,

$$Zn(s) \rightleftharpoons Zn^{2+}(aq) + 2e^-$$

The electrons remain on the metal and so there is a difference of potential between the metal and the solution. It is known as the electrode potential of the metal–metal ion system and is a measure of the tendency of the metal to provide electrons, i.e. to act as a reducing agent. Any attempt to measure this potential difference will entail the introduction of another electrode which will exert its own electrode potential. It is not possible to determine absolute values of electrode potentials, but by employing a standard reference electrode it is possible to establish an order of oxidizing and reducing tendency for electrode systems.

Hydrogen Electrode and Standard Redox Potentials

The hydrogen electrode is used as the reference electrode. It consists of hydrogen gas, at one atmosphere pressure, bubbling over platinum coated in electrically deposited platinum and dipping into a solution which is 1 M with respect to hydrogen ions.

Cells are constructed with the hydrogen electrode as one half and a metal dipping into a 1 M solution of its ions as the other half. The two halves are connected by a salt bridge which is made from a piece of filter paper soaked in potassium nitrate solution (Figure 16.1).

The maximum potential difference, known as the e.m.f., of a cell is then determined by means of a potentiometer or a high-resistance voltmeter. The e.m.f. is the algebraic sum of the electrode potentials of the two half-cells. As the hydrogen half-cell is the reference system, its electrode potential is taken to be zero and the e.m.f. of the cell is said to be the standard electrode potential of the other half-cell. In the cell shown in Figure 16.1 the two equilibria involved are, in the hydrogen half-cell

$$H_2(g) \rightleftharpoons 2H^+(aq) + 2e^- \quad (1)$$

and in the zinc half-cell

$$Zn(s) \rightleftharpoons Zn^{2+}(aq) + 2e^- \quad (2)$$

The zinc has the greater tendency to form ions. Therefore, equilibrium (2) moves to the right and electrons pass round the external circuit from the zinc to the platinum and equilibrium (1) absorbs electrons by moving to

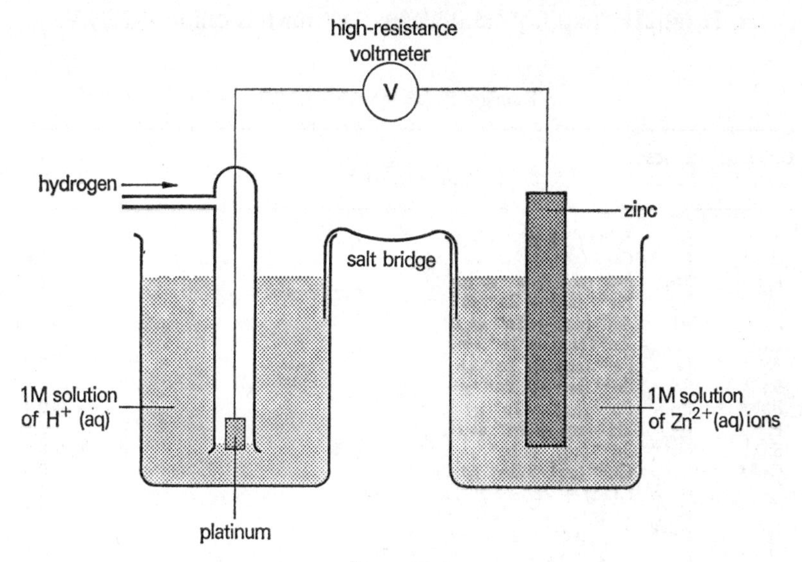

Figure 16.1.

the left. The e.m.f. of the cell is found to be 0.76 V. Conventionally the current is said to flow from the positive terminal of a cell but in fact electrons flow in the opposite direction, that is, from the negative. The zinc is the negative in the above cell and the zinc–zinc ion system is said to have a *standard electrode potential*, $E^\ominus$, of −0.76 V.

When a copper half-cell is substituted for the zinc, the electrons flow from the hydrogen half-cell and the copper is the positive half-cell. The e.m.f. of the cell is 0.34 V and the copper–copper(II) ion system is said to have a standard electrode potential of +0.34 V.

All the half-cell reactions are redox reactions and so it is common for standard electrode potentials to be referred to as standard redox potentials. A table of standard redox potentials is given on page 250. The order in which the systems appear in this table is known as the electrochemical series.

To avoid drawing diagrams or writing lengthy descriptions, cells are represented by inserting the appropriate formulae in the following scheme.

electrode|solution|solution|electrode
salt
bridge

The sign of the e.m.f. of a cell is taken as the sign of the right-hand electrode as it is written. The two cells which have been considered so far are represented as follows:

$Pt, H_2(g)|2H^+ (aq)|Zn^{2+}(aq)|Zn(s)$ $E^\ominus$ for this cell = −0.76 V
$Pt, H_2(g)|2H^+ (aq)|Cu^{2+}(aq)|Cu(s)$ $E^\ominus$ for this cell = +0.34 V

Standard Redox Potentials $E^\ominus$(V)

Oxidized species	*Reduced species*	*Half-cell reaction*	$E^\ominus$(V)
K^+	K	$K^+(aq) + e^- \rightleftharpoons K(s)$	−2.92
Ca^{2+}	Ca	$Ca^{2+}(aq) + 2e^- \rightleftharpoons Ca(s)$	−2.87
Na^+	Na	$Na^+(aq) + e^- \rightleftharpoons Na(s)$	−2.71
Mg^{2+}	Mg	$Mg^{2+}(aq) + 2e^- \rightleftharpoons Mg(s)$	−2.38
Al^{3+}	Al	$Al^{3+}(aq) + 3e^- \rightleftharpoons Al(s)$	−1.67
Zn^{2+}	Zn	$Zn^{2+}(aq) + 2e^- \rightleftharpoons Zn(s)$	−0.76
Fe^{2+}	Fe	$Fe^{2+}(aq) + 2e^- \rightleftharpoons Fe(s)$	−0.44
Sn^{2+}	Sn	$Sn^{2+}(aq) + 2e^- \rightleftharpoons Sn(s)$	−0.14
Pb^{2+}	Pb	$Pb^{2+}(aq) + 2e^- \rightleftharpoons Pb(s)$	−0.13
H^+	H_2	$2H^+(aq) + 2e^- \rightleftharpoons H_2(g)$	0.00
Sn^{4+}	Sn^{2+}	$Sn^{4+}(aq) + 2e^- \rightleftharpoons Sn^{2+}(aq)$	+0.15
Cu^{2+}	Cu	$Cu^{2+}(aq) + 2e^- \rightleftharpoons Cu(s)$	+0.34
I_2	I^-	$I_2(aq) + 2e^- \rightleftharpoons 2I^-(aq)$	+0.54
O_2	H_2O_2	$O_2(g) + 2H^+(aq) + 2e^- \rightleftharpoons H_2O_2(l)$	+0.68
Fe^{3+}	Fe^{2+}	$Fe^{3+}(aq) + e^- \rightleftharpoons Fe^{2+}(aq)$	+0.77
Hg^{2+}	Hg_2^{2+}	$2Hg^{2+}(aq) + 2e^- \rightleftharpoons Hg_2^{2+}(aq)$	+0.91
Br_2	Br^-	$Br_2(aq) + 2e^- \rightleftharpoons 2Br^-(aq)$	+1.09
IO_3^-	I_2	$2IO_3^-(aq) + 12H^+(aq) + 10e^- \rightleftharpoons I_2(aq) + 6H_2O(l)$	+1.19
$Cr_2O_7^{2-}$	Cr^{3+}	$Cr_2O_7^{2-}(aq) + 14H^+(aq) + 6e^- \rightleftharpoons 2Cr^{3+}(aq) + 7H_2O(l)$	+1.33
Cl_2	Cl^-	$Cl_2(aq) + 2e^- \rightleftharpoons 2Cl^-(aq)$	+1.36
MnO_4^-	Mn^{2+}	$MnO_4^-(aq) + 8H^+(aq) + 5e^- \rightleftharpoons Mn^{2+}(aq) + 4H_2O(l)$	+1.52
F_2	F^-	$F_2(g) + 2e^- \rightleftharpoons 2F^-(aq)$	+2.84

Ion–ion Systems

A half-cell may be constructed from an inert electrode (platinum) dipping into a solution containing two types of ions, one of which is the reduced form and the other the oxidized form of a redox equilibrium. For example, the system represented by the equilibrium

$$Fe^{2+}(aq) \rightleftharpoons Fe^{3+}(aq) + e^-$$

is capable of providing or absorbing electrons and can be used in a half-cell. By putting such half-cells with the hydrogen half-cell it is possible to determine a set of values of standard redox potentials for ion–ion systems (see table on page 250).

As previously stated, standard electrode potentials refer to the potentials obtained when molar solutions of ions are used. If the concentrations of the ions concerned are not molar, then the electrode potential of that half-cell (E) will not be equal to the standard electrode potential ($E^{\ominus}$). The relationship between the two electrode potentials is given by the Nernst equation, which is

$$E = E^{\ominus} + \frac{RT}{zF} \ln [\text{ion}]$$

for a metal–metal ion half-cell and

$$E = E^{\ominus} + \frac{RT}{zF} \ln \frac{[\text{oxidized species}]}{[\text{reduced species}]}$$

where R is the gas constant in J K^{-1}, T is the temperature in Kelvin, z is the change in charge involved, F is the Faraday constant in coulombs, and the concentrations are measured in mol dm^{-3}.

Applications of Standard Redox Potentials

1. *Predicting an e.m.f. value of a cell*

The positions, with respect to hydrogen, of two half-cells in the series may be used to predict the e.m.f. of a cell. For example, the systems $Zn^{2+}(aq)|Zn(s)$ and $Pb^{2+}(aq)|Pb(s)$ are 0.63 V apart in the series. A cell constructed from these two half-cells will be capable of producing an e.m.f. of 0.63 V. The systems $Zn^{2+}(aq)|Zn(s)$ and $Cu^{2+}(aq)|Cu(s)$, being on opposite sides of the hydrogen system, are 1.1 V apart in the series.

As previously mentioned the sign of the e.m.f. of the cell is that of the right-hand electrode as the cell is written, thus

$$Zn(s)|Zn^{2+}(aq)|Pb^{2+}(aq)|Pb(s) \quad E^{\ominus} = +0.63 \text{ V}$$

whereas

$$Pb(s)|Pb^{2+}(aq)|Zn^{2+}(aq)|Zn(s) \quad E^{\ominus} = -0.63 \text{ V}$$

The above discussion may be summarized by the statement:

$$E^{\ominus}_{(\text{cell})} = E^{\ominus}_{(\text{right-hand half-cell})} - E^{\ominus}_{(\text{left-hand half-cell})}$$

2. *Predicting whether or not a redox reaction is likely to occur*

Simple displacement reactions show that a metal which is higher in the series will displace a metal which is lower in the series from a solution containing its ions. Such observations are consistent with redox potentials:

a system which is high in the series (high negative $E^\ominus$) is a strong reducing agent and one which is low (high positive $E^\ominus$) will be a strong oxidizing agent. Thus, by referring to the series, one might predict that the reactions

$$Zn(s) + Cu^{2+}(aq) \rightarrow Zn^{2+}(aq) + Cu(s)$$

and

$$Cl_2(g) + 2Br^-(aq) \rightarrow Br_2(l) + 2Cl^-(aq)$$

are likely to occur and that the reaction

$$Br_2(l) + 2F^-(aq) \rightarrow F_2(g) + 2Br^-(aq)$$

is not likely to occur.

In general terms, if the reduced form of a system which is higher in the series is mixed with the oxidized form of a system which is lower in the series then a reaction is likely to occur.

As mentioned on pages 193 and 207 the fundamental factor which determines whether or not a reaction is energetically favourable is the change in free energy. This is related to the e.m.f. of the cell under standard conditions by the equation,

$$\Delta G^\ominus = -zFE^\ominus$$

where z is the number of moles of electrons transferred and F is the Faraday constant. As with all predictions of feasibility based on energy changes for the complete reaction, no indication of the rate is implied. The following are some reactions which could be predicted as feasible from the standard redox potentials of the half reactions and which give observable results in a short time.

(a) Chlorine and bromine oxidize iron(II) salts in solution.

$$2Fe^{2+}(aq) + Cl_2(aq) \text{ or } Br_2(aq) \rightarrow 2Fe^{3+}(aq) + 2Cl^-(aq) \text{ or } 2Br^-(aq)$$

The halogen colour is discharged and the green solution changes to yellow.

(b) Iodine (in potassium iodide solution) oxidizes tin(II) chloride (in moderately concentrated hydrochloric acid).
The brown colour of iodine is discharged, leaving a colourless liquid.

$$I_2(aq) + Sn^{2+}(aq) \rightarrow 2I^-(aq) + Sn^{4+}(aq)$$

(c) The metals with more negative $E^\ominus$ values are oxidized by the hydrogen ion of dilute mineral acids.

$$\left.\begin{array}{l} Zn(s) + 2H^+(aq) \rightarrow Zn^{2+}(aq) + H_2(g) \\ Mg(s) + 2H^+(aq) \rightarrow Mg^{2+}(aq) + H_2(g) \end{array}\right\} \text{Dilute sulphuric and hydrochloric acid}$$

$$2Al(s) + 6H^+(aq) \rightarrow 2Al^{3+}(aq) + 3H_2(g)\} \text{ Dilute hydrochloric acid only}$$

(d) Potassium permanganate in acidic solution oxidizes iron(II) salts in acidic solution.

$$5Fe^{2+}(aq) + MnO_4^-(aq) + 8H^+(aq) \rightarrow 5Fe^{3+}(aq) + Mn^{2+}(aq) + 4H_2O(l)$$

The purple colour of the permanganate is discharged and the green iron(II) solution turns to a yellow iron(III) solution.

(e) Potassium dichromate in acidic solution oxidizes iron(II) salts in acidic solution.

$$6Fe^{2+}(aq) + Cr_2O_7^{2-}(aq) + 14H^+(aq) \rightarrow 6Fe^{3+}(aq) + 2Cr^{3+}(aq) + 7H_2O(l)$$

The orange-yellow dichromate solution is reduced to a green solution of a chromium(III) salt. The green iron(II) solution turns yellow, but this change will be hidden by the other.

(f) Tin(II) chloride (in moderately concentrated hydrochloric acid) reduces an iron(III) salt in acidic solution.

$$Sn^{2+}(aq) + 2Fe^{3+}(aq) \rightarrow Sn^{4+}(aq) + 2Fe^{2+}(aq)$$

The solution changes colour from yellow to green.

17. Rates of Chemical Reactions

The rate of a reaction may be expressed in terms of the rate of decrease in concentration of a reactant or the rate of increase in concentration of a product. As will be discussed later, investigations into rates of reactions may yield useful information which can, for example, help to elucidate reaction mechanisms.

Experimental Methods

1. *Recording the time for a reaction to reach a certain stage*

Examples of this method are the classic iodine clock experiment, in which the time for the iodine-starch colour to appear is recorded, and also the acid-thiosulphate reaction, in which the time for the sulphur precipitate to reach a certain density is recorded. Such experiments can give roughly quantitative information on how rates vary with temperature and concentrations of reactants, but they are of limited value as they only provide an average rate for the period of time over which the experiment has run. Techniques which provide values for initial rates of reactions are of more value.

2. *The progress of a reaction is followed by chemical analysis*

Known volumes are withdrawn from the reaction mixture at regular intervals of time. Each sample is quickly run into an excess of a reagent which will arrest the reaction or alternatively the reaction is stopped by rapid cooling. The sample is then analysed, usually by titration, to determine the concentration of either a reactant or a product. Examples of this technique are:

(a) *Acid catalysed hydrolysis of an ester*

$$CH_3CO_2C_2H_5(l) + H_2O(l) \rightarrow CH_3CO_2H(aq) + C_2H_5OH(aq)$$

The concentration of the ethanoic acid can be determined by titration with standard alkali, suitable allowance being made for the mineral acid which was added to catalyse the reaction.

(b) *Reaction between iodine and propanone (acetone)*

$$I_2 + CH_3COCH_3 \rightarrow CH_3COCH_2I + H^+ + I^-$$

In this case samples can be withdrawn and, after neutralizing the hydrogen ions, each sample can be titrated with standard sodium thiosulphate solution in order to estimate the iodine left.

3. *The progress of a reaction is followed by a physical method*

This group of methods has the advantage that they do not require the removal of portions of the reaction mixture. Examples are:

(a) The volume of a gaseous product is continuously recorded by collecting the gas in a graduated syringe. The decomposition of hydrogen peroxide can be investigated by this method.

$$2H_2O_2(l) \rightarrow 2H_2O(l) + O_2(g)$$

(b) The loss in mass of a reaction mixture due to the loss of a gaseous product.

$$CaCO_3(s) + 2H^+(aq) \rightarrow Ca^{2+}(aq) + H_2O(l) + CO_2(g)$$

(c) The rate of change in concentration of a coloured reactant can be followed by a colorimeter which, by means of a light-sensitive cell, translates the intensity of light transmitted by a coloured solution into an electrical measurement. The instrument is firstly calibrated by finding the ammeter readings for solutions containing known concentrations of the coloured reactant. The progress of the reaction may then be followed by the ammeter readings. The reaction between iodine and propanone, and that between permanganate and oxalate are both suitable for this method.

(d) The change in conductivity of a liquid, due to the change in the number of ions present, may be used to follow a reaction. For example, during the hydrolysis of a halogenoalkane the conductance of the reaction mixture will increase.

$$\underset{\displaystyle CH_3}{CH_3\overset{|}{C}BrCH_3}(l) + H_2O(l) \rightarrow CH_3\,\underset{\displaystyle CH_3}{\overset{|}{C}OHCH_3}(l) + H^+(aq) + Br^-(aq)$$

In addition to the above physical methods, optical activity, refractive index and volume changes of liquids have been used for appropriate reactions.

Factors Which can Affect the Rate of a Chemical Reaction

These are:

1. concentrations of reactants,
2. pressure,
3. temperature,
4. physical state of reactants,
5. catalysts.

In order to find out how the rate of a reaction depends on one of these conditions all of the other conditions must be kept constant and a series of experiments carried out in which the condition under investigation is varied.

Treatment of Experimental Results

1. *Concentrations*

When the progress of a reaction is followed by determining the concentration of a product at regular intervals of time and the concentration of the product is plotted against time, a graph similar to that in Figure 17.1 will be obtained.

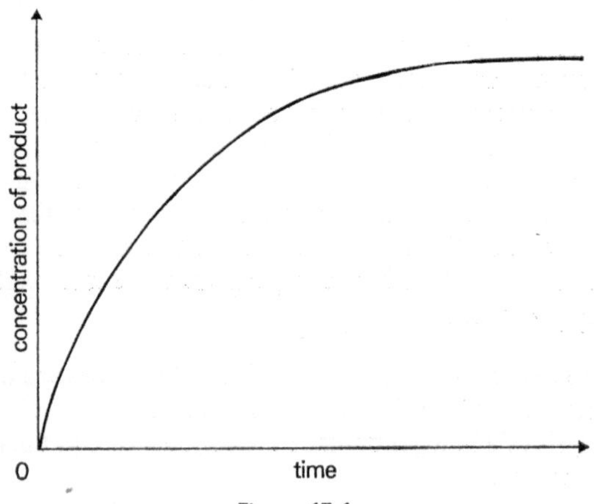

Figure 17.1.

The initial rate of reaction, given by the gradient at $t = 0$, is the most useful as it is only at this point that the concentrations of all the reactants are known. At any time after $t = 0$, some of the reactants will have changed to products. In the general case of

$$A + B \rightarrow \text{products}$$

the dependence of the rate on the concentrations of A and B would be determined by separate investigations. Firstly the concentration of B would be kept constant and a series of experiments carried out with varying concentrations of A. A graph would be plotted for each experiment and the initial rate determined for each starting concentration of A.

Let us say the rate is related to the concentration of A by

$$\text{rate} \propto [\text{A}]^x$$
$$\text{rate} = k\,[\text{A}]^x$$

Taking logarithms,

$$\lg \text{rate} = x \lg[\text{A}] + \lg k$$

This is the equation for a straight line and, when lg[A] is plotted against lg (rate), the gradient will be x and the intercept $\lg k$.

x is known as the *order of the reaction* with respect to A. That is, if rate $\propto$[A] it is first-order with respect to A, or if rate $\propto[\text{A}]^2$ it is second-order with respect to A; k is the velocity or rate constant.

If another series of experiments reveal that the rate is related to the concentration of B by

$$\text{rate} \propto [\text{B}]^y$$

then the order of the reaction with respect to B is y. The overall rate equation is

$$\text{rate} = k[\text{A}]^x\,[\text{B}]^y$$

and the overall order is $x + y$.

First-order reactions

It is not always necessary to carry out a series of experiments to show that a particular reaction is first order. Consider the case of

$$\text{A} \rightarrow \text{products}$$

Assume the concentration of A in the reaction mixture can be determined at various times during the reaction. Let the initial concentration of A be a mol dm^{-3} and the concentration at time t be $(a - x)$ mol dm^{-3}, x mol dm^{-3} having reacted. Then as the reaction is first-order, the rate at which x increases at a particular time will depend on the value of $(a - x)$ at that time, i.e.

$$\frac{\mathrm{d}x}{\mathrm{d}t} = k(a - x)$$

where k is the rate constant. Re-arranged,

$$\frac{\mathrm{d}x}{(a - x)} = k\,\mathrm{d}t$$

Integrating,

$$-\ln (a - x) = kt + c$$

where c is a constant, a value for which is found by considering the case when $t = 0$ and $x = 0$. Thus

$$c = -\ln a$$

$$kt = -\ln(a - x) + \ln a$$

$$kt = \ln \frac{a}{(a - x)}$$

Converting to base 10,

$$kt = 2.303 \lg \frac{a}{(a - x)}$$

Thus if a reaction is first-order, a plot of t against $\lg [a/(a - x)]$ will be a straight line, a value for k being determined by measuring the gradient of the line.

As $\lg [a/(a - x)]$ is a ratio of concentrations, readings such as titration volumes which are proportional to the concentrations may be used instead of the actual concentrations.

Half-life of a first-order reaction

Determination of the time for the concentration of a reactant to fall to half of a particular value (known as the half-life) affords a third method of establishing that a reaction is first-order. The half-life, $t_{\frac{1}{2}}$, is given when

$$(a - x) = \tfrac{1}{2}a$$

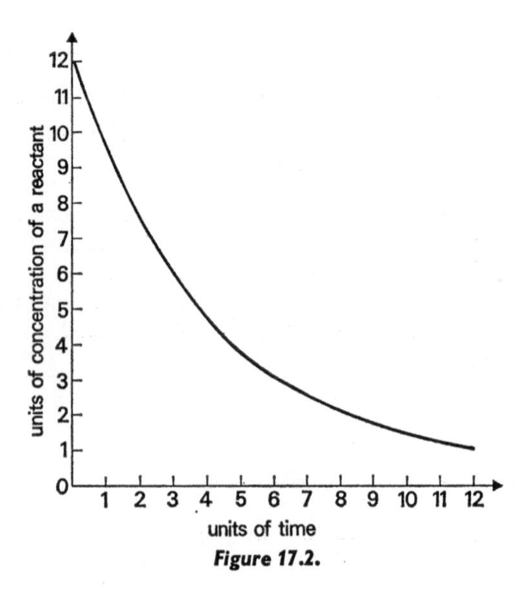

Figure 17.2.

Substituting into

$$kt = 2.303 \lg \frac{a}{(a - x)}$$

we get

$$kt_{\frac{1}{2}} = 2.303 \lg 2$$

From this equation it can be seen that $t_{\frac{1}{2}}$ is independent of the initial concentration of A. Thus a reaction which has a constant half-life throughout its path is a first-order reaction (Figure 17.2). Radioactive decay is an example of a first-order reaction (see page 40).

Second-order reactions

Consider the case of

$$A + B \rightarrow \text{products}$$

Let the initial concentrations of both A and B be a mol dm^{-3}. Then at time t, x mol dm^{-3} of both A and B will have reacted and the rate of the reaction will be given by

$$\frac{dx}{dt} = k(a - x)^2$$

which on integration gives

$$kt = \frac{x}{a(a - x)}$$

Thus for a second-order reaction, a plot of t against $x/a(a - x)$ will be a straight line.

Note that in the case of a second-order reaction, the half-life will depend on the starting concentrations and hence will not be constant throughout the reaction. The half-life $t_{\frac{1}{2}}$ is given when $x = \frac{1}{2}a$. Hence

$$kt_{\frac{1}{2}} = \frac{\frac{1}{2}a}{a(a - \frac{1}{2}a)}$$

$$t_{\frac{1}{2}} = \frac{1}{ka}$$

It must be stressed that the order of a reaction is an experimentally determined quantity. Although it is usually 1 or 2, it is sometimes fractional and, of course, if the rate of a reaction is independent of the concentration of a particular reactant then the reaction is zero-order with respect to that reactant. It is important to realize that the order of a reaction cannot be deduced from the balanced chemical equation for the reaction. The overall order of a reaction can sometimes be modified by carrying out the reaction with one or more reactants present in excess. The rate of the reaction will be independent of changes in concentration of the reactant which is in excess and hence the reaction will be zero-order with respect to that reactant.

Most reactions take place in several stages, and if one of these stages is slower than the others then a measurement of the overall rate of the reaction will be a measurement of the rate of the slowest step. The slow step in a reaction mechanism is known as the **rate-determining step,** The number of molecules or ions involved in the rate-determining step is known as the **molecularity** of the reaction. Unlike the order of a reaction the molecularity must be a whole number, it cannot be fractional or zero.

The order of a reaction provides an indication of the mechanism of the reaction. The following example illustrates this point and may help to clarify the distinction between molecularity and order.

Hydrolysis of halogenoalkanes by alkali

(a) *Bromoethane*

This reaction can be shown experimentally to be first-order with respect to the bromoethane and first-order with respect to the hydroxide ion, thus making the reaction overall second-order.

This result suggests that the rate-determining step involves both of the reactants, i.e.

$$CH_3CH_2Br + OH^- \rightarrow CH_3CH_2OH + Br^-$$

This is a bimolecular reaction and the molecularity and the order ar identical.

If the reaction was carried out with one reactant in a large excess, the reaction would appear to be first-order although still remaining bimolecular.

(b) 2-*bromo*-2-*methylpropane*

In this case the reaction is first-order with respect to the halogenoalkane but zero-order with respect to the hydroxide ion. This experimental result indicates that the rate-determining step does not involve the OH^- ion and the likely mechanism is

$$(CH_3)_3CBr \rightarrow (CH_3)_3C^+ + Br^- \quad \text{(slow)}$$

followed by

$$(CH_3)_3C^+ + OH^- \rightarrow (CH_3)_3COH \quad \text{(fast)}$$

The first stage is the rate-determining step and the reaction is unimolecular.

2. *Temperature*

It can be calculated that the number of collisions between molecules of reactants is so large that if every collision resulted in a reaction, the total reaction would be almost instantaneous. In fact, only those collisions which involve molecules with greater than a certain minimum energy result in a reaction. This minimum energy is known as the **activation energy**.

On average the rate of a reaction doubles for every 10 K rise in temperature. This large increase, which cannot be accounted for by the increase in the number of collisions between the reacting molecules, is due to a doubling of the number of molecules with more than the activation energy. Figure 17.3 shows how the distribution of molecular kinetic energies of a gas changes with a rise in temperature. It is clear that the number of molecules with energy greater than the activation energy, E_a, increases rapidly.

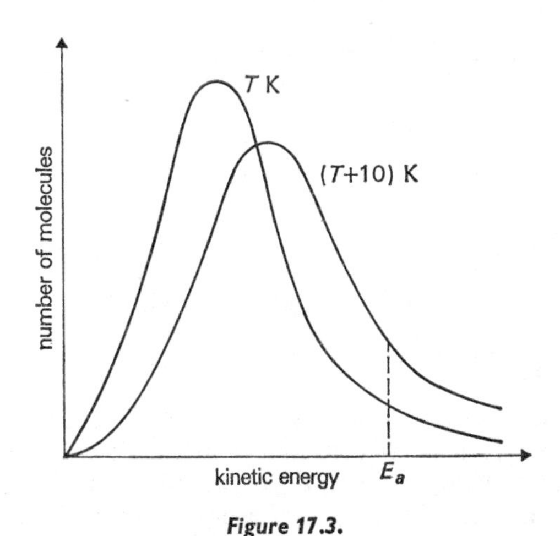

Figure 17.3.

If the rate of a reaction is temperature-dependent, then the rate constant must also vary with temperature. The relationship between the rate constant and temperature, which is known as the **Arrhenius equation**, is

$$k = Ae^{-E_a/RT}$$

Where A is a constant, R is the gas constant, and E_a is the activation energy which can also be taken as a constant. An alternative form of the equation is:

$$\ln k = -\frac{E_a}{RT} + \ln A$$

or

$$\lg k = -\frac{E_a}{2.3RT} + \lg A$$

If a graph of $\lg k$ against $1/T$ is plotted, a straight line is obtained the gradient of which is equal to $-E_a/2.3\ R$. Hence, if the rate constant can be determined at various temperatures, the Arrhenius equation can be used to find the activation energy of the reaction.

3. *Catalysts*

The path of a reaction and the accompanying energy changes may be represented as in Figure 17.4.

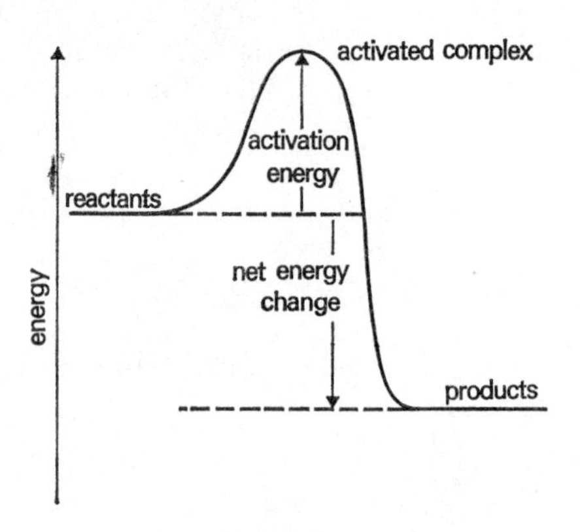

Figure 17.4.

The state of the reactants when they are at the point of maximum energy is sometimes described as the activated complex. To catalyse a reaction it is necessary to provide an alternative pathway in which the activated complex requires a smaller activation energy. This enables a larger proportion of the molecules to achieve the state of activated complex and hence react.

Catalysis is defined as a change in the rate of a chemical reaction brought about by an agent (the *catalyst*) which is left unchanged in mass and in chemical nature at the end of the reaction.

Correspondingly, a **catalyst** is defined as an agent which alters the rate of a chemical reaction and is left unchanged in mass and in chemical nature at the end of the reaction.

The following comments illustrate some points of importance which are not expressly stated in the above definitions but are regarded as additional criteria of catalysis.

(a) The change in the rate of the reaction may be in either direction, i.e., an increase or a decrease. If it is an *increase*, the catalysis is said to be *positive*; if a *decrease*, the catalysis is said to be *negative*. Negative catalysis is also known as *inhibition* and a negative catalyst as an *inhibitor*.

(b) While a catalyst is left unchanged in chemical nature after the reaction, the definition does not exclude physical changes, which are, in fact, quite frequent in catalysis. For example, coarsely powdered manganese(IV) oxide used as catalyst in the decomposition of potassium chlorate becomes much finer in grain.

(c) The definitions above allow *intermediate* chemical changes involving the catalyst, provided that it is regenerated at the close of the catalytic cycle.

(d) In general, a catalyst will operate even when present in very small proportion to the reagents. For example, one ten-millionth of its mass of finely divided platinum will give a measurable increase in the rate of decomposition of hydrogen peroxide. On the other hand, certain agents are classified as catalysts when, to be effective, they need to be present in relatively large amount. For example, in the Friedel-Crafts reaction,

$$C_6H_6(l) + C_2H_5Cl(l) \rightleftharpoons C_6HC_2H_5(l) + HCl(g)$$

the catalyst, anhydrous aluminium chloride, must reach about 30 per cent of the mass of the benzene to be effective.

(e) There is some disagreement as to whether a catalyst can initiate a reaction or merely alter the rate of an existing chemical change. The general opinion favours the second view. On the other hand, this opinion involves the acceptance of the idea that a chemical action must be taken as proceeding even though it cannot be detected over a very long period (perhaps years). For example, a mixture of gaseous hydrocarbons and oxygen, which remains apparently unchanged almost indefinitely at room temperature, can be brought to explosive speed by the catalyst, platinum black, in a few seconds.

The catalyst and equilibrium

In general, a catalyst has no influence on the final state of equilibrium reached by the chemical system. It merely alters the time required to reach the equilibrium, usually shortening the time to a very marked extent. For example, in the reaction

$$N_2(g) + 3H_2(g) \rightleftharpoons 2NH_3(g)$$

the state of equilibrium reached is determined by the choice of temperature, pressure, and the relative proportions of the nitrogen and hydrogen. The introduction of a catalyst, such as reduced iron, merely shortens the time required to reach equilibrium. That is, from the industrial point of view, the economy is in terms of *time*, not in terms of percentage yield from materials. It follows from this experience that a given catalyst must catalyse the forward and back reactions in a reversible system to an equal extent. Otherwise a different equilibrium would be reached on introducing the catalyst. This is also impossible because of energy considerations. Suppose that, in a reversible reaction, a catalyst could move the equilibrium in the direction of energy liberation. Then it would be possible, by alternately inserting and removing the catalyst, to obtain energy indefinitely. This is contrary to the Law of Conservation of Energy.

Classification of catalysis

Two main forms of catalysis are recognized; these are (a) *homogeneous catalysis*, (b) *heterogeneous catalysis*. A third type, mainly of biological interest, is known as *enzyme catalysis*. Enzymes are complex organic catalysts.

1. *Homogeneous catalysis*

All the reagents and the catalyst are in the same physical phase, that is, all are gases or all are liquids (or in solution). Examples of this kind of catalysis are:

(i) Catalysis of the oxidation of sulphur dioxide and steam to sulphuric acid in the Lead Chamber process; the catalyst is nitrogen monoxide.
(ii) Catalysis of the esterification of ethanoic acid by ethanol; the catalyst is dry hydrogen chloride or concentrated sulphuric acid.

2. *Heterogeneous catalysis*

The catalyst is in a different physical phase from the reagents. The most common form of this occurs in *contact catalysis* in which the reagents are in the *gaseous* phase and the catalyst is *solid*. Contact catalysis has great industrial importance; the following are examples of it.

(i) $N_2(g) + 3H_2(g) \rightleftharpoons 2NH_3(g)$, catalysed by finely reduced iron at about 720 K.
(ii) $2SO_2(g) + O_2(g) \rightleftharpoons 2SO_3(g)$, catalysed by platinum (or vanadium pentoxide) at about 720 K, as a stage in sulphuric acid manufacture.
(iii) $4NH_3\ (g) + 5O_2\ (g) \rightarrow 4NO\ (g) + 6H_2O\ (g)$ catalysed by platinum gauze at red heat (or, recently, by iron(III) oxide and bismuth oxide) as a stage in the manufacture of nitric acid.

Theories of catalysis

There are two main theories of catalysis—*the Intermediate Compound Theory* and the *Adsorption Theory*. It is, however, fairly certain that no rigid division can be drawn between these two theories, the suppositions of which may often operate together in the same catalysis.

1. *The Intermediate Compound Theory*

In its simplest presentation, this theory takes the following form. Suppose a reaction, $A + B \rightarrow AB$, is slow and is catalysed by a catalyst, C. The theory supposes that C first combines with one of the reagents, A, to form the compound, AC. This compound then reacts with the reagent, B, to form the compound, AB, releasing the catalyst C, which may then start on another round of the catalytic cycle, i.e.

$$A + C \rightarrow AC$$
$$AC + B \rightarrow AB + C$$

These two reactions together are much faster than the reaction $A + B \rightarrow AB$, to which, in sum, they correspond.

It is obvious that, in being alternately combined and released, the catalyst probably passes through varying states of oxidation. Consequently, elements of variable valency, capable of assuming varying oxidation states, are frequently found in intermediate compound catalysts, especially transition elements such as Mn, Co, and V. Examples of intermediate compound catalysis are quoted below. They conform to the general theory stated above, with, in some cases, rather more complication.

(a) Catalysis of the Lead Chamber process, by nitrogen monoxide.

$$\left.\begin{aligned} NO(g) \quad\quad + \tfrac{1}{2}O_2(g) &\rightarrow NO_2(g) \\ H_2O(l) + SO_2(g) + NO_2(g) &\rightarrow H_2SO_4(l) + NO(g) \end{aligned}\right\}$$

Total result: $H_2O(l) + SO_2(g) + \frac{1}{2}O_2(g) \rightarrow H_2SO_4(l)$

(b) Catalysis of the decomposition of potassium chlorate, by manganese dioxide (according to McLeod).

$$\left.\begin{aligned} 2MnO_2(s) + 2KClO_3(s) &\rightarrow 2KMnO_4(s) + Cl_2(g) + O_2(g) \\ 2KMnO_4(s) &\rightarrow K_2MnO_4(s) + MnO_2(s) + O_2(g) \\ K_2MnO_4(s) + Cl_2(g) &\rightarrow 2KCl(s) + MnO_2(s) + O_2(g) \end{aligned}\right\}$$

Total result: $2KClO_3(s) \rightarrow 2KCl(s) + 3O_2(g)$

It must be remembered, however, that definite evidence of these mechanisms is very difficult to obtain. They are usually plausible rather than proved. Even though each stage can be shown to occur as a single reaction, it does not necessarily follow that the succession of reactions shown above is the actual mechanism of the catalysis. Actually, in case (b) above the oxygen always contains a little chlorine and, if only a relatively little manganese dioxide is used, it can be shown to contain some permanganate after the reaction. This gives some support to the suggested mechanism.

2. *The Adsorption Theory*

This theory applies mainly to catalysis of a reaction between gaseous reagents by a solid catalyst, i.e. to contact catalysis. The theory supposes that the reagents are adsorbed on to the surface of the catalyst, i.e. brought into some form of loose association with the outermost one or two atomic or molecular layers of the catalyst. After the reaction, the products are released (*desorbed*) from the catalyst surface, leaving it free to adsorb more of the reacting substances.

The forces involved in adsorbing the reagents are probably not identical with those concerned in ordinary chemical combination, but are similar to those operating to hold together the atoms (or molecules) of the catalyst. Contact catalysts of this kind are usually most reactive when in a very fine state of division, a fact which points to the surface as the active part of the catalyst.

The factors involved in adsorption catalysis are not fully understood. The adsorption of reacting materials very close together on the catalytic surface is equivalent to a marked concentration of them and accounts partly for the catalytic effect. It is also known that adsorption catalysts are not equally active over their entire surfaces. Spots of exceptional activity occur and are often associated with discontinuities, e.g. cracks and irregularities, and crystal boundaries. It should also be noted that adsorption catalysts also tend to be specific in their action so that the same reagents give different products under the action of different catalysts. An example of this is carbon monoxide and hydrogen:

$$CO(g) + H_2(g) \xrightarrow[\text{Cu}]{\text{heated}} \underset{\text{methanal}}{HCHO(g)}$$

$$CO(g) + 2H_2(g) \xrightarrow[Cr_2O_3]{\text{heated}} \underset{\text{methanol}}{CH_3OH(g)}$$

These facts cannot be explained on a theoretical basis.

The phenomena of catalyst promotion and catalyst poisoning are well known in connection with contact catalysts.

A **catalyst promoter** is an added material which enhances the activity of a main catalyst, but does not necessarily catalyse the reaction if alone. Most mixed catalysts consist of a main catalyst and a promoter. The action of a promoter is not usually very spectacular, but it gives a useful increase in the catalytic activity. For example, the activity of the main catalyst, iron, in Haber's process for ammonia synthesis is substantially promoted by alumina and potassium oxide; also, the production of methanol from carbon monoxide is catalysed more effectively by a mixture of zinc and chromium oxides than by either oxide separately.

There is no adequate theory of the action of catalyst promoters. Possibly some part of their activity lies in the creation of the discontinuities alluded to a little earlier.

Catalyst poisoning occurs when a small proportion of impurity in the reagents gradually suppresses the activity of the catalyst. The impurity is the *catalyst poison.* Well-known examples are the poisoning of platinum as catalyst in the oxidation of sulphur dioxide by arsenic(III) oxide, and the poisoning of the iron catalyst in Haber's process by sulphur compounds, such as H_2S. In some cases, the poisoning occurs because the poison is adsorbed on to the catalyst surface and not released again, so that the surface gradually becomes unavailable for adsorption of the reagents. The poisoning of platinum by arsenic(III) oxide appears to be of this kind. In other cases, the catalyst is chemically changed by reaction with the poison. The poisoning of iron by sulphur compounds falls into this class. The chief safeguard against catalyst poisoning is a thorough purification of the reagents.

Negative catalysis

Catalysis is said to be negative when the catalyst *reduces* the rate of a reaction, often to the point of virtual suppression. Negative catalysis is, for obvious reasons, less useful in industry than positive catalysis, which increases the rate of reaction, but it can be valuable for the suppression of unwanted reactions. For example, about 2 per cent of ethanol in trichloromethane acts as negative catalyst to the oxidation of trichloromethane to poisonous products by the air:

$$4CHCl_3(l) + 3O_2(g) \rightarrow 4COCl_2(g) + 2H_2O(l) + 2Cl_2(g)$$

Also a little benzene-1, 4-diol will suppress the oxidation of benzaldehyde to benzoic acid by the air.

$$2C_6H_5CHO(l) + O_2(g) \rightarrow 2C_6H_5COOH(s)$$

In some cases at least, negative catalysts are believed to operate by interfering with chain-reactions by which the overall reaction progresses. For example, the combination of hydrogen and chlorine is believed to proceed by the chain-reaction

$$Cl_2 \rightarrow Cl\cdot + Cl\cdot \text{ (activated by light)}$$
$$H_2 + Cl\cdot \rightarrow HCl + H\cdot$$
$$H\cdot + Cl_2 \rightarrow HCl + Cl\cdot$$

Nitrogen trichloride is negatively catalytic to the reaction by absorbing free chlorine atoms:

$$NCl_3 + Cl\cdot \rightarrow \tfrac{1}{2}N_2 + 2Cl_2$$

Autocatalysis

Autocatalysis is the catalysis of a reaction by one of its own products. Some examples are:

1. catalysis of the reaction between an oxalate and potassium permanganate by the manganese(II) ion, Mn^{2+}.

   $$2MnO_4^-(aq) + 16H^+(aq) + 5C_2O_4^{2-}(aq) \rightarrow 2Mn^{2+}(aq) + 8H_2O(l) + 10CO_2(g)$$

   The earliest stages of the oxidation are slow, but the accumulation of the catalyst, Mn^{2+}, makes the later stages rapid.

2. catalysis of hydrolysis of an ester by H^+ derived from the acidic product of the hydrolysis, e.g.

   $$CH_3CO_2C_2H_5(l) + H_2O(l) \rightleftharpoons CH_3CO_2H(aq) + C_2H_5OH\ (aq)$$

   $$CH_3CO_2H(aq) \rightleftharpoons CH_3CO_2^-(aq) + H^+(aq)$$

18. Electrolysis and Conductance

Electrolysis

Electrolytes and non-electrolytes

An *electrolyte* can be defined as a compound which, when molten or in solution, can conduct electric current with decomposition.

A *non-electrolyte* is a compound which will not conduct electric current.

The mechanism of electrolysis

> **Electrolysis** can be defined as the decomposition of a compound, molten or in solution, by the passage of electric current.

For electrolysis, a source of direct current is required. The current is conveyed to the electrolyte, from some source such as lead accumulators, by **electrodes.** The electrode connected to the *positive* pole of the source of current is called the **anode**; the electrode connected to the *negative* pole is called the **cathode.** The conventional current enters the electrolyte by the anode and leaves by the cathode.

The mechanism of electrolysis can be described by reference to the case of molten sodium chloride. This salt is completely ionized even in the solid state, and the ions arrange themselves into a rigid cubic lattice. If the temperature of the salt is raised, the ions acquire greater energy until a point is reached at which the lattice is broken down and the salt melts. The ions, Na^+ and Cl^-, are then mobile, and, in the absence of any electrical pressure, move at random in the liquid. If two inert conductors are placed in the liquid, say a platinum wire as cathode and a graphite rod as anode in this case, the cations, Na^+, begin to migrate towards the oppositely charged pole, the cathode; correspondingly the anions, Cl^-, begin to migrate towards the anode. This movement of ions in both directions corresponds to the passage of electric current.

When a sodium ion, Na^+, reaches the cathode, it acquires from the cathode an electron and is converted to a sodium atom. Correspondingly, on reaching the anode, a chlorine ion, Cl^-, gives up an electron to the anode and is converted to a chlorine atom. The chlorine atoms liberated can then combine in pairs to form molecules.

At the cathode

$$Na^+(aq) + e^- \rightarrow Na(s)$$

At the anode

$$Cl^-(aq) - e^- \rightarrow \tfrac{1}{2}Cl_2(g)$$

This accounts for the electrolytic decomposition products of the salt. In an aqueous solution of an electrolyte the situation is more complex, because H^+ and OH^- ions are always present from the ionization of water. In aqueous common salt solution, for example, the situation is:

From sodium chloride: $Na^+(aq) \quad Cl^-(aq)$

$$H_2O(l) \rightleftharpoons H^+(aq) + OH^-(aq)$$

Consequently four ionic species are present. The question at once arises—which of them discharge during electrolysis? This question is decided by a number of factors of which the most important are:

1. The electrode potential of the element or group.
2. The relative concentrations of the ions.
3. The nature of the electrode concerned.

These factors are considered briefly below.

1. *The electrode potential of the element or group.*

If all other factors are constant the order of discharge of ions will be related to that of the electrochemical series. This order, excluding the less common elements (and groups) is:

Cations:

most positive electrode potentials discharge first

$Ag^+ \rightarrow Cu^{2+} \rightarrow H^+ \rightarrow Pb^{2+} \rightarrow Sn^{2+} \rightarrow Fe^{2+} \rightarrow$
$Zn^{2+} \rightarrow Al^{3+} \rightarrow Mg^{2+} \rightarrow Na^+ \rightarrow Ca^{2+} \rightarrow K^+$

most negative electrode potentials discharge last

Anions:

$OH^- \rightarrow I^- \rightarrow Br^- \rightarrow Cl^- \rightarrow NO_3^- \rightarrow SO_4^{2-}$

most negative electrode potentials discharge first — most positive electrode potentials discharge last

This situation in which, in identical conditions, the ions are discharged in the order of their *discharge potentials* is complicated by the following considerations.

2. *Relative concentrations of ions*

The general situation here is that the order of discharge potentials tends to be reversed in an ionic pair if the ion which would normally discharge second is present in relatively high concentration. The most important case is that of OH^- and Cl^-. The order of discharge potentials requires the preferential discharge of OH^- ion and the later production of *oxygen*:

At the anode

$$OH^-(aq) - e^- \rightarrow (OH)$$
$$4(OH) \rightarrow 2H_2O(l) + O_2(g)$$

If, however, the concentration of Cl^- is very high, this order is reversed and the product is almost entirely *chlorine.*

At the anode

$$Cl^-(aq) - e^- \rightarrow \tfrac{1}{2}Cl_2(g)$$

In moderate concentrations of Cl^-, mixtures of chlorine and oxygen are produced with the proportion of chlorine decreasing with dilution. If two ions are far apart in the E.C.S., concentration is an irrelevant factor, e.g. SO_4^{2-} cannot discharge before OH^-.

3. *Nature of the electrode*

In some cases, the nature of the electrode can influence ionic discharge. The most important case of this arises with the ion, Na^+. If a common salt solution is electrolysed with a *platinum* cathode, H^+ (from water) discharges first.

$$2H^+(aq) + 2e^- \rightarrow H_2(g)$$

If, however, a *mercury* cathode is used, the energy of discharge of sodium ion is much reduced and it discharges and forms sodium amalgam.

$$\left.\begin{array}{l} Na^+(aq) + e^- \rightarrow Na(s) \\ Na(s) + Hg(l) \rightarrow NaHg(l) \end{array}\right\}$$

Hydrogen is not liberated in these conditions.

Electrolysis of aqueous solutions

Electrolytes with common cations, e.g. sodium chloride, sodium sulphate, sodium hydroxide, tend to behave similarly at the cathode when electrolysed. Similarly, electrolytes with common anions, e.g. concentrated hydrochloric acid and concentrated solutions of sodium chloride and potassium chloride, tend to behave similarly at the anode. These common cathode and anode changes will now be presented for a number of the better-known ions. The results can be paired to produce the effects of electrolysing particular solutions. In general, the electrode is taken to be inert.

1. *Aqueous solutions containing* $H^+(aq)$, *i.e. dilute acids*

The acid supplies $H^+(aq)$ and so does the water of the solution:

$$H_2O(l) \rightleftharpoons H^+(aq) + OH^-(aq)$$

Consequently there is only one cation to consider.

At the cathode

Hydrogen ion discharges by electron gain

$$H^+(aq) + e^- \rightarrow H$$

The hydrogen atoms combine in pairs to form molecules

$$H + H \rightarrow H_2(g)$$

As the H^+ discharges, the acidity tends to *decrease* at the *cathode.* (There is a corresponding increase of acidity at the anode in the case of H_2SO_4 and HNO_3; see SO_4^{2-} and NO_3^- below). If oxygen is discharged at the anode its volume is *half* of that of cathodic hydrogen.

2. *Aqueous solutions of sodium (and potassium) compounds, containing the ion,* Na^+ (*or* K^+)

The sodium compound provides the ion, Na^+; water ionizes as $H_2O(l) \rightleftharpoons H^+(aq) + OH^-(aq)$. The two cations present are Na^+ and H^+.

At the cathode

H^+ ion discharges in preference to sodium ion:

$$\left.\begin{array}{lll} H^+(aq) & + e^- & \rightarrow H \\ H & + H & \rightarrow H_2 \end{array}\right\}$$

The product is hydrogen gas. As the H^+ discharges, the equilibrium of water ($[H^+][OH^-] = K = 10^{-14}$) is disturbed. To restore K_w, water ionizes further. This puts OH^- into excess and, with Na^+ attracted to the cathode, this is equivalent to the production of a solution of sodium hydroxide. If oxygen is the anode product, its volume is *half* of that of the cathodic hydrogen.

Potassium compounds behave in a corresponding way.

3. *Aqueous solutions of copper (and silver) salts, containing the ion,* Cu^{2+} (*or* Ag^+)

The copper salt produces the ion, Cu^{2+}; water ionizes as

$$H_2O(l) \rightleftharpoons H^+(aq) + OH^-(aq)$$

The two cations present are Cu^{2+} and H^+.

At the cathode

Cu^{2+} ion discharges in preference to hydrogen ion. It discharges by electron-gain.

$$Cu^{2+}(aq) + 2e^- \rightarrow Cu(s)$$

Metallic copper is precipitated. The cathode itself is usually a copper rod or plate.

Silver salts behave in a corresponding way, the cathode usually being a silver plate or wire.

$$Ag^+(aq) + e^- \rightarrow Ag(s)$$

4. *Aqueous solutions of soluble hydroxides, notably sodium hydroxide and potassium hydroxide*

The hydroxide produces the ion, OH^-; water ionizes as

$$H_2O(l) \rightleftharpoons H^+(aq) + OH^-(aq)$$

In this case, only one anion, OH^-, is present.

At the anode

OH^- ion discharges by electron loss. The neutralized (OH) group is not stable. It generates water and *oxygen.*

$$OH^-(aq) - e^- = (OH)$$
$$4(OH) \rightarrow 2H_2O(l) + O_2(g)$$

If (as is usual) hydrogen is the corresponding cathode product, its volume is *twice* that of the oxygen.

5. *Aqueous solutions of soluble sulphates (including sulphuric acid)*

The soluble sulphate produces the ion, SO_4^{2-}; water ionizes as $H_2O(l) \rightleftharpoons H^+(aq) + OH^-(aq)$. Consequently, there are two anions present, SO_4^{2-} and OH^-.

At the anode

OH^- discharges, by electron loss, in preference to SO_4^{2-}, and then oxygen is produced as in (4) above. The relation to cathodic hydrogen is the same.

As OH^- discharges, the equilibrium of water

$$([H^+][OH^-] = K_w = 1 \times 10^{-14})$$

is disturbed. To restore K_w, water ionizes further, putting H^+ into excess. This, with incoming SO_4^{2-}, is equivalent to the production of sulphuric acid (or an increase in its concentration). This increase is balanced, in suitable cases, by a decrease in acidic concentration at the cathode (see H^+ earlier).

In electrolysing copper sulphate solution, the anode is often a copper plate. In this case, no discharge of OH^- (or SO_4^{2-}) will occur. Instead, copper atoms of the anode will ionize by electron loss,

$$Cu(s) - 2e^- \rightarrow Cu^{2+}(aq)$$

and the ions will pass into solution. This process requires least energy. The anode slowly dissolves. A corresponding *equal* precipitation of copper occurs at the cathode.

6. *Aqueous solutions of soluble chloride (including hydrochloric acid)*

The soluble chloride produces the ion, Cl^-; water ionizes as

$$H_2O(l) \rightleftharpoons H^+(aq) + OH^-(aq)$$

Consequently, there are two anions present, Cl^- and OH^-.

At the anode

In very dilute solution, OH^- discharges by electron loss in preference to Cl^- Then *oxygen* is produced, as

$$OH^-(aq) - e^- \rightarrow (OH)$$
$$4(OH) \rightarrow 2H_2O(l) + O_2(g)$$

The volume of oxygen is *half* of that of hydrogen produced in the same electrolysis at the cathode.

In very concentrated solution, the above order of discharge is almost completely reversed by concentration effects and almost the sole product is *chlorine*.

$$Cl^-(aq) - e^- \rightarrow \tfrac{1}{2}Cl_2(g)$$

After saturation of the solution, the volume of chlorine is *equal* to that of hydrogen produced in the same electrolysis at the cathode.

At intermediate concentrations of chloride, the above two effects will occur mixed.

7. *Aqueous solutions of nitrates, especially silver nitrate*

The soluble nitrate produces the ion, NO_3^-; water ionizes as

$$H_2O\ (l) \rightleftharpoons H^+(aq) - OH^-(aq)$$

Consequently, there are two anions present, OH^- and NO_3^-.

At the anode

OH^- discharges, by electron loss, in preference to NO_3^-, and *oxygen* is produced.

$$OH^-(aq) - e^- = (OH)$$
$$4(OH) \rightarrow 2H_2O(l) + O_2(g)$$

The volume of the oxygen is *half* of that of hydrogen produced at the cathode in the same electrolysis. As OH^- discharges, the equilibrium of water ($[H^+][OH^-] = K_w = 1 \times 10^{-14}$) is disturbed. To restore K_w, water ionizes further. This puts H^+ into excess and, with incoming NO_3^-, is equivalent to the production of nitric acid (or an increase in its concentration). This increase is balanced, in suitable cases, by a corresponding decrease at the cathode (see H^+ earlier).

In the case of silver nitrate solution, an anode of silver is often used. In that case, OH^- (or NO_3^-) is not discharged. Instead, silver atoms of the anode ionize as $Ag - e^- = Ag^+$. The ions pass into solution, so the

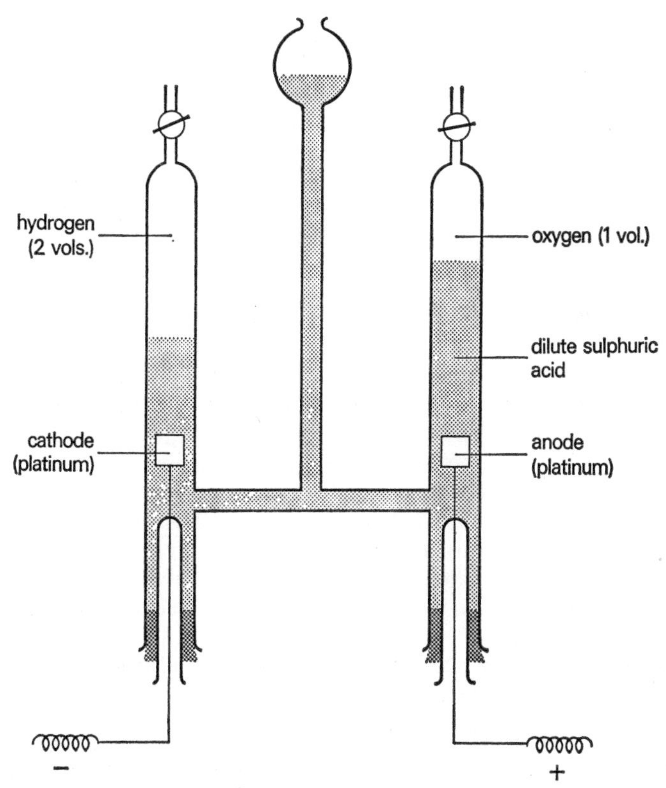

Figure 18.1. Electrolysis of dilute sulphuric acid

anode slowly dissolves. There is a corresponding and *equal* deposition of silver at the cathode.

The apparatus of Figure 18.1 is suitable for the electrolysis of dilute sulphuric acid, sodium (and potassium) sulphate, and sodium (and potassium) hydroxide in solution. The result shown is the same for all these solutions and is equivalent to the electrolysis of water. The cathodic and anodic changes are detailed above.

The same apparatus is suitable for the electrolysis of concentrated hydrochloric acid and concentrated solutions of sodium (and potassium) chloride, but the anode must be changed to a carbon rod (since chlorine produced in this way attacks platinum appreciably). The gaseous products are then hydrogen and chlorine in equal volumes for all the solutions mentioned (after chlorine has saturated the solution).

Copper sulphate and silver nitrate are usually electrolysed with copper plates and silver plates, respectively, as both anode and cathode. The plates are merely inserted (suitably spaced) into the solutions contained in glass containers.

Faraday's Laws of Electrolysis

These laws express the quantitative side of electrolysis. The first law is expressed in the following way:

> The mass of a given element liberated during electrolysis is directly proportional to the magnitude of the steady current used and to the time for which the current passes; that is, to the quantity of electricity consumed during the electrolysis.

The first law can be illustrated by passing known steady currents of electricity through a solution of copper(II) sulphate between copper electrodes for observed times. In each experiment the cathode is weighed before the electrolysis and washed, dried, and weighed after it. The increase of mass of the cathode (i.e. the mass of copper deposited) will be found to be directly proportional to the product (steady current × time).

One mole of singly charged ions will require a transfer of 1 mole of electrons for neutralization. Similarly, 1 mole of doubly charged ions will require 2 moles of electrons. The transfer of 1 mole of electrons corresponds to the passing of approximately 96 500 coulombs of electricity, which is known as the Faraday constant. A coulomb is equivalent to the passage of 1 ampere for 1 second.

Faraday's second law may thus be stated in the following way:

> When the same quantity of electricity is passed through different electrolytes, the masses of the different substances liberated are directly proportional to the masses of the substances which require one mole of electrons (1 faraday) for neutralization.

These masses are equal to:

the mass of 1 mole for singly charged ions,
the mass of $\frac{1}{2}$ mole for doubly charged ions and
the mass of $\frac{1}{3}$ mole for ions with three charges.

These masses have been known as the chemical equivalent masses of the respective ions.

The second law can be illustrated by passing a suitable electric current (the exact magnitude of which need not be known) in series for the same time through, say, dilute sulphuric acid (with platinum electrodes), copper(II) sulphate solution (with copper electrodes) and silver nitrate solution (with silver electrodes). The volumes of hydrogen and oxygen liberated in the first case (see page 275) are observed at known room temperature and pressure, converted to s.t.p. and rendered as masses from the relation: 1 dm^3 of hydrogen has a mass of 0.09 g at s.t.p. or 1 dm^3 of oxygen at s.t.p. has a mass of 1.44 g. The copper (or silver) cathode is weighed before the electrolysis, and washed, dried, and weighed after it. The increase of mass is the mass of copper (or silver) liberated. It will be found that if the mass of oxygen liberated is $8.00x$ g, the masses of hydrogen, copper, and silver are $1.008x$, $31.8x$, and $107.88x$ g.

8 g is the mass of 0.5 mole of oxygen atoms, 1.008 g is the mass of 1 mole of hydrogen atoms, 31.8 g is the mass of 0.5 mole of copper atoms, and 107.88 is the mass of 1 mole of silver atoms.

Electrochemical equivalent of an element

This characteristic of an element is defined in the following way:

> The electrochemical equivalent of an element is the mass of the element liberated during electrolysis by the consumption of one coulomb of electricity, i.e. by the passage of a steady current of one ampere for one second.

Some Aspects of Conductance of Electrolytes

Except in certain exceptional circumstances, electrolytes obey the same laws as metallic conductors with respect to electrical resistance (and conductance); that is, at constant temperature, the electrical resistance of a conductor is directly proportional to its length and inversely proportional to its (constant) cross-sectional area. This can be expressed in the equation

$$R = \text{a constant} \times \frac{l}{a}$$

where R is the resistance (expressed in ohms), l is the length of the conductor (in cm) and a its constant cross-sectional area (in cm^2).

With these units, the constant of this equation is called the **specific resistance** of the material of which the conductor is composed. R becomes equal to the constant when both l and a are unity; that is, the specific resistance of the material is the resistance in ohms between opposite faces of a centimetre cube of it.

The reciprocal of the specific resistance is called the **specific conductance** of the material and is measured in reciprocal ohms, that is, in ohm^{-1}. The specific conductance is the characteristic of an electrolyte which is measured (by means of the specific resistance in actual practice), but it is not the most valuable data about the electrolyte, because it varies with dilution in a two-fold way. In the first place, dilution reduces the number of ions (the conducting particles) in unit volume of the electrolyte and this *increases* the resistance of the electrolyte; in the second place, weak electrolytes increase their ionization on dilution, which *decreases* the resistance of the electrolyte. These two effects are confused in the specific conductance values.

To sort out this confusion, a characteristic of the electrolyte is employed called its *equivalent conductance.* This is obtained by multiplying the specific conductance of the electrolyte by the number of cm^3 containing the gramme-equivalent of the electrolyte, that is, the mass of electrolyte which requires the passage of 1 Faraday for neutralization. The effect of this is to cancel out the simple dilution factor in conductance as concentration varies, and, for weak electrolytes, to exhibit the effect of variation in ionization. The equivalent conductance of an electrolyte is usually denoted by Λ_v, where v is the number of cm^3 containing 1 g-equivalent of the electrolyte. Consequently,

$$\Lambda_v = (\text{specific conductance}) \times v$$

As we shall see a little later, the equivalent conductance of an electrolyte at infinite dilution (i.e. when 1 g-equivalent of it is dissolved in an infinitely large volume of solution) is an important characteristic of it, and is denoted by the symbol, Λ_∞.

The concept of **molar conductance** is also in use and is exactly similar to that of equivalent conductance, except that it relates to 1 mole of the electrolyte; that is, the molar conductance (μ) of an electrolyte is its specific conductance multiplied by the number of cm^3 containing 1 mole of the electrolyte. For electrolytes such as Na^+Cl^-, K^+Cl^-, $Ag^+NO_3^-$, in which both ions are univalent, equivalent and molar conductance have the same value. For $Ba^{2+}(Cl^-)_2$, $Zn^{2+}SO_4^{2-}$, $Pb^{2+}(NO_3^-)_2$ and similar electrolytes, the equivalent mass is *half* the mass of 1 mole and equivalent conductance is *half* of the molar conductance. A brief account of the experimental determination of specific resistance, from which specific equivalent and molar conductance can be calculated, is given later.

Equivalent conductance and ionization

It has already been pointed out that, because of the dilution factor involved in it, equivalent conductance is independent of the effect of simple dilution in reducing the conducting power of an electrolyte. It is affected by two other factors – by the velocity of the ions and by the extent to which the electrolyte is ionized, i.e. its degree of ionization. At first it was assumed that the velocity of a given ion is constant at constant temperature. In this case the equivalent conductance of an electrolyte is directly related to its degree of ionization. Since, for all electrolytes, ionization becomes complete at infinite dilution, the relation was thought to hold:

$$\text{degree of ionization } (\alpha) \text{ of electrolyte at dilution, } v = \frac{\Lambda_v}{\Lambda_\infty}$$

when v is the volume (in cm^3) containing 1 g-equivalent of the electrolyte, and Λ_v and Λ_∞ are the equivalent conductances of the electrolyte at dilution, v, and at infinite dilution.

For weak electrolytes (that is, electrolytes which are only slightly ionized at the usual working dilutions), this relation still holds and is a very valuable method of determining degree of ionization. Examples of its working are given later when some other relevant factors have been considered. For strong electrolytes, however, which are fully ionized (or very nearly so) in all 'dilute' solutions, the above relation is not now considered valid. In strong electrolytes of moderate concentration, the ions are comparatively close together and each tends to interfere with the progress of oppositely charged ions through the solution, giving the effect of a mutual slowing down. This reduces the observed value of equivalent conductance below its 'true' value. Dilution increases the average distance separating ions. Consequently they tend to interfere less with each other and equivalent conductance increases until, at infinite dilution, it reaches a constant maximum value. This change was formerly ascribed to increased ionization; the present view is that it is rather the result of increased velocity of the ions, as their effect on each other lessens. With weak electrolytes, relatively few ions are present and their mutual interference effects are almost negligible at all concentrations.

The above fraction, Λ_v/Λ_∞, is now called the **conductance ratio** of an electrolyte. For *weak electrolytes only*, it is also (with fair accuracy) a measure of their degree of ionization.

Change of equivalent conductance with dilution

The figures in the table on page 280 are typical of the changes of equivalent conductance with dilution in the case of a strong electrolyte (sodium chloride) and a weak electrolyte (ethanoic acid).

The following features are notable about these figures. It will be seen that, in both cases, equivalent conductance *increases* as the solution becomes more dilute. In the case of sodium chloride (the strong electrolyte), however, the increase is relatively small (from 74.2 to 106.3, or about 43 per cent), while in the case of the weak electrolyte, the relative increase is much greater, from 1.32 to 41.0 or about 3000 per cent. Further, in the case of the strong electrolyte the value of equivalent conductance rises very little between 0.01 M and 0.001 M and is reaching a constant value at about 0.001 M. In the case of the weak electrolyte, the equivalent conductance is still rising at 0.001 M dilution and shows no tendency towards constancy. These features are characteristic of the behaviour of strong and

Molarity	*NaCl*	CH_3CO_2H
M	74.2	1.32
0.5 M	80.8	2.01
0.2 M	87.5	3.24
0.1 M	91.8	4.60
0.05 M	95.5	6.48
0.01 M	101.7	14.3
0.002 M	105.3	30.2
0.001 M	106.3	41.0

weak electrolytes, though it must be remembered that an absolutely sharp distinction between them cannot be drawn.

The general position is indicated approximately by the graphs in Figure 18.2.

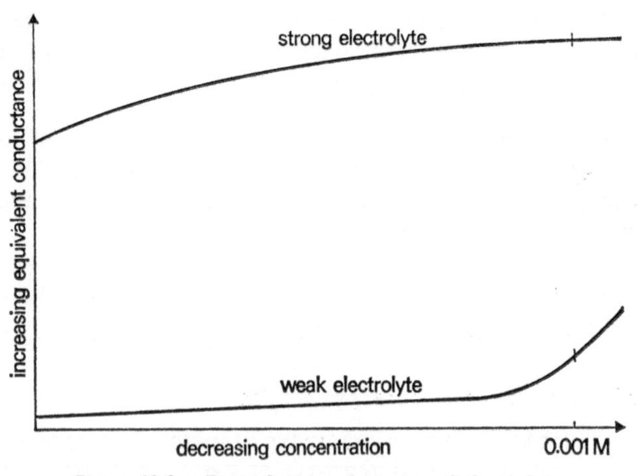

Figure 18.2. Equivalent conductance of electrolytes

Equivalent conductance at infinite dilution

This characteristic of an electrolyte can be determined experimentally for a strong electrolyte as follows. For a few of its solutions, the specific conductance is first determined. Experimental details are given on page 281.

The specific conductance is then multiplied by the number of cm^3 of solution containing 1 g-equivalent of the electrolyte. If a graph is plotted of equivalent conductance against the square root of concentration, a straight line is obtained for low concentrations. If it is extrapolated to zero concentration, i.e. infinite dilution, the value of equivalent conductance at infinite dilution can be read off.

For weak electrolytes, this method is less suitable. A method based on the following facts is used. The values for equivalent conductance of certain potassium salts and the corresponding sodium salts are (at infinite dilution):

	Potassium	*Sodium*	*Difference*
Chloride	130.1	109.0	21.1
Nitrate	126.5	105.3	21.2
Sulphate	133.0	111.9	21.1

It will be seen that varying the cation produces, within experimental limits, a constant difference in the equivalent conductances. Similarly, the KCl–KNO_3 difference ($130.1 - 126.5 = 3.6$) is the same as the $NaCl$–$NaNO_3$ difference ($109.0 - 105.3 = 3.7$) and the K_2SO_4–KNO_3 difference ($133.0 - 126.5 = 6.5$) is the same as the Na_2SO_4–$NaNO_3$ difference ($111.9 - 105.3 = 6.6$). From general results such as these, Kohlrausch reached the conclusion that the equivalent conductance of an electrolyte at infinite dilution is made up of a contribution from the cation and another from the anion, and that these contributions are constant and independent of each other. That is, the equivalent conductance of an electrolyte at infinite dilution is made up of the sum of two independent quantities called the *mobilities* of the two ions concerned. This is known as the *Law of Independent Migration of Ions.*

Experimental determination of specific conductance and equivalent conductance of an electrolyte

In principle, these determinations require the measurement of the resistance of the electrolyte in a cell like that shown in Figure 18.3 between similar parallel electrodes of known area at a known distance apart. In practice, this rather difficult measurement of exact dimensions is avoided by relying on accurate determination of the specific conductance of potassium chloride solutions made by earlier workers. For example, the specific conductance of 0.02 M KCl at 298 K is 0.002 77 ohm^{-1} and of 0.1 M KCl, at the same temperature, 0.0129 ohm^{-1}. From accepted figures such as these, it is possible to calibrate a given cell and avoid measurement of electrode dimensions and distances. This is done by obtaining its *cell constant*. For example, the resistance of 0.02 M KCl at 298 K in a given cell (measured as below) was found to be 550.0 ohm, i.e. the conductance of the electrolyte was $1/550.0 = 0.001\,82$ ohm^{-1}. From these figures, the cell constant for this particular cell is $0.002\,77/0.001\,82 = 1.52(3)$. That is, all conductances found by this cell must be multiplied by 1.52(3), and then yield true values for specific conductance.

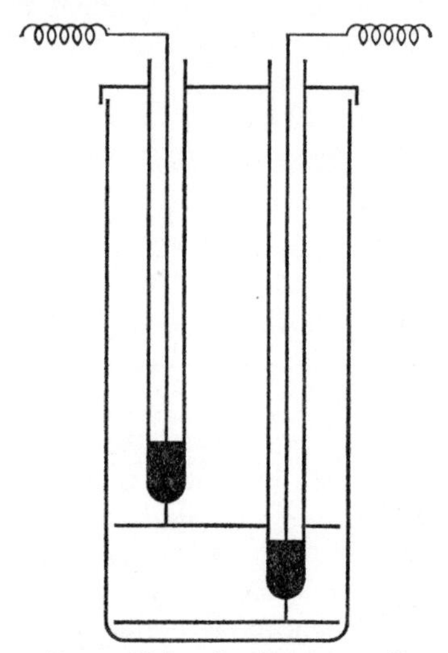

Figure 18.3. Conductance cell

In measuring the resistance of electrolytes, certain precautions are necessary. Among them are the following:

1. The solution must be made up in specially prepared 'conductivity water'. For ordinary measurements, distilled water is redistilled containing potassium permanganate, in apparatus of resistance glass, with ground joints, or corks protected by tin foil.

2. Direct current must not be used because it decomposes the electrolyte. This introduces the double error of altering the concentration of the electrolyte and depositing decomposition products on the electrodes, which will tend to set up a back e.m.f. and so increase the apparent resistance.

3. To avoid polarization, the electrodes are usually covered with platinum black by electrolysing platinic chloride solution in the cell before using it for measurements. Direct current is passed in each direction for half-minute intervals till the electrodes are blackened. (Such electrodes are unsuitable for use with electrolytes on which platinum shows catalytic activity.)

4. Since the conductance of an electrolyte varies with temperature, the cell should be placed in a thermostat (usually at 298 K).

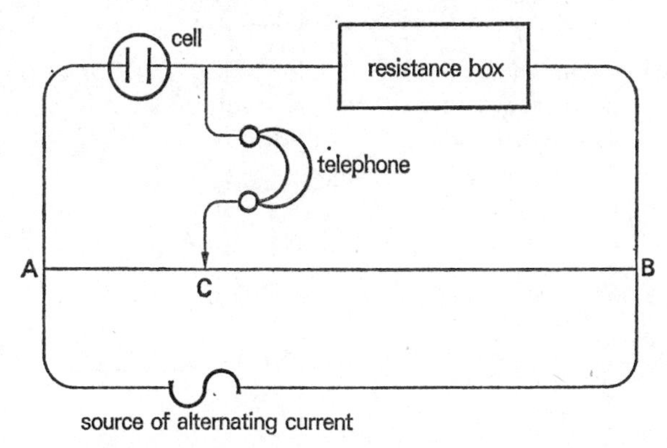

Figure 18.4. Determination of resistance of an electrolyte

For a determination, the lay-out is arranged as in Figure 18.4. AB is a resistance wire of uniform cross-section. The value of the resistance in the resistance box is adjusted by trial till a *minimum* sound is obtained in the telephone with the contact C near the middle of AB. Then

$$\text{resistance of electrolyte in cell} = (\text{resistance in box}) \times \frac{AC}{BC}$$

The reciprocal of the resistance gives the conductance of the electrolyte; this when multiplied by the cell constant, gives the true specific conductance. A sample calculation is given below.

Electrolyte: 0.1 M zinc sulphate solution (at 298 K).

With 70 ohms in the resistance box, AC = 50.8 cm and BC = (100 − 50.8) = 49.2 cm.

That is, resistance of the electrolyte is

$$70 \times \frac{50.8}{49.2} = 72.2 \text{ ohm}$$

From this,

$$\text{conductance of electrolyte} = \frac{1}{72.2} \text{ ohm}^{-1}$$

The cell constant (from the calculation above) being 1.52, the specific conductance of the electrolyte is

$$\frac{1.52}{72.2} = 0.0211 \text{ ohm}^{-1}$$

Since 0.1 M means that 1 mole of the electrolyte is dissolved in 10 000 cm^3 of solution, the molar conductance of zinc sulphate at this dilution is

$$0.0211 \times 10\,000 = 211$$

The g-equivalent of zinc sulphate (Zn^{2+} SO_4^{2-}) is *half* of its formula mass and so is contained in 5000 cm^3 of solution. That is, the equivalent conductance of zinc sulphate at this dilution is

$$0.0211 \times 5000 = 105.5$$

Questions on Chapter 1

1. Summarize the propositions which together constitute the Atomic Theory of Dalton. Show that they lead to the conclusions expressed in the Laws of Constant Composition and Multiple Proportions. In the case of *one* of these laws, outline a piece of experimental work by which it can be illustrated.

2. Outline the work of Landolt in illustrating the Law of Conservation of Matter. Relate the law to the Atomic Theory of Dalton. To what extent has this law been modified in the light of recent discoveries? Explain why pure samples of sodium chloride conform to the Law of Constant Composition though chlorine has two isotopes of masses 35 and 37 on the scale of O = 16.

3. Three oxides of nitrogen contain respectively 46.7%, 30.4%, and 63.6% of nitrogen. Show that these figures illustrate the Law of Multiple Proportions. Relate this law to Dalton's Atomic Theory.

4. Water, sulphur dioxide, and magnesium oxide contain 88.9%, 50.0%, and 40.0% of oxygen respectively. Magnesium sulphide contains 57.1% of sulphur and hydrogen sulphide 94.1% of sulphur. Show that these figures illustrate the Law of Reciprocal Proportions. Explain how this law can be deduced from Dalton's Atomic Theory.

5. 85.0 cm^3 of hydrogen, measured dry at 16 °C and 750 mmHg pressure, were liberated by the action of 0.200 g of a certain metal on excess of dilute acid. Calculate the mass of metal which would liberate 22.4 dm^3 of hydrogen at s.t.p. If the specific heat capacity of the metal is 0.46 $J\,g^{-1}\,K^{-1}$, what are the relative atomic mass and valency of the metal? [1 dm^3 of hydrogen at s.t.p. weighs 0.09 g.]

6. State what is meant by *isomorphism*. State Mitscherlich's Law of Isomorphism. Briefly discuss its limitations. At a certain period in the nineteenth century, the mass of zirconium which combined with 8.00 g of oxygen was known (correctly) to be 22.8 g. The accepted valency of the element was 3. Criticize this situation in the light of the fact that zirconium forms a compound isomorphous with the silicon compound known to be K_2SiF_6 and containing about 32% of zirconium. [K = 39, F = 19.]

7. Given the standard $^{12}C = 12.00$, the relative atomic masses of hydrogen, silver, nitrogen, and chlorine are very important as a basis for fixing other relative atomic masses. Describe *two accurate* experiments which have been performed to assist in the determination of the relative atomic masses of these four elements.

8. A certain metallic element forms (a) an oxide containing 23.1% of oxygen, (b) a chloride containing 43.3% of chlorine, (c) two sulphides containing 37.6% and 25.6% of sulphur. The specific heat capacity of the element is 0.146 $J\,g^{-1}\,K^{-1}$. What deductions can you draw from these facts? [O = 16, S = 32, Cl = 35.5.]

9. The two oxides of titanium contain (a) 66.7% and (b) 60.0% of titanium. The specific heat capacity of the metal is 0.54 $J\,g^{-1}\,K^{-1}$. Calculate the relative atomic mass of titanium as an average derived from these data. What is the percentage composition of the chloride of titanium corresponding to oxide (a)? [Cl = 35.5.]

10. Describe an experiment to demonstrate the composition of hydrogen chloride by volume. State its result. Show that this result requires the hydrogen molecule to contain at least two atoms. What further evidence can be adduced to prove that the hydrogen molecule is actually diatomic?

11. State (a) Gay-Lussac's Law of Gaseous Volumes, (b) Avogadro's Hypothesis. Illustrate (a) by stating the volume compositions of ammonia and steam. By applying Avogadro's Hypothesis to these compositions, and assuming the diatomicity of the hydrogen, nitrogen, and oxygen molecules, deduce the molecular formulae of ammonia and steam.

12. Assuming the diatomicity of the hydrogen molecule, prove the relation between the relative molecular mass and vapour density of a gas. Calculate the relative molecular mass of an elementary gas, X, from the data:

Weight of evacuated globe = 211.470 g

Weight of globe filled with hydrogen = 211.830 g

Weight of globe filled with X = 219.380 g

(Temperature and pressure remain constant.)

Questions on Chapter 2

1. Calculate the mass of each of the following: (a) 2 moles of calcium atoms, (b) 0.5 mole of nitrogen molecules, (c) 0.1 mole of sodium chloride, NaCl, (d) 0.2 mole of hydrated copper(II) sulphate, $CuSO_4 . 5H_2O$, (e) 2.5 moles of ethanol, C_2H_5OH.

2. Calculate the number of moles in each of the following. Clearly state the nature of the species concerned. (a) 3.9 g of benzene, C_6H_6, (b) 41.4 g of lead, (c) 10.1 g of potassium nitrate, KNO_3, (d) 10 kg of magnesium nitride, Mg_3N_2, (e) 8.6 g of gypsum, $CaSO_4 . 2H_2O$.

3. Calculate the molarity of each of the following solutions:
(a) sodium bromide, NaBr, containing 1.03 g of NaBr in 50 cm^3 of solution, (b) hydrogen ion, in a solution containing 9.8 g of sulphuric acid, H_2SO_4, in 200 cm^3 of solution, (c) magnesium sulphate, $MgSO_4$, containing 24 g of $MgSO_4$ in 100 cm^3 of solution, (d) potassium permanganate, $KMnO_4$, containing 7.9 g of $KMnO_4$ in 1000 cm^3 of solution, (e) ammonia, NH_3, containing 11.2 dm^3 (measured at s.t.p.) of NH_3(g) dissolved in 250 cm^3 of solution.

4. A compound containing only carbon, hydrogen, and oxygen gave the following results on analysis: C, 40.0%; H, 6.7%; O, 43.3%. Find its empirical formula. If its relative molecular mass is 60, what is its molecular formula? The compound is found to release carbon dioxide from sodium carbonate solution. Can you suggest a structural formula for the compound?

5. Barium chloride crystals are known to be hydrated. 12.2 g of the crystals were heated to constant weight. The anhydrous barium chloride weighed 10.4 g. Deduce the formula of the hydrated crystals.

6. Calculate the number of moles of each reactant and each product in the reactions described below. Write an equation for each reaction, showing how you have used the information provided. (Gas volumes are measured under usual room conditions, when one mole of gas molecules occupies 24 dm^3.) (a) 6.62 g of lead(II) nitrate, $Pb(NO_3)_2$, were heated to constant weight. 4.46 g of lead(II) oxide, PbO, were left. The other products of the reaction were 1.84 g of dinitrogen tetraoxide, N_2O_4, and 480 cm^3 of oxygen gas. (b) 13 g of zinc reacted with 200 cm^3 of molar copper(II) sulphate, $CuSO_4$. 12.8 g of copper were precipitated and 200 cm^3

of a molar solution of zinc sulphate, $ZnSO_4$, were left. (c) 75 cm^3 of 2 M sodium hydroxide, NaOH, reacted with 50 cm^3 of molar phosphoric(V) acid (orthophosphoric acid), H_3PO_4. The reaction produced 16.4 g of sodium phosphate, Na_3PO_4.

Questions on Chapter 4

1. What do you understand by the term *isotopy*? Illustrate it by reference to the element *chlorine*. Explain why, in spite of the existence of chlorine isotopes, the relative atomic mass of this element appears constant at 35.5 in all ordinary samples.

2. Give an account of *two* different types of experimental evidence which support the view that certain elements exhibit isotopy. Explain this phenomenon in general terms. Is there any observed connection between the atomic number of an element and the occurrence of isotopy in the element? Analysis showed that 2.743 g of ordinary lead chloride, $PbCl_2$, required 2.128 g of silver for complete precipitation as silver chloride; corresponding figures for two samples of the chloride from lead associated with radioactivity were (a) 1.613 g required 1.255 g of silver, (b) 3.748 g required 2.901 g of silver. Calculate the relative atomic mass of lead from each source and explain the results. [Ag = 107.9, Cl = 35.46.]

3. Discuss the isotopy of oxygen and consider its relevance to the relative atomic mass scale. Do you consider that it would be more suitable to adopt a scale based upon F = 19.00? Give your reasons.

4. Account for the existence of three isotopes of hydrogen. Why has it been thought desirable to allot separate chemical names and symbols to them? Describe *one* method of preparation of deuterium oxide (heavy water) and contrast its physical properties with those of protium oxide (ordinary water).

5. Explain why separation of isotopes of the same element must, in general, depend on physical methods. Outline *three* such methods. Why, in spite of isotopy, is the order of relative atomic masses of elements almost the same as the order of atomic numbers?

6. Briefly explain the principles which underlie the use of nuclear energy for the production of electricity. Outline the problems involved in the disposal of the waste products from nuclear power stations.

7. Give an account of the usefulness of isotopes in industry and medicine.

Questions on Chapter 6

1. Explain briefly the basis on which Mendeléeff s Periodic Table was originally drawn up about 1870. Quote the table in its modern form up to the element *calcium*. Using the portion of the table quoted, illustrate the meaning of periodicity of properties by reference to (a) valency towards hydrogen or chlorine, (b) valency towards oxygen, (c) one other property of the elements.

2. Why is it now considered correct that potassium (K = 39.1) should follow argon (Ar = 39.9) in the Periodic Table? What factors explain the lower relative atomic mass of potassium? Mention one other similar case in the table.

3. Relate the classification of elements in the Periodic Table, from He to Ca, to the electronic arrangements shown in the atoms of these elements.

4. Give the names and *full electron structures* of two elements in each case from the following sections of the Periodic Table: (a) the s block, (b) the p block, other than noble gases (rare gases), (c) the d block. (An example of a full electron structure is: Ar, $1s^2 2s^2 2p^6 3s^2 3p^6$; see Chapter 5.)

Questions on Chapter 7

1. Explain what is meant by the terms *electrovalency*, *covalency*, and *dative covalency*. Illustrate them by reference to the compounds calcium chloride, phosphorus trichloride oxide, $POCl_3$, and trichloromethane, $CHCl_3$. Briefly relate the properties of these compounds to the kinds of valency they exhibit.

2. Discuss the valency types shown in the compounds calcium oxide, ammonium chloride, and tetrachloromethane, CCl_4. Explain why ammonium chloride can precipitate silver chloride from silver nitrate solution but tetrachloromethane cannot do this.

3. Discuss in electronic terms (a) the formation of ions of the type, MO_4^{n-}, illustrating by reference to ClO_4^- and PO_4^{3-}, (b) the existence of a stable compound, $BCl_3 \cdot NH_3$, (c) the valency types shown in the formation of ammonium sulphate.

4. What is meant by the term *atomic number* of an element? Explain why this characteristic of an element is closely related to the chemical properties shown by the element. Illustrate by reference to the elements *potassium* and *chlorine*.

5. At a certain period, the compounds magnesium chloride, phosphorus trichloride, phosphorus trichloride oxide, and ammonium chloride were represented by the structures:

Cl—Mg—Cl, Cl—P(—Cl)—Cl,

Cl—P(Cl)(Cl)=O, and N(H)(H)(Cl)(H)(H)

Criticize these formulae in the light of modern ideas about valency and suggest suitable modifications.

6. Discuss the desirability of applying the term *molecule* to particles present in crystalline sodium chloride, ammonia gas, ammonium chloride crystals, and chlorine gas.

7. Account for the shapes of the following species: (a) BCl_3, (b) NH_3, (c) CH_4, (d) SF_6. Why does boron trichloride form an addition compound with ammonia?

8. What is meant by the term *delocalization*? Give examples of materials in which it is thought to occur.

Questions on Chapter 8

1. Account for the properties of the allotropes of carbon in terms of their structures.

2. Explain the following terms: (a) hexagonal close packing, (b) body-centred cubic structure, (c) co-ordination number, (d) molecular lattice. Give an illustrative example in each case.

3. Give a brief account of the use of X-ray diffraction in determining crystal structures. Describe as fully as you can the structure of sodium chloride.

Questions on Chapter 9

1. 'The hydrogen bond is essential to life.' Discuss.

2. Explain the nature of the forces which hold the following materials together in the solid state: (a) pure silver, (b) sodium chloride, (c) diamond, (d) ice, (e) iodine.

3. Consider, as widely as you can, the consequences of water losing its ability to hydrogen bond with itself and with other materials.

Questions on Chapter 10

1. What do you understand by the term *colloid*? Discuss the relation between 'true' solution, colloidal solution, and suspension. Give an account of *two* distinct methods of producing colloidal solutions with *one* illustrative example for each. Discuss the process of *dialysis* as a means of purifying hydrosols.

2. What do you understand by the terms *dialysis* and *peptization* as used in aqueous colloid chemistry? Give *one* example of the use of each process. Compare 'true' and colloidal solutions with respect to *three* different properties.

3. Give a brief account of (a) the Brownian movement, (b) electrophoresis, (c) protection, (d) coagulation in colloid chemistry.

4. What do you understand by *lyophilic* and *lyophobic* sols? Mention *two* examples of each with water as dispersion medium. Tabulate *four* contrasting properties of these two types of sol.

Questions on Chapter 11

1. Describe how you would find the vapour density of trichloromethane by the method of V. Meyer. Illustrate the calculation by the following results: 0.119 g of trichloromethane used; 25.4 cm^3 of air displaced at 17 °C and 745 mmHg pressure; vapour pressure of water at 17 °C is 15 mmHg. [One dm^3 of hydrogen at s.t.p. weighs 0.09 g.]

2. Describe a determination of the vapour density of ether by the method of Dumas. Use the following figures to illustrate the calculation:

Weight of bulb and air = 31.782 g
Weight of bulb and ether vapour after sealing = 32.032 g
Weight of bulb full of water (+ broken tip) = 210.0 g

Laboratory temperature = 20 °C
Temperature of sealing of bulb = 73 °C; pressure, 755 mmHg

[1 dm^3 of dry air and 1 dm^3 of dry hydrogen at s.t.p. weigh 1.293 g and 0.09 g respectively.] Discuss the advantages and disadvantages of Dumas' method.

3. In a V. Meyer determination, 0.156 g of a volatile liquid displaced 47.0 cm^3 of air measured over water at 14 °C and 762 mmHg pressure. [Vapour pressure of water at 14 °C is 13 mmHg.] Calculate the vapour density and relative molecular mass of the liquid.

4. An element X forms a number of gaseous or volatile compounds. The following data are representative:

Compound	*% of X by mass in the compound*	*Vapour density of compound*
1	57.1	14
2	72.7	22
3	50.0	32
4	53.3	15
5	36.3	22
6	60.0	40

What is the probable relative atomic mass of X?

5. When vaporized at 130 °C in a V. Meyer apparatus, 0.100 g of methanoic acid displaced 39.2 cm³ of air measured over water at 12 °C and 750 mmHg. Calculate the relative molecular mass of the acid as indicated by these figures. If the correct molecular formula of the acid is H_2CO_2, how do you account for the calculated value obtained above? [Vapour pressure of water = 11 mmHg at 12 °C. One dm³ of hydrogen at s.t.p. weighs 0.09 g. H = 1, C = 12, O = 16.]

6. Outline briefly *two* different uses for the process of gaseous diffusion in science, giving one example of each use. At a certain temperature and under identical conditions, the volumes of oxygen and nitrogen dioxide effusing from the same apparatus in the same time were in the proportion of 5 : 3. Calculate the relative molecular mass of nitrogen dioxide at this temperature. [O = 16.0.]

7. State Boyle's Law and Charles' Law. What is the mathematical formulation of Boyle's Law? Why do real gases not obey Boyle's Law exactly? Give a qualitative account of van der Waal's attempt to express the behaviour of gases more precisely.

8. What was formerly meant by a *permanent* gas? Methane, CH_4, was placed in this class; carbon dioxide was not. Interpret these facts in the light of later, more correct ideas. A gas, X, has a critical temperature of 430 K and a critical pressure of about 78 atm; corresponding figures for gas, Y, are 155 K and about 50 atm. State, with reasons, suitable methods of liquefaction for X and Y. What do the terms *critical temperature* and *critical pressure* mean?

Questions on Chapter 12

1. Describe the preparation of a good semi-permeable membrane. Explain the meaning of the term *semi-permeable* and describe how you would use the membrane to demonstrate the existence of osmotic pressure in a given solution of cane sugar. State the laws of osmotic pressure and point out the analogy between the behaviour of gases and of unionized solids in dilute solution.

2. Describe a reasonably accurate method of measuring the osmotic pressure of a given sugar solution. A solution containing 4.00 g of a certain sugar in 100 cm³ of aqueous solution has an osmotic pressure of 2.76 atm at 16 °C. Calculate the relative molecular mass of the sugar. [G.M.V. is 22.4 dm³ at s.t.p.]

3. (a) A solution of a non-electrolyte of relative molecular mass 342 has an osmotic pressure of 0.630 atm at 12 °C. Calculate the concentration of the solution in g dm^{-3}.

(b) Calculate the osmotic pressure in atm of a solution of urea (formula mass 60) at 24 °C, which contains 2.00 g of urea in 100 cm³ of solution.

(c) Calculate the temperature of a laboratory in which a solution containing 5.00 g of urea (formula mass 60) in 160 cm³ of solution gave an osmotic pressure of 12.30 atm. [G.M.V. is 22.4 dm³ at s.t.p.]

4. Describe in detail how you would determine the formula mass of urea by measurement of the elevation of the boiling point of water. Illustrate the calculation by the following data: 0.300 g of urea in 30 g of water elevate the boiling point of the solvent by 0.310 °C. [K_b = 1.68 °C per 1000 g of water.]

5. Describe how you would determine the relative molecular mass of naphthalene by observations on the freezing point of benzene, in which naphthalene is readily soluble. [The freezing point of benzene is about 4 °C.] Illustrate the calculation from the data: 6.50 g of naphthalene in 200 g of benzene depressed the freezing point of the solvent by 1.250 °C. [K_f = 5.10 °C per 1000 g of benzene.]

6. What do you understand by the terms *saturated solution* and *supersaturation*? Mention the usual conditions under which a *supersaturated* solution can be obtained. Illustrate by reference to *one* example. How would you prepare a saturated solution of potassium nitrate at 50 °C and then use it to determine the solubility of potassium nitrate at that temperature?

7. Describe in outline *two* methods of finding the solubilities of very sparingly soluble salts, such as silver chloride or lead sulphate.

8. A salt, A, forms no compound with water but, at −25 °C, forms a eutectic of composition 60% water and 40% A. The solubility of A increases, but not very rapidly, with rise of temperature from −25 °C. Draw, with the temperature axis vertical, a rough graph expressing these facts over the range of about −30 °C to +30 °C. Describe what happens during the cooling of various liquids involving water and A, so as to show all the typical behaviour possible. What is an alternative name for the eutectic in this case? Why is it not considered to be a compound in spite of its constant melting point and composition at constant pressure?

9. Two metals, A and B, B having the higher melting point, form a compound containing about 40% of A. This compound forms a eutectic with both A and B, the one with A having the lower melting point. Sketch (with temperature axis vertical) a rough graph to express these facts. Consider what happens during the cooling of various liquids containing A and B, choosing enough cases to illustrate all the typical behaviour possible.

10. Explain the following terms, giving *one* example of each: (a) eutectic, (b) cryohydrate, (c) deliquescence, (d) efflorescence. Why does the addition of sufficient common salt to an icy path produce a liquid on the path? What is the relation of this fact to the use of ice-salt in freezing mixtures to give about −15 °C.?

11. State the Distribution (or Partition) Law. Give an account of *one* set of experiments by which it can be illustrated. 18 g of a compound X distribute themselves between water and an equal volume of an immiscible solvent Y so that 2 g of X are in the water. Calculate to the nearest integer the percentage of X left in water if 1000 cm³ of water containing 1 g of X are extracted by (a) one litre of Y, (b) half a litre of Y twice, (c) one-third of a litre of Y three times, so that one litre of Y is used in each case. Comment briefly on the experimental implications of these figures.

12. State the Partition Law as it relates to a compound in the same molecular state in both solvents. Discuss briefly the variations produced in the statement of the law by (a) ionization, (b) polymerization. The distribution of an acid between benzene and water shows the following figures:

Concentration in water	*Concentration in benzene*
0.0450	0.504
0.0549	0.750
0.0639	1.018
0.0810	1.634

If the acid is monomeric in water, show that it is dimerized in benzene.

13. Describe in essential outline, methods suitable for finding the solubility of (a) a gas slightly soluble in water, (b) a gas very soluble in water. Assuming air to contain 79% of nitrogen, 21% of oxygen, and 0.04% of carbon dioxide (by volume) and that all the gases obey Henry's Law, calculate the composition of air boiled out from water. [The absorption coefficients are: nitrogen, 0.02, oxygen, 0.04, carbon dioxide, 1.80.]

14. Give an account of the principles involved in the process of steam distillation. To what type of compound can this process by suitably applied? Calculate the relative molecular mass of an organic compound, X, from the data: when the compound was distilled in steam, the liquid boiled at 99.3 °C and, of the distillate, 90 cm^3 were water and 14 cm^3 were X. [Barometric height = 760 mmHg; vapour pressure of water at 99.3 °C = 740 mmHg. Density of X = 1.22 g cm^{-3}.]

15. The distillation of a methanol water mixture can produce the pure alcohol, but distillation of a water-ethanol mixture gives a distillate with, at best, about 96% of the alcohol. Explain this difference.

16. Hydrogen chloride forms a *constant-boiling mixture* with water. Explain the meaning of this statement, illustrate it graphically and discuss the results of distilling liquids containing water and hydrogen chloride. Choose enough examples to illustrate all possible types of behaviour.

Questions on Chapter 13

1. Define the terms: *standard enthalpy of formation* and *standard enthalpy of combustion* of a compound. Describe in essential outline how the enthalpy of combustion of a solid compound such as glucose, $C_6H_{12}O_6$, can be determined. Assuming this result, explain how you could obtain the *enthalpy* of formation of glucose, given that water and carbon dioxide are exothermic compounds with standard enthalpies of formation of *a* kJ mol^{-1} and *b* kJ mol^{-1}, respectively.

2. Define the terms *enthalpy of solution* of a compound and *enthalpy of neutralization.* Describe in essential outline how you would determine *one* of these, using a named example.

3. State Hess's Law of Heat Summation. Show that it is essentially an aspect of the general Law of Conservation of Energy. Devise a set of experiments, based on the conversion of barium oxide to a dilute solution of barium chloride, to illustrate Hess's Law.

4. Calculate the standard enthalpy of formation of ethanol, C_2H_5OH, given that its combustion is an exothermic process and the heat of combustion is 1421 kJ mol^{-1}, and that carbon dioxide and water are exothermic compounds with enthalpies of formation of 405 and 284 kJ mol^{-1}, respectively.

5. Define the term *enthalpy of neutralization.* Calculate the enthalpy of neutralization of caustic soda by nitric acid from the data: 250 cm^3 of 0.1 M caustic soda solution were added to 250 cm^3 of 0.1 M nitric acid in a lagged calorimeter. Both solutions were initially at 13.70 °C and the final temperature was 14.32 °C. The water-equivalent of the calorimeter system was 50 g and the specific heat capacities of the solutions can be taken as the same as that of water.

6. The combustion of carbon disulphide is exothermic and the enthalpy of combustion of the compound is 1108 kJ mol^{-1}. Given that carbon dioxide and sulphur dioxide are exothermic compounds, with enthalpies of formation of 405 and 293 kJ mol^{-1}, respectively, calculate the enthalpy of formation of carbon disulphide. Comment on the stability of this compound at various temperatures, considering the result obtained in the light of Le Chatelier's Principle.

7. Discuss the measurements which must be made in order to determine the lattice energy of sodium chloride.

8. What factors influence the solubility in water at room temperature of an ionic compound?

Questions on Chapter 14

1. Discuss briefly what is meant by the term *chemical equilibrium*. In illustration, consider the reaction:

$$CH_3COOH(l) + C_2H_5OH(l) \rightleftharpoons CH_3COOC_2H_5(l) + H_2O(l)$$

occurring in the liquid phase. What is the effect, in the light of Le Chatelier's Principle, of (a) increasing the relative concentration of the alcohol, (b) increasing the relative concentration of water? Why does the superficially similar reaction:

$$HNO_3(aq) + NaOH(aq) \rightarrow NaNO_3(aq) + H_2O(l),$$

proceed rapidly and almost completely from left to right?

2. If a mixture of 1 mole of ethanol and 1 mole of glacial ethanoic acid is allowed to reach equilibrium at 25 °C, two-thirds of the acid is esterified. Calculate the molar composition of the mixtures obtained at equilibrium at this temperature starting from (a) 1 mole of the acid and 8 moles of the alcohol, (b) 1 mole each of ethanoic acid, alcohol and water. This reaction proceeds with very little heat change. What would you expect to be the effect of temperature change on the equilibrium?

3. In an experiment of Bodenstein, 20.57 moles of hydrogen and 5.22 moles of iodine were allowed to react to equilibrium at about 405 °C. At this point, the mixture contained 10.22 moles of hydrogen iodide. Show from the data that the equilibrium constant of the reaction:

$$H_2(g) + I_2(g) \rightleftharpoons 2HI(g)$$

is about 61. If, in a similar experiment, 2 moles of hydrogen and 6 moles of iodine are used, calculate the molar composition of the equilibrium mixture. [Use $K = 61$.] Outline the experimental method required to obtain such a set of observations.

4. At a certain high temperature, the equilibrium constant of the reaction:

$$N_2(g) + O_2(g) \rightleftharpoons 2NO(g),$$

is 8×10^{-4}. If air is a mixture of nitrogen:oxygen in the proportions of 4 : 1 by volume, calculate the percentage of nitrogen oxide by volume in the gas produced when air reacts to equilibrium at this temperature.

5. The equilibrium constant of the reaction:

$$CO_2(g) + H_2(g) \rightleftharpoons CO(g) + H_2O(l)$$

is 1.6 at 1000 °C. What is the percentage by volume of each constituent in the mixture at equilibrium (at 1000 °C) produced from a mixture of (a) 75 cm^3 of hydrogen and 75 cm^3 of carbon dioxide, (b) 50 cm^3 each of carbon dioxide and carbon monoxide and 100 cm^3 of hydrogen?

6. Ammonium hydrogen sulphide being a solid, in what circumstances can the Law of Mass Action be applied to the equilibrium:

$$NH_4HS(g) \rightleftharpoons NH_3(g) + H_2S(g)?$$

If the system is in equilibrium at *t* °C. with the total pressure at 400 mmHg of mercury, what is the partial pressure of ammonia if hydrogen sulphide is added till its partial pressure is 500 mmHg without change of volume at t °C? Under what conditions would the partial pressure of hydrogen sulphide be 640 mmHg at t °C in the same volume?

7. State Le Chatelier's Principle. Consider its application in predicting the results of varying:

(a) Temperature, pressure, and relative concentrations of reagents in the reactions:

$$N_2(g) + 3H_2(g) \rightleftharpoons 2NH_3(g);$$

A kJ evolved.

(b) As for (a) in the reaction:

$$N_2(g) + O_2(g) \rightleftharpoons 2NO(g);$$

B kJ absorbed.

(c) Temperature and pressure in a system of ice-water in equilibrium.

(d) Temperature in a system of potassium nitrate-water in equilibrium (the dissolution of the salt being endothermic).

(e) Temperature and pressure in a system of oxygen-water in equilibrium (the dissolution of oxygen being exothermic).

Questions on Chapter 15

1. What is meant by the 'abnormal' behaviour of electrolytes in aqueous solution with respect to osmotic pressure and its related effects? Outline the ionic theory of Arrhenius as it was first put forward about 1880 and show how it explains the abnormalities. Mention *one* considerable modification of the theory in recent years.

2. Using the weak electrolyte, ethanoic acid, in illustration, obtain a mathematical expression which states the variation of its ionization with concentration. What is the name given to this expression? A molar aqueous solution of a weak monobasic acid was found to freeze at −1.91 °C. Calculate the degree of ionization of the acid and its dissociation constant. [K_f = 1.86 °C per 1000 g of water.]

3. Ammonium hydroxide is 1.4% ionized at 25 °C in 0.1 M solution. Calculate the dissociation constant of this base. Calculate the approximate degree of ionization of the base in 0.01 M solution. What is the hydroxyl ion concentration in this solution and, hence, its pH?

4. If the dissociation constant of formic acid is 2.14×10^{-4} mol dm^{-3} at 25°C, calculate the degree of ionization of the acid in 0.1 M solution at this temperature. Hence calculate its approximate degree of ionization in 0.05 M solution.

5. What is the osmotic pressure of a decimolar solution of ethanoic acid at 25 °C if its dissociation constant at this temperature is 1.8×10^{-5} mol dm^{-3}? Answer in atmospheres. [G.M.V. is 22.4 dm^3 at s.t.p.]

6. Using the 'insoluble' salt, barium sulphate, in illustration, explain what is meant by the term *solubility product.* Write expressions for the solubility product of lead iodide, silver chromate, and calcium phosphate. Assuming the solubility product of barium sulphate to be 1×10^{-10} mol^2 dm^{-6} at about room temperature, calculate (a) the solubility of the salt in g dm^{-3}, (b) its solubility in 0.001 M sulphuric acid, assuming the acid fully ionized. [$BaSO_4$ = 233.]

7. The solubility product of barium carbonate is 7×10^{-9} mol^2 dm^{-6} at 16 °C. What is the greatest mass of barium carbonate that can be dissolved in 1500 cm^3 of water at this temperature? [$BaCO_3$ = 197.]

8. The solubility product of barium oxalate (ethanedioate), BaC_2O_4, at 18 °C is 1.2×10^{-7}mol^2 dm^{-6}. Calculate the mass of barium oxalate that would be precipitated from 1 dm^3 of its saturated solution at 18 °C by dissolving in it 0.67 g of anhydrous sodium oxalate. Assume both salts fully ionized. [Ba = 137, C = 12, O = 16, Na = 23.]

9. Explain, in ionic terms, the following facts: (a) silver chloride is almost insoluble in water but passes readily into solution when excess ammonia is added, (b) iodine is readily soluble in a concentrated solution of potassium iodide but only sparingly soluble in water, (c) lead(II) chloride is appreciably more soluble in concentrated hydrochloric acid than in water, (d) the addition of potassium cyanide solution to a solution of silver nitrate gives a white precipitate which redissolves in excess of potassium cyanide solution.

10. State, in ionic terms, what is meant by (a) a neutral aqueous liquid, (b) the hydrogen ion index, pH, of an aqueous solution. (c) neutralization in aqueous solution. Calculate the pH of: (i) 0.01 M HCl, (ii) 0.001 M KOH, (iii) an aqueous solution containing 7.3 g of HCl per dm^3, (iv) an aqueous solution containing 4 g of NaOH per dm^3, (v) 0.01 M ethanoic acid in which the degree of ionization is 0.044. [Assume the electrolytes fully ionized except where it is stated otherwise. H = 1, Cl = 35.5, O = 16, K = 39, C = 12.]

11. Outline the development in meaning of the term *acid.* Explain the terms *strong* and *weak* as applied to acids and summarize the characteristic properties of an acid. What is the relation between an acid and one of its salts?

12. What is meant by the *basicity* of an acid? Illustrate by reference to sulphuric acid and describe *two* methods by which you could show this acid to be dibasic.

13. Describe in outline *two* methods by which the strength of two acids can be compared. Briefly indicate the limitations of the methods. The heats of neutralization of the following acids by caustic soda are: hydrochloric, 57.3 kJ mol^{-1}; sulphuric, 114.6 kJ mol^{-1}; ethanoic, 56.0 kJ mol^{-1}. Comment on these figures.

14. Discuss, in ionic terms, the nature and mode of operation of a typical acid–alkali indicator. Show by approximate graphs the change of pH during titrations with strong and weak acids and bases. Hence show what indicators are suitable for the following titrations: (a) NaOH-HCl, (b) KOH-$H_2C_2O_4$, (c) NH_4OH-HCl. Discuss the position regarding choice of indicator for an ethanoic acid–ammonia titration.

15. What do you understand by the term *hydrolysis* as applied to inorganic salts? Discuss the situation prevailing in aqueous solutions of: (a) aluminium chloride, (b) ammonium ethanoate, (c) sodium ethanoate. Explain (i) why a dilute solution of sodium cyanide has an almond smell, (ii) why a solution of ammonium chloride, if swallowed by an individual, tends to turn his blood acidic. Briefly describe *one* method of estimating approximately the pH of any one of the above salt solutions.

16. What is meant by a *buffer solution*? State the purpose of such a solution and, illustrating by one actual case, explain why it is able to fulfil it. Mention appropriate buffer solutions for (a) moderately acidic conditions, (b) moderately alkaline conditions.

17. Describe in outline a method for determining the degree of ionization of a weak electrolyte in aqueous solution.

(a) A 0.1 molar solution of sodium chloride freezes at −0.335 °C. Calculate the osmotic pressure of the solution (in mm of mercury) at 8 °C, assuming that the apparent ionization of the salt is unaffected by the temperature change. [K_f = 1.86 °C per 1000 g of water; G.M.V. is 22.4 dm^3 at s.t.p.]

(b) 2.00 g of potassium chloride dissolved in 65 g of water raise the boiling point of the liquid by 0.400 °C. Calculate the apparent degree of ionization of the salt. [$K_b = 0.52$ °C per 1000 g of water.]

18. Calculate the apparent degree of ionization of calcium nitrate in the given solution from the data (at 760 mmHg): (a) the freezing point of a solution of 2.30 g of cane sugar, relative molecular mass 342, in 134 g of water is −0.0935 °C. (b) The freezing point of a solution containing 0.36 mole of calcium nitrate in 1000 g of water is −1.68 °C.

19. A 0.05 molar solution of magnesium sulphate acts as if the electrolyte is 70% ionized. Calculate the osmotic pressure of this solution at 20 °C in mmHg. [G.M.V. is 22.4 dm^3 at s.t.p.]

Questions on Chapter 16

1. State briefly in electronic terms what is understood by *oxidation* and *oxidizing agent*. Illustrate by reference to the action of chlorine on an iron(II) salt. Show that the older ideas of oxidation as 'combination with oxygen' and 'increase in the proportion of the electronegative part of the molecule' can be reconciled with the electronic interpretation. Illustrate by reference to suitable metallic compounds.

2. Rewrite the 'molecular' equations given below in ionic terms; then, for each molecular or ionic species on the left-hand side of each equation, state whether it is oxidized or reduced or left unchanged, giving reasons for your answer in electronic terms:

(a) $2Na(s) + Cl_2(g) \rightarrow 2NaCl(s)$
(b) $Zn(s) + H_2SO_4(dil.\ aq) \rightarrow ZnSO_4(aq) + H_2(g)$
(c) $2Na_2S_2O_3(aq) + I_2(s) \rightarrow Na_2S_4O_6(aq) + 2NaI(aq)$
(d) $SnCl_2(aq) + HgCl_2(aq) \rightarrow Hg(l) + SnCl_4(aq)$
(e) $2FeSO_4(aq) + Br_2(l) + H_2SO_4(dil.\ aq) \rightarrow Fe_2(SO_4)_3(aq) + 2HBr(aq)$

3. Give *one* example in each case, stating the experimental conditions required, of the *oxidizing action* of chlorine, hydrogen ion, and potassium permanganate, and of the *reducing action* of hydrogen sulphide, tin(II) chloride, and hydriodic acid. In one oxidizing action and one reducing action of the above, interpret the change in electronic terms.

4. The following are all cases of oxidation, in an older meaning of this term. Discuss the application of electronic ideas of oxidation to them.

(a) $2Ca(s) + O_2(g) \rightarrow 2CaO(s)$
(b) $C(s) + O_2(g) \rightarrow CO_2(g)$
(c) $2Na_2SO_3(aq) + O_2(g) \rightarrow 2Na_2SO_4(aq)$
(d) $2H_2S(g) + O_2(g) \rightarrow 2H_2O(l) + 2S(s)$

Discuss the oxidizing behaviour of (a) sulphuric acid, (b) nitric acid on copper and zinc. Wherever you can, interpret the reactions in electronic terms.

5. Describe briefly how *standard redox potentials* may be measured. Explain how they may be used to predict (a) the e.m.f. values of cells, (b) the likelihood of a redox reaction occurring. Give two examples in each case. [Use the data given in the table on page 250.]

Questions on Chapter 17

1. Define the term *catalyst*. Outline the two principal theories of catalysis and illustrate them by two examples in each case. Discuss, giving *one* example in each case, what is meant by (a) catalyst promotion, (b) catalyst poisoning, (c) autocatalysis, (d) negative catalysis (inhibition).

2. Give *three* characteristics of a catalyst. Describe experiments by which you would try to show that manganese(IV) oxide is catalytic for the decomposition of potassium chlorate by heat. Mention a possible mechanism for this catalysis. Are there any experimental facts supporting the mechanism you suggest?

3. Distinguish between the terms *order* and *molecularity* of a reaction. Describe, for a reaction of your choice, how the order of a reaction may be determined.

4. How does the rate constant of a reaction depend on the temperature? What is meant by *activation energy*? How may the activation energy of a reaction of your choice be found?

5. Dilute hydrochloric acid reacts with a solution of sodium thiosulphate to produce, initially, an opaque suspension of sulphur. Describe a series of experiments to determine how the rate of this reaction varies with (a) concentration of the reactants, (b) temperature.

6. Write an essay entitled 'Catalysis in Industry'. Include economic considerations where appropriate.

Questions on Chapter 18

1. What are the products of electrolysis of the following in aqueous solution with inert electrodes: (a) sulphuric acid, (b) silver nitrate, (c) copper(II) sulphate? Give suitable equations. What difference is made if (b) is performed with a silver anode and (a) with a copper anode? Give a sketch of apparatus suitable to demonstrate (a).

2. State Faraday's Laws of Electrolysis and describe, in outline, experiments by which they can be illustrated. (a) 0.396 g of copper is deposited from copper(II) sulphate solution in 40 min by a steady current of half an ampere. Calculate the mass of copper deposited by passage of 1 mole of electrons. [Assume that one Faraday = 96 000 C.] (b) An electric current is passed in series through dilute sulphuric acid and silver nitrate solution for the same time. 112 cm^3 of hydrogen are liberated in the acid (measured wet at 16 °C and 750 mmHg.) Calculate the mass of silver deposited at the cathode in the other cell. [Ag = 108, H = 1.0, V.P. of water at 16 °C is 14 mmHg; 1 dm^3 of hydrogen at s.t.p. has a mass of 0.09 g.]

3. Electric current is passed in series for the same time through solutions of copper(II) sulphate, silver nitrate, and dilute sulphuric acid. If 0.315 g of copper is deposited in the first cell, calculate (a) the mass of silver deposited in the second, (b) the volume of oxygen liberated in the third at 15 °C and 740 mmHg pressure. [O = 16, Cu = 63.5, Ag = 108. G.M.V. of a gas at s.t.p. is 22.4 dm^3.]

4. Discuss the principal factors which determine the products obtained at the cathode and anode in the electrolysis of aqueous solutions containing ions such as Na^+, Cu^{2+}, Ag^+, SO_4^{2-}, Cl^-, and the ions derived from water. Give suitable examples.

5. Explain what is meant by (a) specific conductance, (b) molar conductance, (c) equivalent conductance of an electrolyte. Explain why (c) can be determined experimentally for a strong electrolyte such as potassium chloride, but not for a weak electrolyte such as ethanoic acid. Describe in outline the determination of the specific resistance of 0.05 M sodium chloride solution. If the value is 210 ohms, calculate the equivalent conductance of this electrolyte at this dilution.

6. What do you understand by Kohlrausch's Law of Independent Migration of Ions? The equivalent conductivities of the following electrolytes at infinite dilution are: AgCl, 119.8;

$AgNO_3$, 116.2; Ag_2SO_4, 122.8; KCl, 130.1; KNO_3, 126.5; K_2SO_4, 133.1. Show that these figures are in accordance with the law. Describe, in outline, how any one of these figures can be determined.

7. Why cannot the equivalent conductance at infinite dilution of ethanoic acid be determined directly? Indicate briefly *one* method which can be used for this determination.

General Questions

The questions in the following pages are included by permission of the Examining Boards concerned: Joint Matriculation Board (*J.M.B.*); Oxford and Cambridge Schools Examination Board (*O.C.*); Examination Board for Wales (*W.*); London University Entrance and Schools Examination Council (*L.*); London University Intermediate B.Sc. (*Lond. Inter.*); University of Durham (*D.*).

1. Measurement has shown that even the most highly purified distilled water has a small electrical conductivity. How has this been explained? Calculate the pH values in solutions formed by adding 48, 50, and 52 cm^3 of 0.1 M caustic soda separately to 50 cm^3 of 0.1 M hydrochloric acid. The ion product for water may be taken as $K_w = 10^{-14}$ $mol^2\ dm^{-6}$. (*D.*)

2. Show how the electronic theory of valency provides an explanation of the regular variation of valency in passing from group to group of the periodic classification of the elements. Tabulate as completely as the facts allow the following information about the elements Al, C, Ca, Cl, N, Na, Ne, S, rearranging the elements in the order of their groups: (a) the group in the Periodic Table, (b) the formula of the characteristic hydride, (c) the action of the hydride on water, (d) the formula of the oxide characteristic of the groups to which the element belongs, (e) the action of this oxide on water. (*J.M.B.*)

3. Define the expression *allotropic modification* and give three examples of allotropy. In the case of *one* example give an account of a method for converting one form into another, and vice versa, and indicate the chief differences between them. (*Lond. Inter.*)

4. Explain the following terms and illustrate them with examples drawn from your experience in qualitative analysis: (a) solubility product, (b) common ion effect, (c) complex ion formation, (d) amphoteric behaviour. (*L.*)

5. The atom of an element A contains twelve protons. Give your reason for regarding the element as a metal or non-metal and write the electronic formula of its compound with chlorine. The specific heat capacity of A is 1.045 $J\ g^{-1}\ K^{-1}$. What is its approximate relative atomic mass? Outline a method for obtaining additional informa- for the accurate determination of the relative atomic mass. Comment briefly on the adoption of $^{12}C = 12.000$ as the standard for relative atomic mass. (*W.*)

6. Explain fully what you understand by *three* of the following: (a) negative catalyst, (b) the heat of formation of a metallic oxide, (c) the vapour pressure of a liquid, (d) the valency of nitrogen in ammonium chloride. (*O.C.*)

7. List the three main fundamental particles which are constituents of atoms. Give their relative charges and masses. Explain concisely from the standpoint of atomic structure: (a) the difference in chemical properties between a metal and a non-metal, (b) the difference in valency between sodium and magnesium, (c) isotopes, (d) oxidation and reduction. Illustrate your answer by reference to suitable examples. (*J.M.B.*)

8. Trace the developments in the classification of the elements which culminated in the work of Lothar Meyer and Mendeléeff. Indicate briefly how modern views of atomic structure have provided a more satisfactory basis of classification than that of Mendeléeff. (*D.*)

9. Of what value is the knowledge of the specific heat capacity of an element in deciding its relative atomic mass? Illustrate your answer by reference to the cases of (a) copper, (b) argon. (*Lond. Inter.*)

10. Calculate the pH value of a solution of (a) 0.01 M sodium hydroxide, (b) 0.01 M ethanoic acid. The dissociation constant of ethanoic acid is 1.8×10^{-5} mol^2 dm^{-6} at room temperature. Explain the observation that when equivalent amounts of (a) and (b) are mixed the pH value of the resulting solution is greater than 7. What indicator would you use for titrating ethanoic acid against sodium hydroxide? Give reasons for your choice of indicator. (*W.*)

11. What is meant by the electrochemical series? Arrange the elements Ag, Fe, H, K in order of their positions in this series. Explain why (a) a steel knife becomes coated with copper when dipped into a copper(II) sulphate solution, (b) galvanizing with zinc affords more lasting protection to sheet iron than tinning, (c) when sodium amalgam is added to water the sodium alone reacts with the water, (d) copper will dissolve in dilute nitric acid but not in dilute sulphuric acid. (*J.M.B.*)

12. Define the terms *velocity constant* and *equilibrium constant*. What is meant by the common ion effect? Two metals D and E form chlorides DCl_2 and ECl_2 which are both soluble in water and sulphides DS and ES which are both insoluble in water. The solubility product of DS is *very much less* than that of ES. Suggest a method by which the insolubilities of the sulphides may be made the basis of the separation of the metals when they are present as chlorides in an aqueous solution. Explain the theory underlying the method. (*J.M.B.*)

13. Comment on, illustrate, or explain the following statements: (a) hydrogen chloride does not obey Henry's Law, (b) metals can displace hydrogen from sodium hydroxide, (c) a strong electrolyte does not obey Ostwald's Dilution Law, (d) some allotropes differ in chemical properties, others in physical properties only. (*O.C.*)

14. What are the distinctive characteristics of the metallic and non-metallic elements? Discuss their distribution in the periodic classification of the elements. By reference to any *one* group of the Periodic Table, illustrate the gradation of the above characteristics with increasing relative atomic mass. (*W.*)

15. Explain the following statements: (a) it is incorrect to refer to concentrated sulphuric acid as *strong* sulphuric acid, (b) on dissolving in water hydrogen chloride ionizes and dissociates but potassium chloride only dissociates, (c) water is acid to phenolphthalein but alkaline to methyl orange, (d) it is inadmissible to speak of the solubility product of ethanoic acid. (*J.M.B.*)

16. Explain and illustrate *four* of the following terms: (a) atomic number, (b) isotope, (c) isobar, (d) complex ion, (e) covalent bond. (*L.*)

17. What do you understand by a reversible reaction? Discuss how a system in equilibrium is affected by changes in the quantities of the various substances present, and illustrate your answer by reference to (a) the reaction between arsenic(III) oxide and iodine, (b) one other reaction. (*Lond. Inter.*)

18. How do the volumes of (a) ideal gases, (b) real gases vary with temperature and pressure? Illustrate your answer by referring (graphically or otherwise) to the real gases hydrogen, oxygen, and carbon dioxide, and indicate the two chief reasons for the failure of real gases to obey the ideal, gas laws. What do the terms *critical temperature* and *critical pressure* signify? (*J.M.B.*)

19. An element X forms a sulphate which contains 64.06% of X and gives an alum with iron(III) sulphate. Find the relative atomic mass of X, explaining the arguments used in your calculation. What other experiments would you suggest to confirm this relative atomic mass? (*L.*)

20. Define *osmosis* and *osmotic pressure of a solution.* Show how you can demonstrate the existence of the former, and how the latter can be measured for an aqueous solution. (*O.C.*)

21. Explain, with illustrative examples, and distinguish between *three* of the following pairs of chemical terms:

(a) allotrope and isotope, (b) isomorphism and isomerism, (c) dialysis and electrophoresis, (d) transition point and eutectic point. (*L.*)

22. Suggest explanations for the following: (a) The addition of a decimolar monobasic acid to an equal volume of a decimolar monoacidic base does not necessarily produce a neutral solution. (b) Aqueous solutions of copper(II) sulphate and sodium ethanoate have opposite effects on litmus. (c) Determination of the relative molecular mass of ethanoic acid in benzene solution by the cryoscopic method gives a result which does not conform to the formula $C_2H_4O_2$. (*L.*)

23. What do you understand by the term *colligative property*? Describe a method for the determination of one such property and show how this method might be used for the determination of the degree of ionization of an inorganic acid. The depression of freezing point observed for an aqueous 1 M solution of HCl is practically identical with that of a 1 M solution of H_2SO_4 in water. What inference can you draw from this about the state of sulphuric acid in water? (*L.*)

24. Devise experiments to demonstrate that: (a) aqueous mercury(II) chloride is less ionized than aqueous sodium chloride, (b) ethanoic acid is a stronger acid than phenol, (c) iron(III) chloride exists as double molecules in the vapour phase, (d) phosphorus pentachloride dissociates on heating. (*L.*)

25. State Raoult's law of vapour pressure lowering and explain qualitatively why a solution of a non-volatile solute in a volatile solvent has a higher boiling point than the solvent. At 25 °C a saturated aqueous solution of calcium hydroxide, $Ca(OH)_2$, possesses a vapour pressure of 23.765 mmHg. At the same temperature, the vapour pressure of water alone is 23.790 mmHg. Assuming the calcium hydroxide to be completely ionized in solution, what is its solubility in moles per dm³ at 25 °C? (*L.*)

26. It is known that ammonia distributes itself between equal volumes of trichloromethane and water in the molar ratio of 1 : 26 respectively. In an experiment 25 cm³ of molar aqueous ammonia were mixed with 25 cm³ of 0.1 M aqueous copper(II) sulphate. The mixture was shaken with 50 cm³ of trichloromethane for several minutes and allowed to settle. The trichloromethane layer was separated off and repeatedly shaken with several portions of distilled water. The aqueous extracts were combined and required 11.4 cm³ of 0.05 M hydrochloric acid for neutralization, using methyl red as indicator. (a) What information can you deduce from the above results about any compound formed between copper(II) sulphate and ammonia? Suggest what its structure would be. (b) Account, as far as possible, for the greater solubility of ammonia in water than in trichloromethane. (*L.*)

27. State Avogadro's Hypothesis.

20.0 cm^3 of a mixture of two gaseous hydrocarbons was mixed with 80.0 cm^3 of oxygen. After explosion in a eudiometer, the volume of the gas mixture was found to have decreased to 60.0 cm^3. Reaction with potassium hydroxide solution led to a reduction in volume of 30.0 cm^3. After reaction with alkaline pyrogallol solution, a further decrease in volume by 30.0 cm^3 was measured. (a) Find the composition of the hydrocarbon mixture. (b) Suggest a method suitable for the separation of the gases in the mixture. (*L.*)

28. Give a concise account of the principles underlying the following methods of separation of chemical substances: (a) steam distillation, (b) fractional distillation, (c) fractional crystallization, (d) diffusion of gases through a membrane. (*L.*)

29. A current of 0.2 amperes is passed for 60 minutes through 100 cm^3 of an acid solution of 0.5 M copper(I) chloride, the cathode being platinum and the anode being carbon. What mass of copper is deposited at the cathode and what would be the new concentration of the solution in moles of copper(I) chloride per dm^3? What ionic equilibria are involved in this reaction? If a solution of 0.5 M copper(II) chloride in 100 cm^3 was used with copper electrodes, how long would be required to deposit the same mass of copper using the same current? What is the new concentration of the solution in moles of copper(II) chloride per dm^3 in this case? [Cu = 63.5, 1 faraday = 96 500 ampere seconds.] (*L.*)

30. State Raoult's Law and explain concisely its application in the determination of relative molecular masses by the boiling point method. Outline the experimental procedure which you would adopt for the determination of the relative molecular mass of a non-electrolyte which is soluble in water. Draw a sketch of the apparatus that you would employ. In an experiment, a 5% solution of glucose, $C_6H_{12}O_6$, in water was found to give the same boiling point elevation as a 3.3% aqueous solution of the simple carbohydrate erythrose. If the composition of simple carbohydrates may be represented by the general formula $C_nH_{2n}O_n$, what is the molecular formula of erythrose? [C = 12.0, H = 1.00, O = 16.0.] (*L.*)

31. Discuss and explain the following phenomena: (a) When an aqueous solution of ethanoic acid is titrated with a standard solution of sodium hydroxide, using methyl orange or methyl red as the acid/base indicator, no satisfactory titration end-point can be observed. (b) When aqueous ammonia solution is cautiously added to a solution of copper(II) sulphate, a blue precipitate results which is soluble in excess of the ammonia solution to give a deep blue solution. (c) When dry hydrogen chloride gas is dissolved in dry methylbenzene or dry benzene, the solution (i) fumes in air, (ii) is found to be a *non*-conductor of electricity. When this fuming solution is shaken with water, it is observed that (iii) the fuming ceases, (iv) the aqueous layer conducts electricity, (v) the aqueous layer turns blue litmus red. (*L.*)

32. (a) Explain what is meant by the term 'ionic product of water' and discuss how the pH scale is related to it. (b) In an attempt to determine the ionic product of water by a conductivity procedure, Kohlrausch and Heydweiler, in 1894, redistilled water 42 times under reduced pressure and obtained water of specific conductivity 4.1×10^{-8} ohm^{-1} cm^{-1} at 18 °C. Given that the ionic (equivalent) conductances at 18 °C of H^+ and OH^- are 316 ohm^{-1} cm^2 mol^{-1} and 176 ohm^{-1} cm^2 mol^{-1}, respectively, calculate the value of the ionic product of water at this temperature. Briefly explain why it was necessary for Kohlrausch and Heydweiler to purify their water in so elaborate a manner. (*L.*)

33. Describe the phenomenon of

osmosis, illustrating your answer with one appropriate example. Define the term 'osmotic pressure' and describe, in outline *only*, a method for the determination of the osmotic pressure of a solution of sugar. Calculate the osmotic pressure at 25 °C of an aqueous solution of cane sugar, $C_{12}H_{22}O_{11}$, containing 3.42 g dm^{-3}. [C = 12, H = 1, O = 16; G.M.V. at s.t.p.. = 22.4 dm^3.] (*L.*)

34. A student has measured the pH values of a number of dilute aqueous salt solutions and reports the following results:

Solute	*pH value*
(i) Ammonium sulphate	4.5
(ii) Sodium cyanide	11
(iii) Sodium hydrogen-carbonate	7.5
(iv) Iron(III) chloride	4

(a) Explain the experimental results obtained by the student. (b) The pH values given above were obtained by means of indicator paper. Give a brief account of the theory of acid/base indicators. (*L.*)

35. X-ray diffraction is one of the most powerful techniques for the elucidation of the structure of crystalline solids. Give a brief outline of the principles on which this technique is based. Crystalline solids may be classified, according to the nature of their structural units, as ionic, molecular or atomic crystals. For each of these classes give an appropriate example, state how the particles of the substances chosen are held together in the solid state, and discuss briefly the relationship between the structure and properties of these materials. (*L.*)

36. The following table lists the logarithmic values of the successive ionization energies (in kJ mol^{-1}) of the element sodium.

Number of electrons removed	*Logarithm of ionization energy*
1	2.69
2	3.66
3	3.84
4	3.97
5	4.12
6	4.22
7	4.30
8	4.41
9	4.46
10	5.15
11	5.30

(a) Plot a graph of the logarithm of ionization energies against the number of electrons removed. (b) Explain how from your graph information about the electronic structure of a sodium atom may be obtained. (c) Discuss how a knowledge of successive ionization energies of the elements can help the chemist in explaining the chemical properties of elements. (*L.*)

37. The solubility product of magnesium hydroxide, $Mg(OH)_2$, is 1.25 × 10^{-10} mol^3 dm^{-9}. (a) Calculate the solubility (in g dm^{-3}) of magnesium hydroxide in water. (b) Suppose you are given 1 dm^3 of a 0.01 M aqueous ammonia solution. What is the maximum quantity (in moles) of magnesium hydroxide that can be dissolved in it? [$K_{dissociation}$ = 1.8 × 10^{-5} mol dm^3 for NH_3(aq)]. (c) How does the solubility product of $Mg(OH_2)$ compare with the solubility products of the other alkaline earth metal hydroxides? [Mg = 24.3, H = 1, O = 16.] (*L.*)

38. 'Covalent substances tend to be volatile, of low melting point, soluble in non-aqueous solvents but not in water, and non-electrolytes; whereas electrovalent substances tend to have the opposite properties.' Discuss this statement critically, referring particularly to silicon dioxide, SiO_2, hydrogen chloride,

HCl, sodium chloride, NaCl, trichloromethane, $CHCl_3$, and ammonium chloride, NH_4Cl. (*L.*)

39. The kinetic theory leads to the equation

$$pV = \frac{1}{3}nm\bar{c}^2, \text{ for an ideal gas.}$$

(a) What do the letters p, V, n, m, and $\bar{c}$ in this equation stand for? Give one *consistent* set of units in which they may be expressed. (b) Draw a rough sketch to show how the speeds of the molecules of a gas are distributed at some temperature, T_1. On the same sketch, show the distribution of the speeds at some higher temperature, T_2. Discuss briefly the bearing this shift in distribution has on the rates of reactions in gases. (c) State Avogadro's Law and show that it necessarily follows for an ideal gas from the equation above. (d) Real gases deviate to a greater or less extent from 'ideal' behaviour. Discuss *qualitatively* those assumptions made in deducing the ideal gas equation which are not true for real gases and the effect they have on the behaviour of the latter. Under what conditions does the behaviour of a gas like carbon dioxide tend to that of an ideal gas? (*L.*)

40. The collision theory and the transition state theory represent two different scientific models which the chemist uses in order to interpret experimental reaction rates. Give a concise account of each of these theories, and discuss the major similarities and/or differences that exist between them. (*L.*)

41. It is customary to classify mixtures of two miscible liquids according to whether or not they obey Raoult's Law. (a) State Raoult's Law and show diagrammatically how (i) the vapour pressure and (ii) the boiling point, of an ideal binary mixture vary with composition. (b) Give one example each of binary mixtures showing positive and negative deviations from Raoult's Law and give approximate vapour pressure/composition diagrams for these. Discuss the reason(s) for these deviations in terms of molecular interactions. (*L.*)

42. What properties must a substance possess if it is to be suitable for use as an acid-base indicator? A certain indicator has an acid dissociation constant of 10^{-5} mol dm^3; in a solution of pH 1, the indicator is blue, whereas when the pH is 13 the colour is yellow. Explain carefully (a) how this indicator could be used to find the pH of a sample of slightly cloudy pond-water, and (b) why the indicator would be suitable for titrating hydrochloric acid with either ammonia solution or sodium hydroxide solution but not for titrating ethanoic acid with sodium hydroxide solution or with ammonia solution. (*L.*)

Scholarship Level

1. 'Oxidation and reduction are electron transfer reactions.' Discuss this statement. Name *two* compounds which can function as both oxidizing and reducing agents. Give an illustrative reaction in each case. (*W.*)

2. Give an account of variable valency as exhibited by the elements of the first two periods of the periodic system and discuss the explanation of the phenomena given by the modern electronic theory. What properties do you associate with (a) electrovalent linkage, (b) covalent linkage? Comment on the types of chemical bond present in (c) HCl, (d) KCl, (e) NH_4Cl, (f) CO, (g) HNO_3. (*L.*)

3. How would you remove the chemically active constituents of air in order to show that part of it is chemically inert? What part did the discovery of the noble gases play in the development of modern chemical ideas? (*O.C.*)

4. Describe and explain what takes

place when an electric current passes between platinum electrodes immersed in: (a) an aqueous solution of copper(II) sulphate, (b) a hot aqueous solution of potassium chloride, (c) fused sodium ethanoate, (d) a concentrated aqueous solution of potassium hydrogensulphate at 0 °C, (e) a solution of silver cyanide in potassium cyanide solution. (*J.M.B.*)

5. Discuss the following, with illustrative examples where necessary: (a) the determination of relative atomic mass by Cannizzaro's method, (b) the laws of conservation of mass and of energy, (c) the rule of Dulong and Petit, (d) the reducing power of a metal in relation to its position in the electrochemical series. (*L.*)

6. Discuss the application of the electronic theory of valency to account for the properties of the following: methane, sodium chloride, ammonia, phosphorus pentachloride. It is often found difficult to assign to a molecule or an ion a single electronic formula which adequately represents its properties. Cite such an example and comment very briefly on the present interpretation of its structure. (*W.*)

7. Give a brief account of the present theory of atomic structure and show how it accounts for (a) the periodic system of the elements, (b) the existence of isotopes, (c) the principal types of valence bonds occurring in chemical compounds. (*J.M.B.*)

8. Give a definition of catalysis by describing *three* characteristics of a catalysed reaction. Show, by describing *three* experiments each illustrating a different catalysed reaction, how you could demonstrate the characteristics you have mentioned. Describe briefly *two* suggestions as to the mechanism of catalysis. Give *one* reaction to illustrate each type of mechanism you mention. (*J.M.B.*)

9. What do you understand by the electropositive character of an element? Show by appropriate examples how the electropositive character is related to the occurrence and mode of extraction of an element and to the properties of its oxide and chloride. What connection has the electropositive character with the position of an element in the Periodic Table? (*L.*)

10. Explain, briefly, the theory governing each of the following processes, and in each case give a practical illustration: (a) ether extraction, (b) distillation under reduced pressure, (c) fractional recrystallization to constant melting point, (d) drying in a desiccator. (*L.*)

11. The equilibrium constant, K_p, for the thermal dissociation of ammonium chloride according to

$$NH_4Cl(s) \rightleftharpoons NH_3(g) + HCl(g)$$

is 0.36 atm^2, at 327 °C.
(a) Outline an experimental procedure which you would adopt for the determination of the equilibrium constant for this reaction. (b) State the expression for the equilibrium constant, K_p, of the above reaction. Suppose that exactly 0.2 moles of $NH_4Cl(s)$ were introduced into an evacuated container of 10 dm^3 capacity and then heated to 327 °C. What is the quantity (in moles) of NH_4Cl that would remain as a solid? (c) Consider a closed reaction vessel containing an equilibrated mixture of ammonia, hydrogen chloride, and solid ammonium chloride. Explain whether the addition of more $NH_4Cl(s)$ to this system would increase, decrease or leave unchanged the partial pressure of ammonia. (*L.*)

12. The 'strength' of acids is customarily expressed in pK-units, where pK denotes the negative logarithm to the base 10 of the dissociation constant of an acid. For an investigation of the strength of hydrofluoric acid in aqueous solution, the following procedure was adopted: 25.00 cm^3 of a dilute hydrofluoric acid solution was added to 50.00 cm^3 of a 1.000 M sodium hydroxide solution contained in a laboratory flask.

The excess sodium hydroxide was then back-titrated with 1.000 M sulphuric acid solution, requiring 6.25 cm^3 of the latter. In a separate experiment the pH of the hydrofluoric acid solution was found to be 1.66. (a) From the above information, calculate the dissociation constant and the pK value of hydrofluoric acid. (b) Briefly comment upon the procedure adopted for the titration. (c) Suggest a method that might be used for the determination of the pH of the above solution. (d) How does the strength of hydrofluoric acid compare with that of the other hydrogen halide acids? Suggest reasons for any differences you may infer. (*L.*)

13. Discuss and explain the following observations: (a) The boiling point of methane is considerably lower than that of the corresponding silicon hydride (SiH_4, monosilane), whereas the boiling points of ammonia and of water are higher than those of phosphine and of hydrogen sulphide respectively. (b) Aniline is a weaker base than ammonia, but ethylamine is a stronger base than ammonia. (c) 1 M aqueous solutions of hydrogen chloride, hydrogen bromide, and hydrogen iodide have pH values of 0.09, 0.06, and 0.02 respectively, whereas the pH of a 1 M aqueous solution of hydrogen fluoride is 1.7. (*L.*)

14. Outline the essential assumptions made about the nature of a gas in the kinetic theory of gases and discuss critically the extent to which these assumptions are justified. Show how the kinetic theory of gases accounts for (a) Boyle's Law, (b) Avogadro's Law, and (c) Dalton's Law of Partial Pressures. Under what conditions would you expect these laws to apply strictly to real gases? How have any of these laws been modified to explain the behaviour of real gases? (*L.*)

15. What is meant by the term 'lattice energy'? Outline the theoretical principles underlying the methods of evaluating lattice energies from (a) (experimental) thermochemical data, (b) information about the solid structure of an ionic compound, especially the packing arrangement of the ions and interionic distances. Discuss why method (b) appears to be unsatisfactory when applied to substances such as silver chloride and zinc sulphide. (*L.*)

16. The Arrhenius equation, $k = Ae^{-\frac{E_a}{RT}}$ describes the variation of the rate constant, k, of a chemical reaction with temperature. In this equation, e is the base to the natural logarithm, R the gas constant, and A and E_a are constants for a particular reaction. (a) Explain concisely the physical significance of the two constants A and E_a. (b) The decomposition of N_2O_5 was studied over a range of different temperatures and the following data obtained:

Temperature, T (K)	$1/T$ (K^{-1})	*log k* (k in s^{-1})
298	3.36×10^{-3}	−4.76
308	3.25×10^{-3}	−4.18
318	3.14×10^{-3}	−3.60
328	3.05×10^{-3}	−3.12
338	2.96×10^{-3}	−2.62

Plot a graph of log k against $1/T$ and from it obtain a value for E_a, using the modified Arrhenius equation

$$\log k = \log A - \frac{E_a}{2.3\ RT},$$

where $R = 8.3\ J\ K^{-1}\ mol^{-1}$. Also calculate a value for A. (*L.*)

17. Discuss briefly the procedure by which *three* of the following could be determined experimentally: (a) The number of molecules in a mole. (b) The first ionization potential of an element. (c) The atomic number of a given element. (d) The standard redox potential for the system Fe^{2+}/Fe^{3+}. (*L.*)

18. 'Chemical bonding, irrespective of whether it is ionic or covalent, is basically the result of electrostatic interactions between electrons and nuclei. The essential difference between the two types of bonding lies in the distribution of electrons.' Explain and discuss this statement. (*L.*)

19. The direction of change of a chemical system at constant temperature and pressure may be said to be governed by two factors: the enthalpy change and the entropy change for the given reaction. (a) Explain what you understand by the term 'entropy'. (b) Discuss, with suitable examples of your own choice, the way in which the enthalpy and entropy of a reaction determine the direction of the change. (*L.*)

20. The atomic spectrum of hydrogen consists of a number of lines arranged in series (called e.g., 'Balmer' series, 'Paschen' series, etc.). Describe how, in principle, these spectra could be observed and recorded. What information about the hydrogen atom can be obtained from these line spectra? Discuss briefly (a) the relation between the atomic spectrum of an element and its ionization energy; (b) the information about the chemical properties of an element that can be obtained from its ionization energies. (*L.*)

21. At room temperature the electrode potential for the system $Fe^{3+}(aq)$, $Fe^{2+}(aq)|Pt$ is given by the equation

$$E = E^{\ominus} + 0.06 \log_{10}\frac{[Fe^{3+}]}{[Fe^{2+}]} \qquad (1)$$

For this system, $E^{\ominus} = 0.77$ V. Similarly for the silver electrode, $Ag^{+}(aq)|Ag(s)$

$$E = E^{\ominus} + 0.06 \log_{10}[Ag^{+}] \qquad (2)$$

and $E^{\ominus} = 0.80$ V. (a) Calculate the electrode potential of the half-cell $Fe^{3+}(aq)$, $Fe^{2+}(aq)|Pt$ when the iron(III) ion concentration is 1.0 M and the iron(II) ion concentration is 0.6 M (b) What concentration of silver ions (Ag^{+}) in contact with metallic silver would give the same electrode potential as you have calculated in (a)? (c) What would happen if a solution 1.0 M with respect to Fe^{3+}, 0.6 M with respect to Fe^{2+}, and containing silver ions of the concentration you have found in (b), was in contact with metallic silver? Explain. (d) Use equations (1) and (2) to derive a value for the equilibrium constant *K* for the reaction

$$Fe^{3+}(aq) + Ag(s) \rightleftharpoons Fe^{2+}(aq) + Ag^{+}(aq).$$

(*L.*)

Answers to Numerical Questions (pages 285–306)

Chapter 1

3. 4:2:1
5. 56.0 g; 56.0; 2
6. Valency 4; relative atomic mass 91.2
8. Relative atomic mass 186; valencies 4 and 7
9. 48.0; 31.0% Ti
12. 43.9

Chapter 2

1. (a) 80 g; (b) 14 g; (c) 5.85 g; (d) 49.9 g; (e) 115 g
2. (a) 0.05, molecules; (b) 0.2, atoms; (c) 0.1, formula units; (d) 100, formula units; (e) 0.05, formula units
3. (a) 0.2; (b) 1.0; (c) 2.0; (d) 0.05; (e) 2.0
4. CH_2O; $C_2H_4O_2$; CH_3COOH
5. $BaCl_2 . 2H_2O$

Chapter 4

2. 207.4; 206.2; 207.9

Chapter 11

1. 58.9
2. 36.9

3. 39.3
4. 16
5. 60.8; 32% dimerized
6. 89

Chapter 12

2. 344
3. (a) 9.22 g dm^{-3}; (b) 8.1 atm; (c) 14.8 °C
4. 60
5. 133
13. N_2, 63.5; O_2, 33.7; CO_2, 2.9%
14. 127

Chapter 13

4. −242 kJ mol^{-1}
5. −56.8 kJ mol^{-1}
6. +117 kJ mol^{-1}

Chapter 14

2. (a) ester and water 0.97, acid 0.03, alcohol 7.03 mole; (b) ester 0.54, water 1.54, acid and alcohol 0.46 mole
3. HI 3.88, H_2 0.06, I_2 4.06 mole
4. 1.2%
5. (a) CO and H_2O, 28%; CO_2 and H_2, 22%; (b) CO_2, 10.5; H_2, 35.5; CO, 39.5; H_2O, 14.5%
6. NH_3; 80 mm; 62.5 mm

Chapter 15

2. 0.027; 7.5×10^{-4} mol dm^{-3}
3. 1.99×10^{-5} mol dm^{-3}; 4.4%; pH 10.6
4. 0.045; 0.063
5. 2.48 atm
6. (a) 2.3×10^{-5} g dm^{-3} (b) 2.3×10^{-7} g dm^{-3}
7. 0.025 g
8. 0.077 g
10. (i) 2; (ii) 11; (iii) 0.7; (iv) 13; (v) 3.4
17. (a) 3154 mmHg; (b) 0.86
18. 0.75
19. 1552 mmHg

Chapter 18

2. (a) 31.7 g; (b) 0.996 g
3. (a) 1.07 g; (b) 60.2 cm^3
5. 95.2 ohm^{-1}

General Questions

1. 2.69; 7.0; 11.39
10. (a) 12; (b) 3.4
19. 85.6 (Rb)
25. 1.94×10^{-2} mol dm^{-3}
32. $K_w = 0.695 \times 10^{-14}$ mol^2 dm^{-6}
37. (a) 0.0184 g dm^{-3}

Scholarship Questions

11. (b) 0.078 mole
12. (a) $K_{diss} = 3.24 \times 10^{-4}$ mol dm^{-3}; pK = 3.49
21. (a) 0.783 V; (b) 0.521 mol dm^{-3}; (d) 3.17 mol dm^{-3}

TABLE OF RELATIVE ATOMIC MASSES TO FOUR SIGNIFICANT FIGURES

(Scaled to the relative atomic mass $^{12}C = 12$ exactly)

Values quoted in the table, unless marked * or †, are reliable to at least ±1 in the fourth significant figure. A number in parentheses denotes the atomic mass number of the isotopes of longest known half-life.

At. No.	Name	Symbol	Relative Atomic Mass	At. No.	Name	Symbol	Relative Atomic Mass
1	Hydrogen	H	1.008	53	Iodine	I	126.9
2	Helium	He	4.003	54	Xenon	Xe	131.3
3	Lithium	Li	6.941*†	55	Caesium	Cs	132.9
4	Beryllium	Be	9.012	56	Barium	Ba	137.3
5	Boron	B	10.81†	57	Lanthanum	La	138.9
6	Carbon	C	12.01	58	Cerium	Ce	140.1
7	Nitrogen	N	14.01	59	Praseodymium	Pr	140.9
8	Oxygen	O	16.00	60	Neodymium	Nd	144.2
9	Fluorine	F	19.00	61	Promethium	Pm	(145)
10	Neon	Ne	20.18	62	Samarium	Sm	150.4
11	Sodium	Na	22.99	63	Europium	Eu	152.0
12	Magnesium	Mg	24.31	64	Gadolinium	Gd	157.3
13	Aluminium	Al	26.98	65	Terbium	Tb	158.9
14	Silicon	Si	28.09	66	Dysprosium	Dy	162.5
15	Phosphorus	P	30.97	67	Holmium	Ho	164.9
16	Sulfur	S	32.06†	68	Erbium	Er	167.3
17	Chlorine	Cl	35.45	69	Thulium	Tm	168.9
18	Argon	Ar	39.95	70	Ytterbium	Yb	173.0
19	Potassium	K	39.10	71	Lutetium	Lu	175.0
20	Calcium	Ca	40.08†	72	Hafnium	Hf	178.5
21	Scandium	Sc	44.96	73	Tantalum	Ta	180.9
22	Titanium	Ti	47.90*	74	Wolfram (Tungsten)	W	183.9
23	Vanadium	V	50.94	75	Rhenium	Re	186.2
24	Chromium	Cr	52.00	76	Osmium	Os	190.2
25	Manganese	Mn	54.94	77	Iridium	Ir	192.2
26	Iron	Fe	55.85	78	Platinum	Pt	195.1
27	Cobalt	Co	58.93	79	Gold	Au	197.0
28	Nickel	Ni	58.70	80	Mercury	Hg	200.6
29	Copper	Cu	63.55	81	Thallium	Tl	204.4
30	Zinc	Zn	65.38	82	Lead	Pb	207.2†
31	Gallium	Ga	69.72	83	Bismuth	Bi	209.0
32	Germanium	Ge	72.59*	84	Polonium	Po	(209)
33	Arsenic	As	74.92	85	Astatine	At	(210)
34	Selenium	Se	78.96*	86	Radon	Rn	(222)
35	Bromine	Br	79.90	87	Francium	Fr	(223)
36	Krypton	Kr	83.80	88	Radium	Ra	(226)
37	Rubidium	Rb	85.47	89	Actinium	Ac	(227)
38	Strontium	Sr	87.62†	90	Thorium	Th	232.0
39	Yttrium	Y	88.91	91	Protactinium	Pa	(231)
40	Zirconium	Zr	91.22	92	Uranium	U	238.0†
41	Niobium	Nb	92.91	93	Neptunium	Np	(237)
42	Molybdenum	Mo	95.94*	94	Plutonium	Pu	(244)
43	Technetium	Tc	(97)	95	Americium	Am	(243)
44	Ruthenium	Ru	101.1	96	Curium	Cm	(247)
45	Rhodium	Rh	102.9	97	Berkelium	Bk	(247)
46	Palladium	Pd	106.4	98	Californium	Cf	(251)
47	Silver	Ag	107.9	99	Einsteinium	Es	(254)
48	Cadmium	Cd	112.4	100	Fermium	Fm	(257)
49	Indium	In	114.8	101	Mendelevium	Md	(258)
50	Tin	Sn	118.7	102	Nobelium	No	(259)
51	Antimony	Sb	121.8	103	Lawrencium	Lr	(260)
52	Tellurium	Te	127.6				

* Values so marked are reliable to ±3 in the the fourth significant figure.

† Values so marked may differ from the atomic weights of the relevant elements in some naturally occurring samples because of a variation in the relative abundance of the isotopes.

Index